A Practical Introduction to Electrical Circuits

A Practical Introduction to Electrical Circuits represents a fresh approach to the subject which is compact and easy to use, yet offers a comprehensive description of the fundamentals, including Kirchhoff's laws, nodal and mesh analysis, Thevenin and Norton's theorems, and maximum power transfer for both DC and AC circuits, as well as transient analysis of first- and second-order circuits. Advanced topics such as mutual inductance and transformers, operational amplifier circuits, sequential switching, and three-phase systems reinforce the fundamentals. Approximately one hundred solved examples are included within the printed copy. Extra features online include over two hundred additional problems with detailed, step-by-step solutions, and 40 self-service quizzes with solutions and feedback.

A Practical Introduction to Electrical Circuits

John E. Ayers

CRC Press
Taylor & Francis Group
Boca Raton London New York

CRC Press is an imprint of the
Taylor & Francis Group, an **Informa** business

Cover: Web Large Image (Public).

First edition published 2024
by CRC Press
6000 Broken Sound Parkway NW, Suite 300, Boca Raton, FL 33487-2742

and by CRC Press
4 Park Square, Milton Park, Abingdon, Oxon, OX14 4RN

CRC Press is an imprint of Taylor & Francis Group, LLC

ISBN: 9781032528151 (hbk)
ISBN: 9781032528168 (pbk)
ISBN: 9781003408529 (ebk)

DOI: 10.1201/9781003408529

Typeset in Times
by codeMantra

Access the support material *www.routledge.com/9781032528168*

Contents

Preface

The concept for this book developed over three decades, as I fine-tuned my delivery of the electrical circuits course at the University of Connecticut. Having talked with thousands of students taking the course, I learned what they really wanted and needed to master the subject. They wanted a text that was comprehensive yet compact, concise, and easy to use. They needed solved examples, with a wide cross-section of topics, which attacked the problems from start to finish, and which explained the process in a step-by-step fashion. They needed a source that was practical in its description of analysis techniques, components, design, and methods of measurement.

A Practical Introduction to Electrical Circuits represents a fresh approach that delivers on all counts. The printed book is compact and easy to use, yet offers a comprehensive description of Kirchhoff's laws, nodal and mesh analysis, Thevenin and Norton's theorems, and maximum power transfer for both DC and AC circuits, as well as transient analysis of first- and second-order circuits. Advanced topics such as mutual inductance and transformers, operational amplifier circuits, sequential switching, and three-phase systems reinforce the fundamentals. Approximately one hundred solved examples are included within the printed copy. Extra features online (www.routledge.com/9781032528151) include detailed solutions to over two hundred additional problems and 40 self-service quizzes with solutions and feedback. The ebook+ version includes multiple additional tools for learning, including 12 laboratory exercises, 39 animated presentations on the full range of topics, and 10 example exams with detailed solutions.

I would like to express my sincere gratitude to the thousands of electrical circuits students who inspired this project, as well as my many teaching assistants for their fruitful discussions and suggestions. Most of all, I would like to thank my family for their unending support and patience throughout this endeavor.

John E. Ayers
July 25, 2023
Storrs, CT

Author

John Ayers has been engaged in teaching and academic research for over three decades at the University of Connecticut, where he is a University Teaching Fellow. Ayers considers electrical circuits to be one of his passions. He earned the BSEE from the University of Maine in 1984 and the MSEE and PhDEE from Rensselaer Polytechnic Institute in 1987 and 1990, respectively. His industrial experience includes Fairchild Semiconductor, National Semiconductor, Phillips Laboratories, and Epitax Engineering. Ayers has authored nine books as well as hundreds of research papers, and is the recipient of eight awards for excellence and innovation in teaching.

1 Beginning Concepts and Resistive Circuits

1.1 WHY CIRCUIT THEORY IS IMPORTANT

Circuit theory is an important tool for the design and analysis of electrical systems, which include digital systems and computers; biomedical instruments such as electrocardiographs and blood glucose monitors; imaging systems such as radars, sonars, and ultrasound systems; sensor systems including air quality monitors, seismographs, and strain gauges on transportation systems; communication systems such as the internet, radio, television, and telephones; power systems including nuclear reactors, wind turbines, and photovoltaic panels; and vehicle systems including engine monitors and controls, anti-lock braking and traction control systems, and self-driving systems. Electrical circuit theory is therefore applicable to every subfield of engineering, including the electrical, computer, biomedical, civil, mechanical, automotive, aerospace, and environmental engineering fields.

1.2 ASSUMPTIONS OF CIRCUIT THEORY

Circuit theory is useful for the analysis of devices, circuits, and systems provided that three assumptions are satisfied.

The first assumption of circuit theory is that of a **lumped parameter system**. In engineering terms, this assumption means that the wavelength λ associated with electrical signals is much greater than the physical size of the system, L.

$$\lambda \gg L \tag{1.1}$$

Electrical signals propagate at the speed of light, c, so an electrical signal with frequency f has an associated wavelength:

$$\lambda = \frac{c}{f}. \tag{1.2}$$

For example, at the frequency of 60 Hz used for power distribution in the United States,

$$\lambda = \frac{c}{f} = \frac{3 \times 10^8 \, \text{m/s}}{60 \, \text{s}^{-1}} = 5 \times 10^6 \, \text{m}. \tag{1.3}$$

This is equal to 5000 km, or roughly the width of the continental United States. The first assumption applies for a system with a physical size much smaller than this; if we use "much smaller" to mean a factor of ten, this would correspond to roughly the width of Arizona. Clearly, the first assumption is easily satisfied at low frequencies!

DOI: 10.1201/9781003408529-1

The second assumption is that there is **zero net electrical charge on each component in the system**. This is approximately true for all of the situations we will encounter, although there is one device (the capacitor) which can contain equal and opposite charges within.

The third assumption is that there is **zero magnetic coupling between components**. Although this will generally hold for the circuits we consider, there are important devices which have internal magnetic coupling (the inductor and the transformer).

Going forward, we will assume that these three assumptions apply, so we can use circuit theory for our analysis, rather than resorting to the more complicated electromagnetic theory based on Maxwell's equations. First, though, we will explore the physical origins of basic electrical phenomena.

1.3 CHARGES, VOLTAGES, AND CURRENTS

Electrical circuit phenomena are associated with electrical charges, which are quantified in Coulombs (abbreviated C). These charges are **discrete** (any electrical charge is an integer multiple of the unit charge q, which is the absolute value of the charge of a single electron, where $q = 1.602 \times 10^{-19}$ C) and **bipolar** (charges may be positive or negative).

The physical separation of charge gives rise to **electric fields** (measured in volts per meter, or V/m) and **potential differences**, measured in volts (V). Figure 1.1 shows in a qualitative way the electric field lines and equipotential contours in two dimensions for the case of a single positive charge and a single negative charge separated by some distance. A positive test charge will be repelled by the positive charge on the left and attracted by the negative charge on the right. Hence, the electric field vectors (shown by arrows) point away from the positive charge and toward the negative

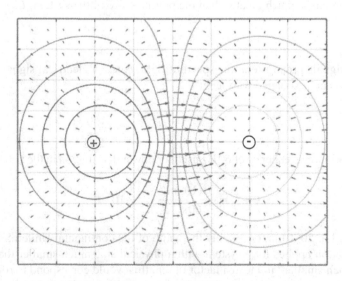

FIGURE 1.1   Separation of charge gives rise to electric fields and potential differences.

charge, in the way that a positive test charge would move. At any point, the electric field is the negative of the gradient of the electric potential.

$$\vec{E} = -\nabla v. \tag{1.4}$$

The equipotential contours curve around the two charges, everywhere perpendicular to the electric field lines. Integration of the electric field $\vec{E}$ between two points a and b gives the electric potential difference v_{ab} between the two points: $v_{ab} = v_a - v_b = -\int_a^b \vec{E}\ \overrightarrow{dr}$. This electric potential difference, usually called the potential difference or voltage, between these points may be related to the incremental amount of work done on an incremental positive test charge to move it from one point to the other:

$$v_{ab} = \frac{dW}{dq}. \tag{1.5}$$

Therefore, if the work is measured in Joules (J) and the charge is measured in Coulombs (C), then the unit of potential difference, Volt (V), can be expressed in other units as $1\ V = 1\ J/C$. In this example, the positive charge on the left is at a higher electric potential than the negative charge on the right, because net work must be done to move a positive test charge from right to left.

The motion of charge gives rise to electrical current. Consider a section of copper wire carrying an electrical current as shown in Figure 1.2. Negatively-charged electrons flow from right to left, giving rise to a conventional current flowing from left to right. (Conventional current is defined with respect to the motion of positive charges.) The electrical current, measured in Amperes (A), is given by the incremental amount of charge dq passing through a cross section of the wire (for example, the cross section shaded in gray) in an incremental amount of time dt:

$$i = \frac{dq}{dt}. \tag{1.6}$$

The unit of electrical current is Ampere (A), and in terms of other units $1\ A = 1\ C/s$.

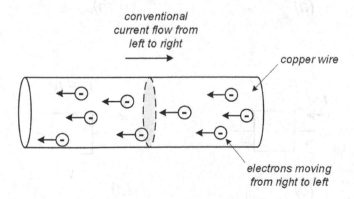

FIGURE 1.2 Motion of charges (negatively-charged electrons in a copper wire) giving rise to an electrical current.

1.4 POWER, ENERGY, AND THE PASSIVE SIGN CONVENTION

Power is the rate at which work is done:

$$p = \frac{dw}{dt}. \tag{1.7}$$

For a two-terminal device with a potential difference across the terminals given by v and an electrical current equal to i flowing into the positive terminal as shown in Figure 1.3a, we can relate the power to the current and voltage as follows:

$$p = \frac{dw}{dt} = \frac{dw}{dq}\frac{dq}{dt} = vi. \tag{1.8}$$

The unit of power is Watt (W). In terms of the other units, $1\ W = 1\ J/s = 1\ VA$.

The sign of the power is important. Here, we will use the **passive sign convention**, so a positive value of power indicates that power is absorbed by the circuit element, whereas negative power indicates that power is extracted from the circuit element. Absorbed power could be dissipated (turned into heat) or stored in the circuit element. Extracted power is either developed (as in a source, such as a battery) or delivered from previously-stored energy (as from a capacitor or an inductor).

To properly use the passive sign convention, we must observe the following rule: if the reference direction of the current is entering the terminal with the plus sign for the voltage reference polarity, we calculate the power by $p = iv$, whereas if the reference direction of the current is entering the terminal with the minus sign for the voltage reference polarity, we should apply $p = -iv$. This is illustrated in Figure 1.3,

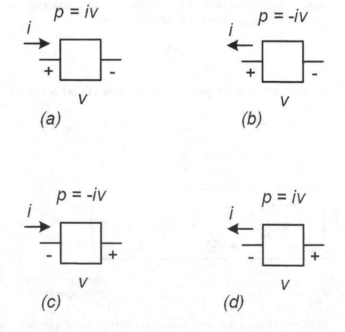

FIGURE 1.3 (a–d) Use of the passive sign convention with assumed reference polarities for the voltage and current.

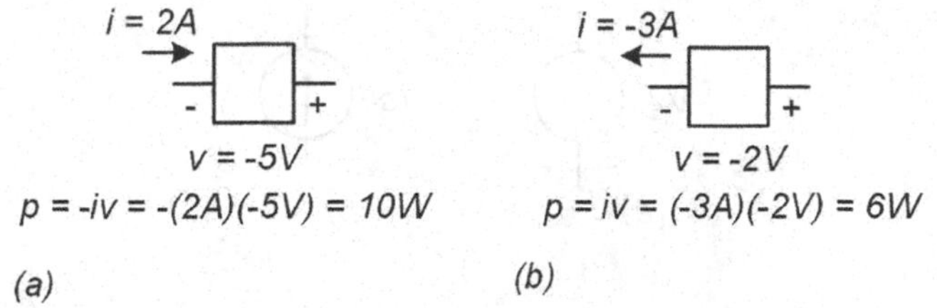

$$i = 2A$$

$$v = -5V$$
$$p = -iv = -(2A)(-5V) = 10W$$

(a)

$$i = -3A$$

$$v = -2V$$
$$p = iv = (-3A)(-2V) = 6W$$

(b)

FIGURE 1.4 (a and b) Examples of application of the passive sign convention.

which shows the four possible permutations for the reference polarities. When applying this rule we must also account for the algebraic signs of the current and the voltage. If the actual current is flowing opposite to the reference direction, we indicate that by assigning the current a negative value. Likewise, if the polarity of the actual voltage is opposite to the reference polarity, we assign a negative value to the voltage. Both of these situations arise commonly, because we often set up reference polarities before we know the actual polarities (i.e., before we have solved the circuit). Two possible situations of this type are illustrated in Figure 1.4.

1.5 INDEPENDENT AND DEPENDENT SOURCES

Sources are two-terminal circuit elements, which supply voltage or current to a circuit. They may be independent (fixed) sources or dependent sources, which are controlled by a voltage or current elsewhere in the circuit.

An independent source may supply voltage or current, so there are two types of independent sources as shown in Figure 1.5a and b. In our notation, an independent source is shown by a circle and the current source is distinguished by an arrow within the circle. Note that a battery can be modeled as an independent voltage source.

We will assume that these sources are ideal. Therefore, the voltage source will provide a fixed voltage regardless of the amount of current we draw, and the current source will output a fixed current regardless of the voltage we apply across its terminals. Any real source has limitations, but we can nonetheless model a real source using an ideal source along with a resistor.

A dependent source may provide voltage or current, and it may be controlled by either a voltage or a current elsewhere in the circuit. This results in four types of dependent sources as shown in Figure 1.5c–f. These are the voltage-controlled voltage source (VCVS), current-controlled voltage source (CCVS), voltage-controlled current source (VCCS), and current-controlled current source (CCCS), respectively.

Dependent sources are important for modeling electronic devices such as transistors and amplifiers. For example, the bipolar junction transistor may be modeled by a CCCS, whereas a field-effect transistor may be modeled by a VCCS.

In our notation, the dependent source has a diamond shape, and a current source is distinguished by the inclusion of an arrow within the diamond. The controlling variable is made clear by the equation of the dependent source, and in Figure 1.5, the controlling variable is referred to as i_x (for a controlling current) or v_x (for a controlling voltage). The coefficient that multiplies the controlling variable has units which are determined by the output and controlling variable for the source. For example,

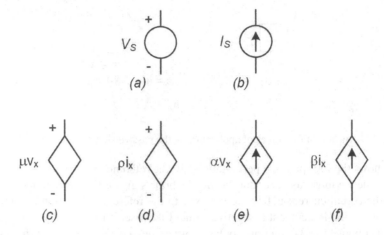

FIGURE 1.5   Independent and dependent sources: (a) independent voltage source, (b) independent current source, (c) VSVS, (d) CCVS, (e) VCCS, and (f) CCCS.

a CCVS has a coefficient μ with units of V/A, whereas a CCCS has a unitless coefficient β (units of A/A). It should be noted that the controlling variable must be labeled in the circuit with its polarity shown.

1.6   THE RESISTOR AND OHM'S LAW

The resistor is a two-terminal circuit element with the symbol as shown in Figure 1.6. It exhibits a linear relationship between voltage and current, given by Ohm's law:

$$v = iR, \tag{1.9}$$

where R is the resistance in units of Ohms, shown by the upper case Greek letter omega ($1\,\Omega = 1\,\text{V/A}$).

Current always flows into the more positive terminal of the resistor. Therefore, according to the passive sign convention, the power absorbed by a resistor is always positive, and the resistor always dissipates power. Referring to the polarity references given in Figure 1.6, we can say that the voltage and current given in the figure must have the same algebraic sign (both positive or both negative). If we change either one of the polarity references, we expect the voltage and current to have opposite signs.

To review, see Presentation 1.1 in ebook+. To test your knowledge, try Quiz 1.1 in ebook+.

1.7   KIRCHHOFF'S CURRENT LAW

Kirchhoff's current law (KCL) is one of the three most important laws of circuit analysis (along with Ohm's law and Kirchhoff's voltage law). Kirchhoff's laws may be understood using some simple circuit definitions. A **node** is a point in the circuit where two or more elements are connected. An **essential node** is one which is connected to more than two elements. A **path** is a trace of elements, none repeated.

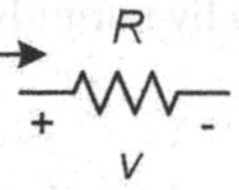

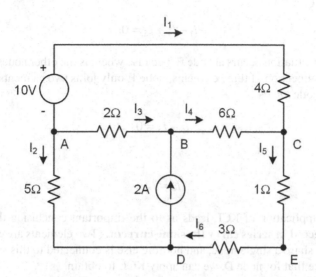

FIGURE 1.6 The resistor.

FIGURE 1.7 A circuit involving sources and resistors for the illustration of KCL.

A **branch** is a path that connects two essential nodes. A **loop** is a path that begins and ends on the same node. A **mesh** is a loop that does not enclose another loop.

One statement of KCL is that the algebraic sum of the currents leaving a node is zero. By "algebraic sum," we mean that we must account for the algebraic signs of the individual currents, some of which may be negative. A second, equivalent statement of KCL is that the algebraic sum of the currents entering a node is zero. However, our convention will be to always add the currents leaving a node, and you are strongly encouraged to use this convention consistently in order to avoid sign errors.

Some examples of the use of KCL may be developed using the circuit diagram of Figure 1.7. For node A, there are three branches connected. Adding the currents leaving this node, we obtain

$$I_1 + I_2 + I_3 = 0. \tag{1.10}$$

Notice that for this to be true for non-zero currents, at least one of the currents involved must be positive and at least one of the currents must be negative. Often we will arbitrarily assign the reference polarities before we know the true directions of any of the currents. This is fine as long as we use the reference polarities consistently after they have been chosen. Once we have solved the circuit, we will find the algebraic signs of the currents and thus discover the true directions of current flow.

We can write additional equations by using KCL for the other nodes in the circuit. For node B,

$$-I_3 - 2\,\text{A} + I_4 = 0, \tag{1.11}$$

and for node C,

$$-I_1 - I_4 + I_5 = 0. \tag{1.12}$$

An important situation occurs at node E, because whereas the other nodes considered involve the connection of three elements, node E only joins two elements. The use of KCL at this node yields

$$I_6 - I_5 = 0 \tag{1.13}$$

or

$$I_6 = I_5. \tag{1.14}$$

This trivial application of KCL leads us to the important conclusion that **two elements connected in series carry the same current**. (Two elements are connected in series if they share a single node, and nothing else is connected to this shared node.) Finally, notice that for node D, we can apply KCL to obtain

$$-I_2 + 2\,\text{A} - I_5 = 0. \tag{1.15}$$

This is not an independent equation, but rather may be obtained by combining the equations for nodes A, B, and C.

1.8   KIRCHHOFF'S VOLTAGE LAW

A third important circuit law is Kirchhoff's voltage law (KVL). **A statement of Kirchhoff's voltage law is that the algebraic sum of the voltages around a loop is zero**. By "loop," we mean a connection of elements, none repeated, which starts and ends at the same point in the circuit. **Our convention will be to proceed clockwise around a loop and add voltage drops**. Operationally, this means that, when proceeding clockwise, we will add a voltage term if we encounter the plus sign first but we will subtract a voltage term if we first arrive at the negative sign. It is strongly recommended that you always use this convention to avoid sign errors. Note that the order of the terms does not matter, so we can start at any point along the loop as long as we end at the same point.

We can develop several applications of KVL using the example circuit of Figure 1.8. For the top loop, proceeding clockwise and adding voltage drops, we obtain

$$-V_A + V_B + 15\,\text{V} - V_C = 0. \tag{1.16}$$

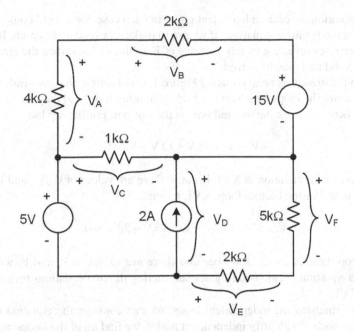

FIGURE 1.8 A circuit with sources and resistors for the demonstration of KVL.

Application of KVL to the loop on the lower left yields

$$-5\,V + V_C + V_D = 0, \tag{1.17}$$

and for the loop on the lower right, we obtain

$$-V_D + V_F - V_E = 0. \tag{1.18}$$

There are additional loops in the circuit, but each of these encloses at least two smaller loops and therefore does not yield an additional independent equation. For example, for the outer loop, we obtain

$$-5\,V - V_A + V_B + 15\,V + V_F - V_E = 0. \tag{1.19}$$

However, this is not a fourth independent equation. Because the outer loop encloses the three smaller loops considered above, its equation may be obtained by simply adding the three previous equations.

1.9 KIRCHHOFF'S LAWS AND THE VALIDITY OF CIRCUIT CONNECTIONS

One application of Kirchhoff's laws is to determine whether a circuit connection is valid. To do this, we write the KVL equation for each independent loop and we write the KCL equation for each independent node. In each case, we should consider combinations of the resulting equations if we are to make an exhaustive search. If neither of these exercises reveals a violation of one of Kirchhoff's laws, then the circuit connection is valid and may be solved.

We can illustrate this analysis using Figure 1.9. This circuit has two **independent loops**; these are the loops which do not enclose another loop, and they are also called **meshes**. There is one on the top and one on the bottom. For the top loop,

$$-V_w + V_x - 18\,\text{V} + 12\,\text{V} - V_y = 0, \tag{1.20}$$

and this reveals no violation of KVL because there are values of V_w, V_x, and V_y which make this true. For the bottom loop, KVL requires

$$V_y - 12\,\text{V} + 10\,\text{V} - V_z + 15\,\text{V} - 20\,\text{V} = 0, \tag{1.21}$$

and here too there is no violation because there are values of V_y and V_z which can satisfy this equation. Neither adding nor subtracting the two equations reveals a violation of KVL.

Having exhausted the independent loops, we can consider the equation for each independent node. To identify independent nodes, we find all of the nodes connected to three or more elements, called the **essential nodes**. We must consider KCL for all but one of these essential nodes. The last essential node will not provide an independent equation, and neither will any of the non-essential nodes connected to only two

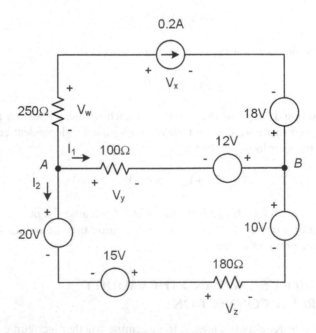

FIGURE 1.9   An example circuit for the determination of its validity.

elements. In the circuit of Figure 1.9, there are two essential nodes labeled "A" and "B." Applying KCL to node A, we obtain

$$0.2\,\text{A} + I_1 + I_2 = 0, \tag{1.22}$$

and this reveals no violation because there are values of I_1 and I_2 which make it true. Having eliminated every possible violation of Kirchhoff's laws, we can say that the circuit of Figure 1.9 is valid and may be solved.

As another example, consider the circuit of Figure 1.10. Here, there are four essential nodes, labeled "A," "B," "C," and "D," and three independent loops, or meshes. To make an exhaustive search for a violation of Kirchhoff's laws, we would write the KCL equation for three of the essential nodes, write the KVL equations for each of the three independent loops, and consider combinations of the equations in each case. The circuit can be considered valid only if the exhaustive search fails to reveal any violation of Kirchhoff's laws, and a single violation of either of Kirchhoff's laws shows that the circuit is not valid. Here, the application of KCL at node A yields

$$1.5\,\text{A} - 1\,\text{A} + 0.5\,\text{A} = 0, \tag{1.23}$$

and this is not true. The circuit is not valid, and we don't need to explore further.

To review, see Presentation 1.2 in ebook+. To test your knowledge, try Quiz 1.2 in ebook+.

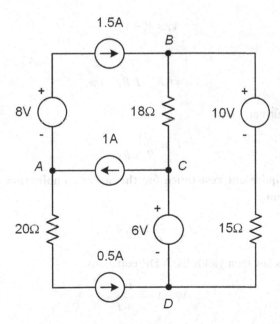

FIGURE 1.10   A second circuit for the determination of its validity.

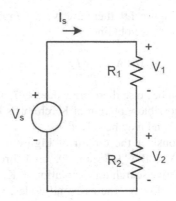

FIGURE 1.11 Two series resistors.

1.10 SERIES RESISTORS AND THE VOLTAGE-DIVIDER RULE

If two or more resistors are connected in series, and the total voltage across the combination is known, we can use the **voltage-divider rule (VDR)** to determine the individual voltages. To see this, consider the connection of two series resistors as shown in Figure 1.11. By KVL,

$$-V_s + V_1 + V_2 = 0. \tag{1.24}$$

Rearranging,

$$V_s = V_1 + V_2. \tag{1.25}$$

By Ohm's law,

$$V_s = I_s R_1 + I_s R_2 = 0. \tag{1.26}$$

Solving for the current,

$$I_s = \frac{V_s}{R_1 + R_2}. \tag{1.27}$$

Therefore, **the equivalent resistance for the series combination of resistors is equal to their sum**:

$$R_{eq} = \frac{V_s}{I_s} = R_1 + R_2. \tag{1.28}$$

The use of Ohm's law then yields the VDR equations:

$$V_1 = V_s \frac{R_1}{R_1 + R_2} \tag{1.29}$$

and

$$V_2 = V_s \frac{R_2}{R_1 + R_2}. \tag{1.30}$$

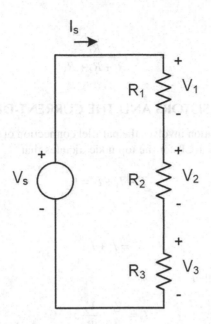

FIGURE 1.12 Three series resistors.

Therefore, a statement of the VDR is that: **when two resistances are connected in series, the fraction of the total voltage appearing across one of the resistors is equal to the fraction of the total resistance contained in this resistor.**

We can extend the VDR to three or more resistors as well, and Figure 1.12 shows the case of three series resistors. In this case,

$$-V_s + I_s R_1 + I_s R_2 + I_s R_3 = 0, \tag{1.31}$$

$$I_s = \frac{V_s}{R_1 + R_2 + R_3}, \tag{1.32}$$

and the equivalent resistance for the series combination is the sum of the individual resistances:

$$R_{eq} = \frac{V_s}{I_s} = R_1 + R_2 + R_3. \tag{1.33}$$

The use of Ohm's law then yields the VDR equations:

$$V_1 = V_s \frac{R_1}{R_1 + R_2 + R_3}, \tag{1.34}$$

$$V_2 = V_s \frac{R_2}{R_1 + R_2 + R_3}, \tag{1.35}$$

and

$$V_3 = V_s \frac{R_3}{R_1 + R_2 + R_3}. \tag{1.36}$$

1.11   PARALLEL RESISTORS AND THE CURRENT-DIVIDER RULE

Another important situation involves the parallel connection of two or more resistors, as shown in Figure 1.13. KCL for the top node dictates that

$$-I_s + I_1 + I_2 = 0 \tag{1.37}$$

or

$$I_s = I_1 + I_2. \tag{1.38}$$

By Ohm's law, we find that

$$I_s = \frac{V_s}{R_1} + \frac{V_s}{R_2}. \tag{1.39}$$

The equivalent resistance for the parallel combination is

$$R_{eq} = \frac{V_s}{I_s} = \left(\frac{1}{R_1} + \frac{1}{R_2} \right)^{-1}. \tag{1.40}$$

In other words, **the equivalent resistance for two parallel resistors is equal to the reciprocal of the sum of their reciprocals**. In the particular case of two parallel resistors, we can rewrite this equation in an equivalent form:

$$R_{eq} = \frac{R_1 R_2}{R_1 + R_2}. \tag{1.41}$$

For two parallel resistors, the equivalent resistance is equal to their product divided by their sum.

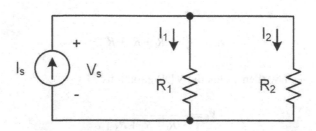

FIGURE 1.13   Two parallel resistors.

When two resistances act in parallel, we can understand how the total current splits between the two parallel resistors using the **current-divider rule (CDR).** For the situation shown in Figure 1.13,

$$V_s = I_s R_{eq} = I_s \frac{R_1 R_2}{R_1 + R_2}. \tag{1.42}$$

The use of Ohm's law leads us to the current-divider equations:

$$I_1 = I_s \frac{R_2}{R_1 + R_2} \tag{1.43}$$

and

$$I_2 = I_s \frac{R_1}{R_1 + R_2}. \tag{1.44}$$

A statement of the CDR is as follows: when two resistors are connected in parallel, the fraction of the total current flowing through one of the resistors is equal to the value of the opposite resistor divided by the sum of the two resistances.

We can generalize the CDR to three or more resistors in parallel as well. Consider the case of four parallel resistors as shown in Figure 1.14a. To find the current, we can combine three of these resistors (R_2, R_3, and R_4) in parallel as shown in Figure 1.14b. This yields the current-divider relation

$$I_1 = I_s \frac{R_2 \parallel R_3 \parallel R_4}{R_1 + R_2 \parallel R_3 \parallel R_4}. \tag{1.45}$$

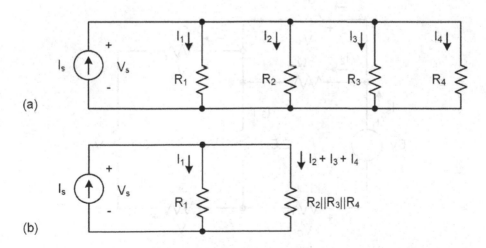

(a)

(b)

FIGURE 1.14 (a) Four resistors connected in parallel and (b) simplified circuit created by combining three of the resistors.

We can determine the other currents in a similar fashion:

$$I_2 = I_s \frac{R_1 \parallel R_3 \parallel R_4}{R_2 + R_1 \parallel R_3 \parallel R_4},$$ (1.46)

$$I_3 = I_s \frac{R_1 \parallel R_2 \parallel R_4}{R_3 + R_1 \parallel R_2 \parallel R_4},$$ (1.47)

and

$$I_4 = I_s \frac{R_1 \parallel R_2 \parallel R_3}{R_4 + R_1 \parallel R_2 \parallel R_3}.$$ (1.48)

1.12 SERIES–PARALLEL COMBINATIONS OF RESISTORS

Often a network involving a single source and a number of resistors may be simplified by combining parallel- and series-connected resistors. If the network can be simplified to a single source and a single resistance, it may then be solved completely by repeated applications of Ohm's law, the VDR, and the CDR. An example is shown in Figure 1.15.

Here, the equivalent resistance connected to the source is

$$R_{eq} = 3\,\Omega \parallel 6\,\Omega + 2.5\,\Omega \parallel (2\,\Omega + 3\,\Omega + 5\,\Omega) + 1\,\Omega$$

$$= 3\,\Omega \parallel 6\,\Omega + 2.5\,\Omega \parallel 10\,\Omega + 1\,\Omega = 2\,\Omega + 2\,\Omega + 1\,\Omega = 5\,\Omega.$$ (1.49)

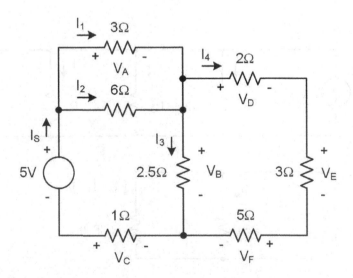

FIGURE 1.15 Circuit involving parallel and series combinations of resistors with a single source.

The source current can be found by Ohm's law:

$$I_s = \frac{5\,\text{V}}{R_{\text{eq}}} = \frac{5\,\text{V}}{5\,\Omega} = 1\,\text{A}. \tag{1.50}$$

The currents I_1 and I_2 may be found by using the CDR:

$$I_1 = I_S \frac{6\,\Omega}{6\,\Omega + 3\,\Omega} = \left(1\,\text{A}\right)\frac{6\,\Omega}{6\,\Omega + 3\,\Omega} = 0.667\,\text{A}, \tag{1.51}$$

and

$$I_2 = I_S \frac{3\,\Omega}{6\,\Omega + 3\,\Omega} = \left(1\,\text{A}\right)\frac{3\,\Omega}{6\,\Omega + 3\,\Omega} = 0.333\,\text{A}. \tag{1.52}$$

The voltage V_A may be found by using Ohm's law:

$$V_A = I_1\left(3\,\Omega\right) = \left(0.667\,\text{A}\right)\left(3\,\Omega\right) = 2.00\,\text{V}. \tag{1.53}$$

We can find I_3 and I_4 by two more applications of the CDR:

$$I_3 = I_S \frac{10\,\Omega}{10\,\Omega + 2.5\,\Omega} = \left(1\,\text{A}\right)\frac{10\,\Omega}{10\,\Omega + 2.5\,\Omega} = 0.800\,\text{A} \tag{1.54}$$

and

$$I_4 = I_S \frac{2.5\,\Omega}{10\,\Omega + 2.5\,\Omega} = \left(1\,\text{A}\right)\frac{2.5\,\Omega}{10\,\Omega + 2.5\,\Omega} = 0.200\,\text{A}. \tag{1.55}$$

The voltages V_B and V_C may be found using Ohm's law:

$$V_B = I_3\left(2.5\,\Omega\right) = \left(0.800\,\text{A}\right)\left(2.5\,\Omega\right) = 2.00\,\text{V} \tag{1.56}$$

and

$$V_C = -I_S\left(1\,\Omega\right) = -\left(1.000\,\text{A}\right)\left(1\,\Omega\right) = -1.00\,\text{V}. \tag{1.57}$$

The remaining voltages may be found through the use of the VDR:

$$V_D = V_B \frac{2\,\Omega}{2\,\Omega + 3\,\Omega + 5\,\Omega} = \left(2.00\,\text{V}\right)\frac{2\,\Omega}{2\,\Omega + 3\,\Omega + 5\,\Omega} = 0.40\,\text{V}, \tag{1.58}$$

$$V_E = V_B \frac{3\,\Omega}{2\,\Omega + 3\,\Omega + 5\,\Omega} = \left(2.00\,\text{V}\right)\frac{3\,\Omega}{2\,\Omega + 3\,\Omega + 5\,\Omega} = 0.60\,\text{V}, \tag{1.59}$$

and

$$V_F = V_B \frac{5\,\Omega}{2\,\Omega + 3\,\Omega + 5\,\Omega} = \left(2.00\,\text{V}\right)\frac{5\,\Omega}{2\,\Omega + 3\,\Omega + 5\,\Omega} = 1.00\,\text{V}. \tag{1.60}$$

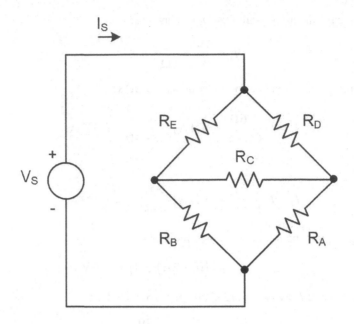

FIGURE 1.16 Wheatstone bridge circuit.

1.13 OTHER CONFIGURATIONS AND DELTA–WYE AND WYE–DELTA TRANSFORMATIONS

Some resistive circuits contain connections, which are neither parallel nor series. They may not be analyzed by simple applications of Ohm's law, the VDR, and the CDR, but instead require additional tools. An example is the Wheatstone bridge circuit as shown in Figure 1.16. This circuit has no pair of resistors, which may be combined in series; this would require the pair of resistors to share a single node with nothing else connected to this shared node. For example, the resistors R_E and R_B may not be combined in series because of the presence of R_C. The circuit also has no pair of resistors which may be combined in parallel; this would require that the resistors be connected to the same two nodes. For example, the resistors R_E and R_D may not be combined in parallel because they share only one node.

A circuit like the Wheatstone bridge may be analyzed by first using a delta-to-wye or a wye-to-delta transformation if this renders a transformed circuit, which may then be analyzed by consideration of series and parallel combinations of resistors. To illustrate this, we will first consider the basic delta-to-wye and wye-to-delta transformations and then apply the delta-to-wye transformation to a Wheatstone bridge.

Consider three delta-connected resistors and three wye-connected resistors, shown in Figure 1.17 on the left and right, respectively. If the resistors R_A, R_B, and R_C are known, we can find values of R_1, R_2, and R_3 such that the delta and wye configurations will be externally equivalent. (By "externally equivalent," we mean that all currents and voltages outsider of the terminals A, B, and C will be unchanged if we

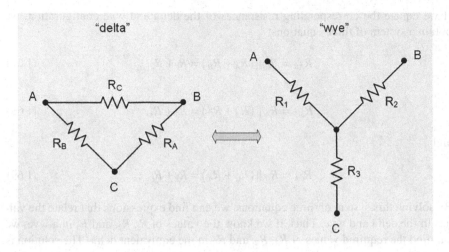

FIGURE 1.17 Delta-to-wye and wye-to-delta transformations.

replace the delta with the wye.) Similarly, if we know the values of R_1, R_2, and R_3, we can find values of R_A, R_B, and R_C such that the delta and wye configurations will be externally equivalent.

If the delta and wye configurations are externally equivalent, then it must be true that the resistance measured between any pair of terminals is the same for the delta and the wye. For the delta, the resistances measured between the pairs of terminals are

$$R_{AB} = R_C \parallel (R_A + R_B), \tag{1.61}$$

$$R_{BC} = R_A \parallel (R_B + R_C), \tag{1.62}$$

and

$$R_{CA} = R_B \parallel (R_C + R_A). \tag{1.63}$$

where R_{AB} is the resistance between the terminals A and B, R_{BC} is the resistance between the terminals B and C, and R_{CA} is the resistance between the terminals C and A. Similarly, for the wye connection, the resistances are given by

$$R_{AB} = R_1 + R_2, \tag{1.64}$$

$$R_{BC} = R_2 + R_3, \tag{1.65}$$

and

$$R_{CA} = R_3 + R_1. \tag{1.66}$$

If we equate the corresponding resistances of the delta and wye configuration, we obtain a system of three equations:

$$R_{AB} = R_C \parallel (R_A + R_B) = R_1 + R_2, \tag{1.67}$$

$$R_{BC} = R_A \parallel (R_B + R_C) = R_2 + R_3, \tag{1.68}$$

and

$$R_{CA} = R_B \parallel (R_C + R_A) = R_3 + R_1. \tag{1.69}$$

By solving this system of three equations, we can find expressions that relate the values in the delta and wye. Thus, if we know the values of R_1, R_2, and R_3 in a wye, we can find the required values of R_A, R_B, and R_C in the equivalent delta. The solution is

$$R_A = \frac{R_1 R_2 + R_2 R_3 + R_3 R_1}{R_1}, \tag{1.70}$$

$$R_B = \frac{R_1 R_2 + R_2 R_3 + R_3 R_1}{R_2}, \tag{1.71}$$

and

$$R_C = \frac{R_1 R_2 + R_2 R_3 + R_3 R_1}{R_3}. \tag{1.72}$$

Similarly, if we know the values of R_A, R_B, and R_C in a delta, we can find the required values of R_1, R_2, and R_3 in the equivalent wye. They are given by

$$R_1 = \frac{R_B R_C}{R_A + R_B + R_C}, \tag{1.73}$$

$$R_2 = \frac{R_C R_A}{R_A + R_B + R_C}, \tag{1.74}$$

and

$$R_3 = \frac{R_A R_B}{R_A + R_B + R_C}. \tag{1.75}$$

Now we turn to a specific example of a Wheatstone bridge (shown in Figure 1.18) and show how it can be solved by making use of a delta-to-wye transformation.

To solve this circuit, we will first undertake a delta-to-wye transformation on the delta comprising R_A, R_B, and R_C. The resistors in the transformed wye section are given by

$$R_1 = \frac{R_B R_C}{R_A + R_B + R_C} = \frac{(175\,\Omega)(150\,\Omega)}{350\,\Omega + 175\,\Omega + 150\,\Omega} = 38.9\,\Omega, \tag{1.76}$$

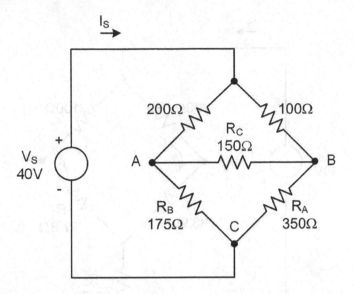

FIGURE 1.18 Wheatstone bridge circuit.

$$R_2 = \frac{R_C R_A}{R_A + R_B + R_C} = \frac{(150\,\Omega)(350\,\Omega)}{350\,\Omega + 175\,\Omega + 150\,\Omega} = 77.8\,\Omega, \qquad (1.77)$$

and

$$R_3 = \frac{R_A R_B}{R_A + R_B + R_C} = \frac{(350\,\Omega)(175\,\Omega)}{350\,\Omega + 175\,\Omega + 150\,\Omega} = 90.7\,\Omega. \qquad (1.78)$$

The transformed circuit therefore takes the form as shown in Figure 1.19, and this circuit may be analyzed readily because it involves only simple combinations of parallel and series branches.

The equivalent resistance for the network is given by

$$R_{eq} = (200\,\Omega + 38.9\,\Omega) \,\|\, (100\,\Omega + 77.8\,\Omega) + 90.7\,\Omega = 192.6\,\Omega. \qquad (1.79)$$

The source current is given by

$$I_S = \frac{V_S}{R_{eq}} = \frac{40\,\text{V}}{192.6\,\Omega} = 0.2077\,\text{A}. \qquad (1.80)$$

We can determine the branch currents and by use of the CDR:

$$I_1 = 0.2077\,\text{A}\left(\frac{100\,\Omega + 77.8\,\Omega}{200\,\Omega + 38.9\,\Omega + 100\,\Omega + 77.8\,\Omega}\right) = 0.0886\,\text{A} \qquad (1.81)$$

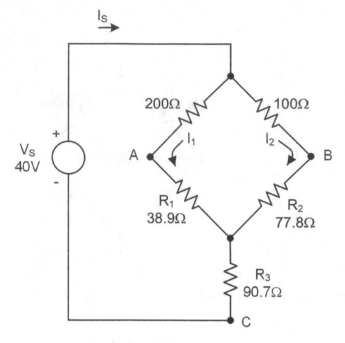

FIGURE 1.19   Transformed Wheatstone bridge circuit.

and

$$I_2 = 0.2077\,\text{A}\left(\frac{200\,\Omega + 38.9\,\Omega}{200\,\Omega + 38.9\,\Omega + 100\,\Omega + 77.8\,\Omega}\right) = 0.1191\,\text{A}. \qquad (1.82)$$

Once we know the three unique branch currents, we can find all of the voltages in the circuit by use of Ohm's law.

To review, see Presentation 1.3 in ebook+. To test your knowledge, try Quiz 1.3 in ebook+. To put your knowledge to practice, try Laboratory Exercise 1.1 in ebook+.

1.14   SUPERPOSITION

A circuit with multiple sources can not always be solved simply by using repeated applications of the CDR and VDR. Usually, an additional tool must be used, and one such tool is the **principle of superposition**.

The principle of superposition states that **when a linear system is driven by multiple independent sources, the overall response is the sum of the individual responses**. In a circuit, the response to be found is a voltage or current. By finding the contributions to this voltage or current associated with each independent source, and summing them, we can find the value of the voltage or current in question.

To see how this works, we can consider the determination of I_x in the circuit of Figure 1.20. Here, there are three independent sources, so we should find the contribution to I_x from each and sum these together. To determine the contribution I_x' associated with the 10 V source, we disable the 2 A and 20 V sources and apply some combination of the CDR, VDR, and Ohm's law. (To disable a voltage source, we set it to zero volts or a short circuit; to disable a current source, we set it to zero amperes or an open circuit.) Then, we find the contribution I_x'' associated with the 2 A source by disabling the other two sources, and finally we find the contribution I_x''' from the 20 V source by disabling the 10 V and 2 A sources. The value of I_x is then found by adding the three contributions: $I_x = I_x' + I_x'' + I_x'''$.

To find the contribution from the 10 V source, we open-circuit the 2 A source and short-circuit the 20 V source, as shown in Figure 1.21.

By the CDR,

$$I_x' = -\underbrace{\left(\frac{10\,\text{V}}{10\,\Omega + 20\,\Omega \,\|\, (2\,\Omega + 5\,\Omega + 3\,\Omega)}\right)}_{\text{total current flowing from the 10 V source}} \times \underbrace{\left(\frac{2\,\Omega + 5\,\Omega + 3\,\Omega}{2\,\Omega + 5\,\Omega + 3\,\Omega + 20\,\Omega}\right)}_{\text{fraction of total current flowing in 20 }\Omega\text{ resistor}} = -0.200\,\text{A}.$$

(1.83)

To find the contribution from the 2 A source, we short-circuit the 10 V and 20 V sources, as shown in Figure 1.22.

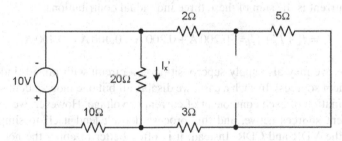

FIGURE 1.20   A circuit for the application of the principle of superposition.

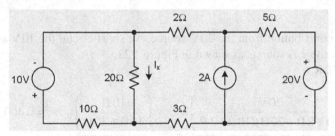

FIGURE 1.21   Determination of the contribution I_x' from the 10 V source.

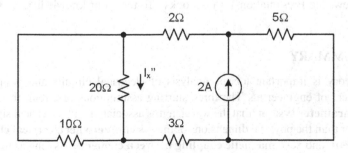

FIGURE 1.22   Determination of the contribution I_x'' from the 2 A source.

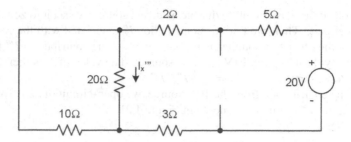

FIGURE 1.23   Determination of the contribution I_x''' from the 20 V source.

Here, it is necessary to apply the CDR twice:

$$I_x'' = 2\,\text{A} \times \underbrace{\left(\frac{5\,\Omega}{5\,\Omega + 2\,\Omega + 20\,\Omega \,\|\, 10\,\Omega + 3\,\Omega} \right)}_{\text{fraction of total current flowing in }2\,\Omega\text{ resistor}} \times \underbrace{\left(\frac{10\,\Omega}{10\,\Omega + 20\,\Omega} \right)}_{\text{fraction of current in }2\,\Omega\text{ which flows in }20\,\Omega} = 0.200\,\text{A}.$$

$$(1.84)$$

To find the contribution from the 20 V source, we short-circuit the 10 V source and open-circuit the 2 A source, as shown in Figure 1.23.

By the CDR,

$$I_x''' = \underbrace{\left(\frac{20\,\text{V}}{5\,\Omega + 2\,\Omega + 20\,\Omega \,\|\, 10\,\Omega + 3\,\Omega} \right)}_{\text{total current flowing from the 10 V source}} \times \underbrace{\left(\frac{10\,\Omega}{10\,\Omega + 20\,\Omega} \right)}_{\text{fraction of total current flowing in }20\,\Omega} = 0.400\,\text{A.} \quad (1.85)$$

The total current is the sum of these three individual contributions:

$$I_x = I_x' + I_x'' + I_x''' = -0.200\,\text{A} + 0.200\,\text{A} + 0.400\,\text{A} = 0.400\,\text{A}. \qquad (1.86)$$

In principle, we may also apply superposition to a circuit with mixed (independent and dependent sources). In such a case, we disable all but one independent source for the determination of each component of current or voltage. However, we must leave all dependent sources active, and this process doesn't lend itself to simple applications of the VDR and CDR. Instead, it is often better to apply the node voltage method, described in the next chapter.

To review, see Presentation 1.4 in ebook+. To test your knowledge, try Quiz 1.4 in ebook+.

1.15   SUMMARY

Circuit theory is important for the analysis of electrical circuits and is applicable to every field of engineering. The three starting assumptions of circuit theory are a **lumped parameter system** (that the wavelengths associated with electrical signals are much larger than the physical dimensions of the system), **zero net electrical charge on each element**, and **zero magnetic coupling between elements**. As long as these hold, we may apply circuit theory instead of resorting to the full electromagnetic theory.

Electrical circuit phenomena are associated with electrical charges, which are **discrete** (in increments of the electronic charge q) and **bipolar** (may be positive or negative). Separation of charge gives rise to electric fields (in V/m) and electric potentials (in V). Motion of charge gives rise to electrical current (in A).

Power, or the rate at which work is done, may be related to the product of the voltage and current. Using the **passive sign convention**, if the current and voltage reference polarities for a two-terminal element are such that the current reference direction enters the terminal with the positive voltage reference, then $p = iv$, where i is the current and v is the voltage. If we reverse either reference polarity, so the reference direction for current is entering the terminal with the negative voltage reference, then $p = -iv$. We must take into account the algebraic signs of both the voltage and current as well, either may be negative. Once we find the sign of the power, a positive value means that the element absorbs power but a negative value means that power is extracted from the element.

There are independent (fixed) and dependent sources. Independent sources may supply voltage or current, leading to two types. Dependent sources may supply voltage or current and may be controlled by either a voltage or a current, leading to four basic types; these are the VCVS, CCVS, VCCS, and CCCS.

The three most important laws for circuit analysis are Ohm's law, KCL, and KVL. The resistor is a linear element which follows Ohm's law, which states that $v = iR$ where R is the resistance in Ω.

Kirchhoff's laws may be understood using some simple circuit definitions. A **node** is a point in the circuit where two or more elements are connected. An **essential node** is one which is connected to more than two elements. A **path** is a trace of elements, none repeated. A **branch** is a path that connects two essential nodes. A **loop** is a path that begins and ends on the same node. A **mesh** is a loop that does not enclose another loop.

KCL states that the sum of the branch currents leaving a node is zero. KVL states that the sum of the voltages around a closed path (a loop or a mesh) is zero. Our convention is to proceed clockwise around such a closed path and to add voltage drops. (If we encounter the plus sign first, we add the voltage term.)

For **resistors in series**, the equivalent resistance is the sum of the individual resistances. The **VDR** may be used to find how the total voltage across series resistances splits between the individual resistances. It states that the fraction of the voltage appearing across a resistor is equal to the fraction of total resistance in that resistor. For **resistors in parallel**, the equivalent resistance is the reciprocal of the sum of the reciprocals of the individual resistances. The **CDR** may be used to find how a total current splits among parallel resistances. It states that the fraction of total current flowing in one parallel resistor is equal to the equivalent resistance for the other resistors divided by the sum of the equivalent resistance for the other resistors and the resistance for the branch in question.

Some configurations or resistors may not be simplified by the use of parallel and series combinations. These configurations generally contain wye-connected or delta-connected combinations of resistors, and an example is the **Wheatstone bridge**. Circuits of this type may be simplified by using **delta–wye or wye–delta transformations**.

The **principle of superposition** is a useful tool for solving some circuit problems. It states that when a linear system is driven by more than one independent source, the total response is the sum of the individual responses associated with each of the independent sources. Here, the total response is a voltage or a current, and we can find it by summing the individual responses, each determined by leaving one independent source active while deactivating all others.

PROBLEMS

Problem 1.1. Consider the electrical circuit of Figure P1.1 containing a battery and two resistors.

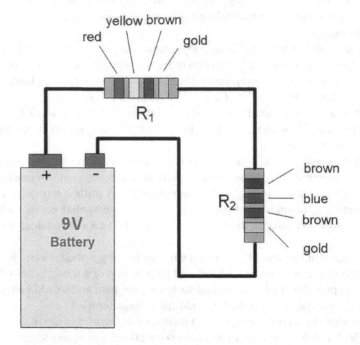

FIGURE P1.1 Circuit involving a battery and resistors.

a. Does conventional current enter or leave the positive terminal of the battery?
b. Do electrons enter or leave the positive terminal of the battery?
c. What is the absolute value of the conventional current flowing in the circuit?
d. What is the power for the battery, and does the battery develop or dissipate power?

Problem 1.2. Consider the three-battery circuit shown in Figure P1.2.

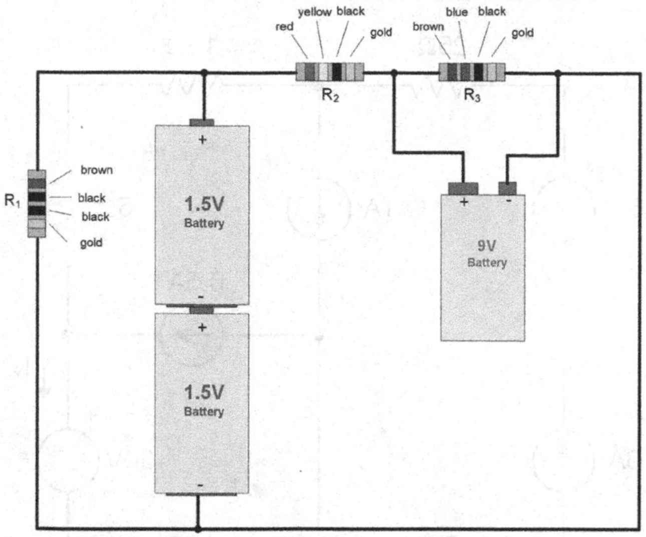

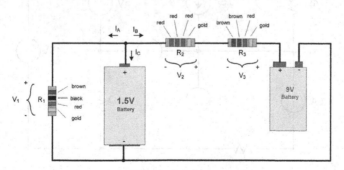

FIGURE P1.2 Circuit involving three batteries and three resistors.

a. Does conventional current enter or leave the top terminal of the upper 1.5 V battery?
b. Do electrons enter or leave the top terminal of the upper 1.5 V battery?
c. Does conventional current enter or leave the positive terminal of the 9 V battery?
d. Do electrons enter or leave the positive terminal of the 9 V battery?
e. Find the power for each of the 1.5 V batteries.
f. Find the power for the 9 V battery.

Problem 1.3. Consider the two-battery circuit shown in Figure P1.3.

FIGURE P1.3 Circuit involving two batteries and three resistors.

a. Find I_A, I_B, and I_C.
b. Determine the voltages V_1, V_2, and V_3.
c. Determine the power for each of the batteries.

Problem 1.4. Determine if the circuit connection of Figure P1.4 is valid. If it is valid, solve for the value of I_X.

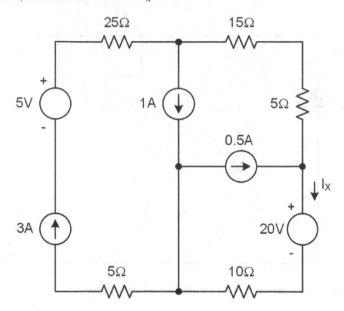

FIGURE P1.4   Circuit with three meshes and four essential nodes to test for validity.

Problem 1.5. Determine if the circuit connection of Figure P1.5 is valid. If it is valid, solve for the value of V_X.

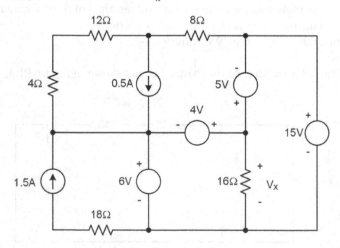

FIGURE P1.5   Circuit with five meshes and five essential nodes to test for validity.

Problem 1.6. Determine if the circuit connection of Figure P1.6 is valid. If it is valid, solve for the value of V_Z.

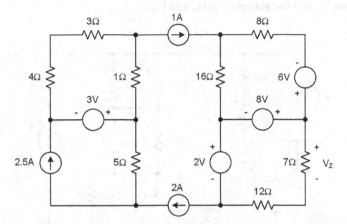

FIGURE P1.6   Circuit with five meshes and eight essential nodes to test for validity.

Problem 1.7. Use the ad hoc application of Kirchhoff's laws and Ohm's law to solve the circuit shown in Figure P1.7. Thus, determine the currents I_A, I_B, I_C, I_D, I_E, I_F, and I_G, and the voltages V_X, V_Y, and V_Z.

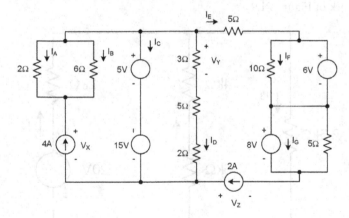

FIGURE P1.7   Circuit involving sources and resistors.

Problem 1.8. Use the ad hoc application of Kirchhoff's laws and Ohm's law to solve the circuit of Figure P1.8. Thus, determine the currents I_A, I_B, I_C, I_D, I_E, I_F, and I_G, and the voltages V_X, V_Y, and V_Z.

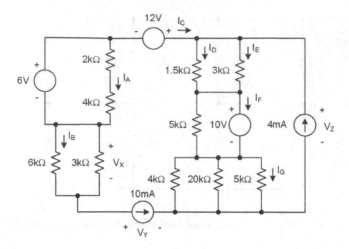

FIGURE P1.8   Complex circuit involving many independent sources and resistors.

Problem 1.9. Determine the currents I_1, I_2, and I_3, and the voltage V_X in the network of Figure P1.9.

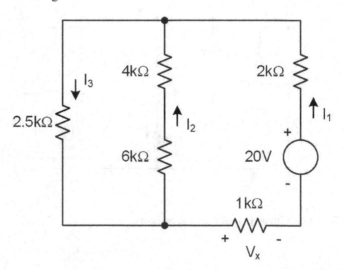

FIGURE P1.9   Circuit containing a single voltage source and five resistances.

Problem 1.10. Find the voltages V_1, V_2, and V, and the current I_X for the circuit of Figure P1.10.

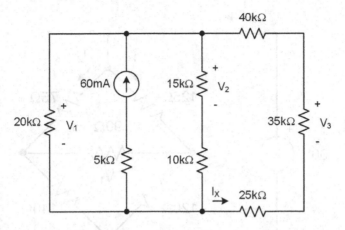

FIGURE P1.10 Circuit containing a single current source and seven resistances.

Problem 1.11. Determine the currents I_A, I_B, and I_C, and the voltage V_X for the circuit shown in Figure P1.11.

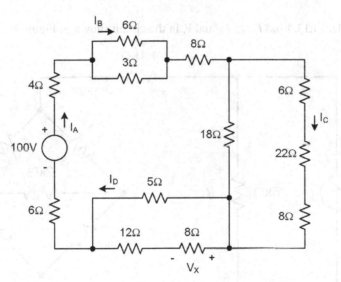

FIGURE P1.11 Circuit containing a single voltage source and twelve resistances.

Problem 1.12. Find I_S, I_1, I_2, and V_1 in the circuit of Figure P1.12.

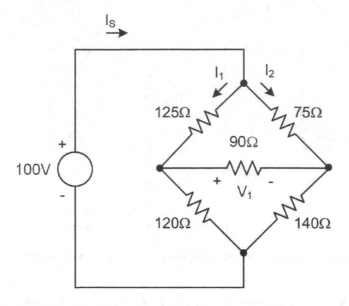

FIGURE P1.12   Wheatstone bridge circuit.

Problem 1.13. Find I_S, I_1, I_2, and V_1 in the circuit shown in Figure P1.13.

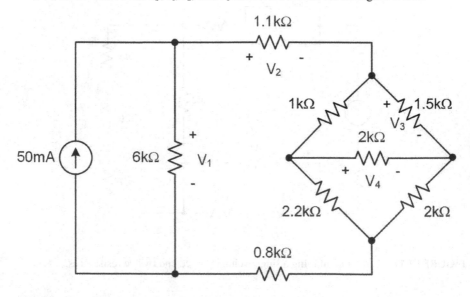

FIGURE P1.13   Resistive circuit containing a Wheatstone bridge.

Problem 1.14. Determine V_A, V_B, V_C, and the power for the voltage source in the circuit of Figure P1.14.

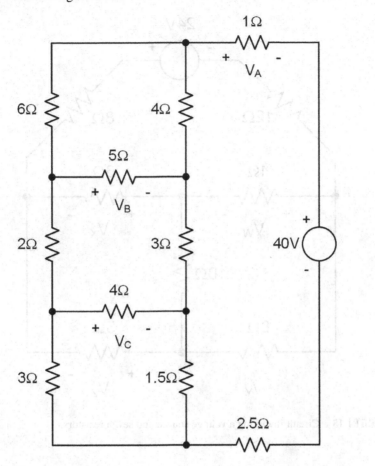

FIGURE P1.14   Circuit involving a voltage source and ten resistors.

Problem 1.15. Calculate the four voltages V_W, V_X, V_Y, and V_Z in the circuit shown in Figure P1.15.

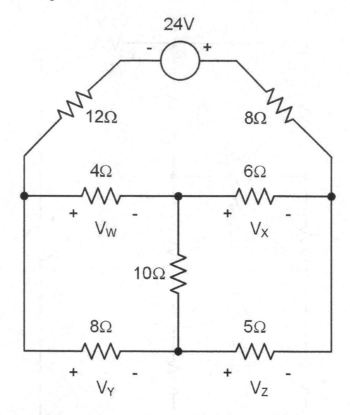

FIGURE P1.15   Circuit involving a voltage source and seven resistors.

Problem 1.16. Determine I_X and the voltages V_1, V_2, and V_3 in the circuit shown in Figure P1.16.

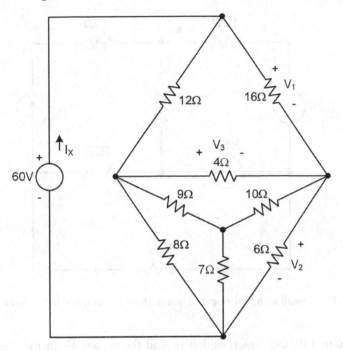

FIGURE P1.16   Circuit with a voltage source and resistors.

Problem 1.17. Use superposition to find the current I_X in the circuit of Figure P1.17.

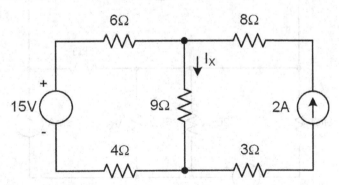

FIGURE P1.17   Circuit with independent sources and resistors.

Problem 1.18. Use superposition to find the current I_X in the circuit shown in Figure P1.18.

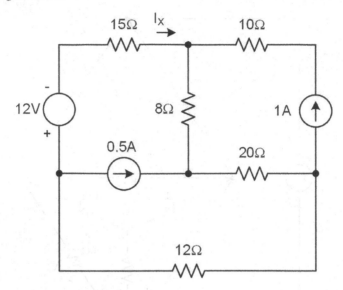

FIGURE P1.18   Circuit with four essential nodes, three sources, and five resistors.

Problem 1.19. Use superposition to find the voltage V_X in the circuit of Figure P1.19.

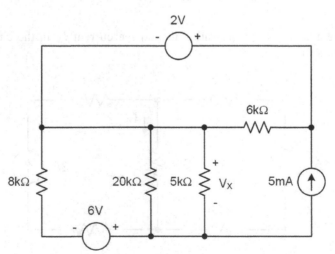

FIGURE P1.19   Circuit with three essential nodes.

Problem 1.20. Use superposition to find the current I_Z in the circuit shown in Figure P1.20, and thereby determine the power for the voltage source.

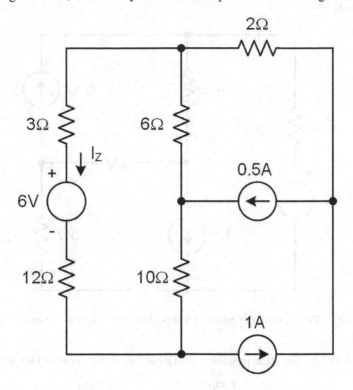

FIGURE P1.20 Circuit with four essential nodes, three sources, and resistors.

Problem 1.21. Use superposition to determine how much the current I_Y in the circuit of Figure P1.21 will change if the 3 A source is changed to 2 A.

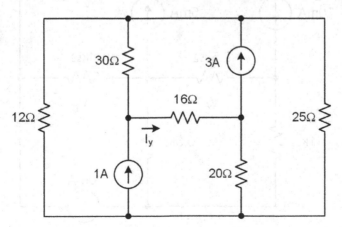

FIGURE P1.21 Circuit with four essential nodes.

Problem 1.22. Use superposition to determine how much the current I_Y in the circuit of Figure P1.22 will change if the voltage source is reversed in polarity.

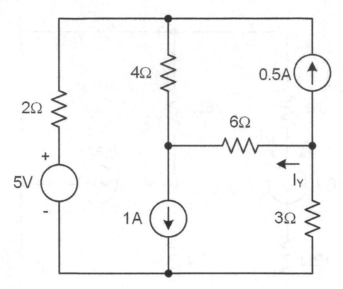

FIGURE P1.22   Circuit with four essential nodes, three sources, and four resistors.

Problem 1.23. Use superposition to find V_A and V_B in the circuit of Figure P1.23.

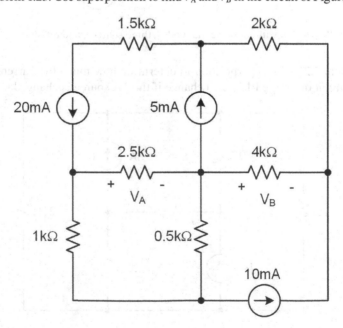

FIGURE P1.23   Circuit with five essential nodes.

Problem 1.24. Use superposition to determine how much the voltage V_Q in the circuit of Figure P1.24 will change if the current source is reversed in polarity.

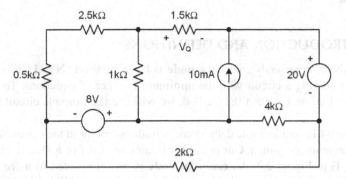

FIGURE P1.24 Circuit with five essential nodes, three sources, and six resistors.

Problem 1.25. Use superposition to calculate how much the voltage V_Z in the circuit of Figure P1.25 will change if the voltage source is changed from 4 V to 8 V.

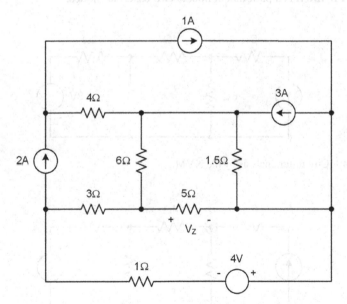

FIGURE P1.25 Circuit with four independent sources and resistors.

2 Nodal Analysis

2.1 INTRODUCTION AND DEFINITIONS

Nodal analysis, commonly called the **node voltage method (NVM)**, is a standard method for solving a circuit with the minimum number of equations. To establish some definitions and explain the method, we will use the example circuit shown in Figure 2.1.

First, we will consider basic definitions. A **node** is a point in the circuit where two or more elements are joined. Our example circuit contains five nodes, indicated and labeled A–E in Figure 2.2. An **essential node** is one connected to more than two elements. Our circuit contains three essential nodes: B, C, and E. Nodes A and D are non-essential because each of them is connected to only two elements. (It is important to note that node E is a single node, even though it is drawn with two "solder blobs." This is because there is only wire between these solder blobs, and the circuit could be redrawn with wires coming together at a single point on the bottom if we tilted the 6 Ω and 3 Ω resistors.) A **path** is a trace of adjoining elements with none repeated. A **branch** is a path that connects two essential nodes.

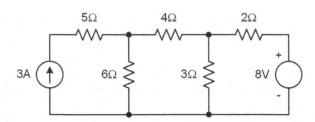

FIGURE 2.1 Circuit for analysis by the NVM.

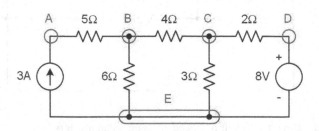

FIGURE 2.2 Identification of nodes in the example circuit.

DOI: 10.1201/9781003408529-2

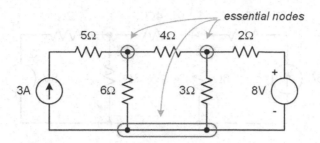

FIGURE 2.3 Identification of essential nodes in the example circuit.

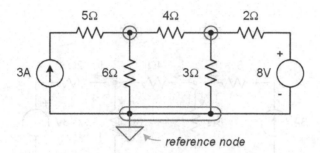

FIGURE 2.4 Choice of a reference node in the example circuit.

2.2 THE BASIC NODE VOLTAGE METHOD (NVM)

To illustrate the node voltage, we will solve the circuit in Figure 2.1. This involves six steps. The first step is to identify the essential nodes as shown in Figure 2.3, and as mentioned previously there are three in this circuit.

The second step is to choose one of the essential modes as the **reference node**, as shown in Figure 2.4. It can be any of the essential nodes, but it will simplify the analysis somewhat if we choose the essential node which is connected to the most elements. Here, we have chosen the bottom node to be the reference node, and this means that the other node voltages will be determined **with respect to this reference node**.

The third step is to number the remaining $(n_e - 1)$ nodes, where n_e is the number of essential nodes. Here we have labeled the top two essential nodes 1 and 2 as shown in Figure 2.5. The node voltages V_1 and V_2 represent the minimum set, which we must determine in order to solve the circuit, and they are both measured with respect to the reference node as shown. For example, V_1 is the voltage across the $6\,\Omega$ resistor, with the voltage reference having the "plus" at node 1.

As a fourth step, we write $(n_e - 1)$ node voltage equations in terms of the $(n_e - 1)$ node voltages. (In this case, we need to solve two node voltage equations to find two node voltages. This is the minimum number of equations that must be solved. We could write additional equations involving the node voltages, but they are not independent equations and are not needed.)

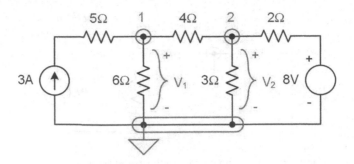

FIGURE 2.5   Numbering of remaining essential nodes.

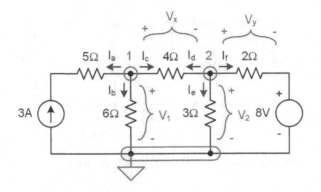

FIGURE 2.6   Branch currents used to write the node equations.

The node equations are written by making use of Kirchhoff's current law as shown in Figure 2.6. For node 1, we sum the branch currents leaving node 1 and set this sum to zero. Thus, $I_a + I_b + I_c = 0$. In a similar fashion for node 2, we write $I_d + I_e + I_f = 0$.

These two node equations must be rewritten in terms of the node voltages and circuit quantities in order to be useful. The branch current I_a is fixed by the current source so $I_a = -3\,\text{A}$. Here, the minus sign applies because the reference direction of I_a is opposite to the direction of the current source. The second branch current I_b may be found by the use of Ohm's law: $I_b = V_1 / 6\,\Omega$. For the determination of the third branch current I_c, we need to apply Kirchhoff's voltage law and Ohm's law with respect to the 4 Ω resistor. The voltage across this resistor, V_x, may be determined by consideration of KVL: $-V_1 + V_x + V_2 = 0$ so $V_x = V_1 - V_2$. Therefore, the branch current I_c is given by $I_c = V_x / 4\,\Omega = (V_1 - V_2)/4\,\Omega$. The node 1 equation may be written as $-3\,\text{A} + V_1 / 6\,\Omega + (V_1 - V_2)/4\,\Omega = 0$.

When developing the node 2 equation we can use similar ideas, but we should note that I_d has the opposite reference polarity compared to I_c, and therefore $I_d = -I_c = (V_2 - V_1)/4\,\Omega$. The determination of I_f requires the use of KVL for the determination of V_y. The basic KVL equation is $-V_2 + V_y + 8\,\text{V} = 0$, which may be rewritten as $V_y = V_2 - 8\,\text{V}$. Then, Ohm's law can be used to find the branch current:

$I_f = V_x / 2\,\Omega = (V_2 - 8\,\text{V}) / 2\,\Omega$. For simplicity, we often state the units as V, A, and Ω, and then write the node equations with only numerical quantities given. Thus, we would write the system of two node equations in V, A, and Ω as

$$N1 \quad -3 + \frac{V_1}{6} + \frac{V_1 - V_2}{4} = 0 \tag{2.1}$$

and

$$N2 \quad \frac{V_2 - V_1}{4} + \frac{V_2}{3} + \frac{V_2 - 8}{2} = 0. \tag{2.2}$$

The next and fifth step in the process is to solve this system of equations for the node voltages (two equations, two unknowns). There are a number of approaches which may be taken, but we will illustrate one of them. For example, we can start by multiplying each equation by the least common denominator (LCD). Here, the LCD for each equation is 12, and multiplying by this we obtain

$$N1 \quad -36 + 2V_1 + 3V_1 - 3V_2 = 0 \tag{2.3}$$

and

$$N2 \quad 3V_2 - 3V_1 + 4V_2 + 6V_2 - 48 = 0. \tag{2.4}$$

Collecting like terms, we can rewrite the node equations as

$$N1 \quad 5V_1 - 3V_2 = 36. \tag{2.5}$$

$$N2 \quad -3V_1 + 13V_2 = 48. \tag{2.6}$$

Next, it is convenient to write this system of equations in matrix form so we can solve by matrix techniques.

$$\begin{bmatrix} 5 & -3 \\ -3 & 13 \end{bmatrix} \begin{bmatrix} V_1 \\ V_2 \end{bmatrix} = \begin{bmatrix} 36 \\ 48 \end{bmatrix}. \tag{2.7}$$

For example, we can solve by use of **Cramer's rule**. The node voltages are given by

$$V_1 = \frac{\begin{vmatrix} 36 & -3 \\ 48 & 13 \end{vmatrix}}{\begin{vmatrix} 5 & -3 \\ -3 & 13 \end{vmatrix}} = 10.93\,\text{V} \tag{2.8}$$

and

$$V_2 = \frac{\begin{vmatrix} 5 & 36 \\ -3 & 48 \end{vmatrix}}{\begin{vmatrix} 5 & -3 \\ -3 & 13 \end{vmatrix}} = 6.21\,\text{V}. \qquad (2.9)$$

In each case, we take the ratio of two determinants. To find V_1, the numerator is the determinant of the original 2×2 matrix modified by replacing the **first** column with the 2×1 matrix, and the denominator is the determinant of the original 2×2 matrix. To find V_2, the numerator is the determinant of the original 2×2 matrix modified by replacing the **second** column with the 2×1 matrix, and the denominator is the determinant of the original 2×2 matrix.

The sixth and final step in solving the circuit is to use these node voltages to find the remaining circuit quantities. **All currents, voltages, and power values in the circuit may be determined once the node voltages have been found**, although we may not need to find each of these quantities. For example, the branch current flowing from top to bottom in the $6\,\Omega$ resistor is

$$I_b = \frac{V_1}{6\,\Omega} = \frac{10.93\,\text{V}}{6\,\Omega} = 1.822\,\text{A} \qquad (2.10)$$

and the power in the $4\,\Omega$ resistor is

$$P_{4\Omega} = \frac{(V_1 - V_2)^2}{4\,\Omega} = \frac{(10.93\,\text{V} - 6.21\,\text{V})^2}{4\,\Omega} = 5.57\,\text{W}. \qquad (2.11)$$

2.3   THE NODE VOLTAGE METHOD AND ALTERNATE NODE NUMBERING

Our choices of reference node and node numbers are arbitrary, but the circuit quantities are unique and not changed by these choices. To show this, we will reconsider the circuit in Figure 2.5 with different choices of reference node and node numbering as shown in Figure 2.7. It should be understood that V_A and V_B will not be equivalent to V_1 and V_2 found in the previous solution; this is because the former are defined differently than the latter. However, fundamental circuit quantities such as I_b and $P_{4\Omega}$ are invariant with respect to our choices of reference node and node numbering.

The node equations for the alternate node numbering in V, A, and Ω are

$$NA \quad 3 + \frac{V_A - V_B}{6} + \frac{V_A}{3} + \frac{V_A + 8}{2} = 0 \qquad (2.12)$$

and

$$NB \quad -3 + \frac{V_B - V_A}{6} + \frac{V_B}{4} = 0. \qquad (2.13)$$

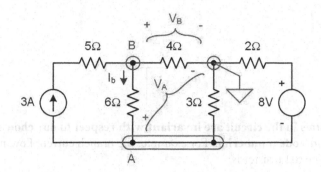

FIGURE 2.7   The electrical circuit of the previous section with the reference node and node labels chosen differently.

Multiplying each by the LCD (6 and 12, respectively), we obtain

$$NA \quad 18 + V_A - V_B + 2V_A + 3V_A + 24 = 0 \tag{2.14}$$

and

$$NB \quad -36 + 2V_B - 2V_A + 3V_B = 0. \tag{2.15}$$

Collecting like terms,

$$NA \quad 6V_A - V_B = -42 \tag{2.16}$$

and

$$NB \quad -2V_A + 5V_B = 36. \tag{2.17}$$

Next, it is convenient to write this system of equations in matrix form so we can solve by matrix techniques.

$$\begin{bmatrix} 6 & -1 \\ -2 & 5 \end{bmatrix} \begin{bmatrix} V_A \\ V_B \end{bmatrix} = \begin{bmatrix} -42 \\ 36 \end{bmatrix}. \tag{2.18}$$

Solving,

$$V_A = \frac{\begin{vmatrix} -42 & -1 \\ 36 & 5 \end{vmatrix}}{\begin{vmatrix} 6 & -1 \\ -2 & 5 \end{vmatrix}} = -6.21\,\text{V} \tag{2.19}$$

and

$$V_B = \dfrac{\begin{vmatrix} 6 & -42 \\ -2 & 36 \end{vmatrix}}{\begin{vmatrix} 6 & -1 \\ -2 & 5 \end{vmatrix}} = 4.71\,\text{V}. \tag{2.20}$$

Basic quantities in the circuit are invariant with respect to our choices of reference node and node numbering. For example, the branch current flowing from top to bottom in the $6\,\Omega$ resistor is

$$I_b = \dfrac{V_B - V_A}{6\,\Omega} = \dfrac{4.71\,\text{V} - (-6.21\,\text{V})}{6\,\Omega} = 1.820\,\text{A} \tag{2.21}$$

and the power in the $4\,\Omega$ resistor is

$$P_{4\Omega} = \dfrac{V_B^2}{4\,\Omega} = \dfrac{(4.71\,\text{V})^2}{4\,\Omega} = 5.55\,\text{W}. \tag{2.22}$$

These values are the same as those determined with the original node labeling in the previous section, apart from round-off differences.

2.4 THE NODE VOLTAGE METHOD WITH FOUR ESSENTIAL NODES

The method described in the previous section can readily extend to cases with higher numbers of essential nodes. To illustrate this, we will consider the circuit in Figure 2.8.

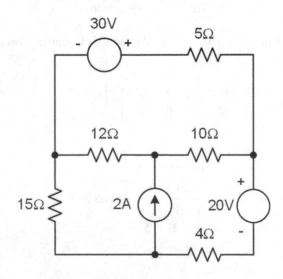

FIGURE 2.8 An electrical circuit with four essential nodes.

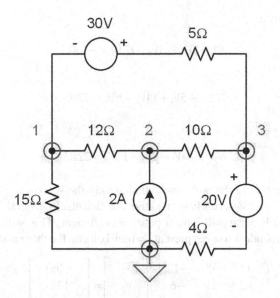

FIGURE 2.9 An electrical circuit with four essential nodes which are labeled.

In Figure 2.9, the four essential nodes have been identified, the reference node has been chosen, and the remaining nodes have been numbered.

The node voltage equations are the following in units of V, A, and Ω.

$$N1 \quad \frac{V_1}{15} + \frac{V_1 - V_2}{12} + \frac{V_1 + 30 - V_3}{5} = 0, \tag{2.23}$$

$$N2 \quad \frac{V_2 - V_1}{12} - 2 + \frac{V_2 - V_3}{10} = 0, \tag{2.24}$$

and

$$N3 \quad \frac{V_3 - 30 - V_1}{5} + \frac{V_3 - V_2}{10} + \frac{V_3 - 20}{4} = 0. \tag{2.25}$$

Multiplying by the LCD in each case, we obtain

$$N1 \quad 4V_1 + 5V_1 - 5V_2 + 12V_1 + 360 - 12V_3 = 0, \tag{2.26}$$

$$N2 \quad 5V_2 - 5V_1 - 120 + 16V_2 - 6V_3 = 0, \tag{2.27}$$

and

$$N3 \quad 4V_3 - 120 - 4V_1 + 2V_3 - 2V_2 + 5V_3 - 100 = 0. \tag{2.28}$$

Collecting like terms,

$$N1 \quad 21V_1 - 5V_2 - 12V_3 = -360, \tag{2.29}$$

$$N2 \quad -5V_1 + 11V_2 - 6V_3 = 120, \tag{2.30}$$

and

$$N3 \quad -4V_1 - 2V_2 + 11V_3 = 220. \tag{2.31}$$

It is useful to note that in the node one equation only the V_1 term has a positive coefficient while the V_2 and V_3 terms have negative coefficients. In general, in the nth node equation, only the V_n term will have a positive coefficient. This will always be true unless there are dependent sources present. In matrix form, the three node equations are

$$\begin{bmatrix} 11 & -5 & -12 \\ -5 & 21 & -6 \\ -4 & -2 & 11 \end{bmatrix} \begin{bmatrix} V_1 \\ V_2 \\ V_3 \end{bmatrix} = \begin{bmatrix} -360 \\ 120 \\ 220 \end{bmatrix}. \tag{2.32}$$

This system of equations may be solved by use of Cramer's rule:

$$V_1 = \frac{\begin{vmatrix} -360 & -5 & -12 \\ 120 & 11 & -6 \\ 220 & -2 & 11 \end{vmatrix}}{\begin{vmatrix} 21 & -5 & -12 \\ -5 & 11 & -6 \\ -4 & -2 & 11 \end{vmatrix}} = 4.72\,\text{V}, \tag{2.33}$$

$$V_2 = \frac{\begin{vmatrix} 21 & -360 & -12 \\ -5 & 120 & -6 \\ -4 & 220 & 11 \end{vmatrix}}{\begin{vmatrix} 21 & -5 & -12 \\ -5 & 11 & -6 \\ -4 & -2 & 11 \end{vmatrix}} = 27.64\,\text{V}, \tag{2.34}$$

$$V_3 = \frac{\begin{vmatrix} 21 & -5 & -360 \\ -5 & 11 & 120 \\ -4 & -2 & 220 \end{vmatrix}}{\begin{vmatrix} 21 & -5 & -12 \\ -5 & 11 & -6 \\ -4 & -2 & 11 \end{vmatrix}} = 26.74\,\text{V}, \tag{2.35}$$

Once the three node voltages are known, we can find all remaining circuit quantities as needed.

To review, see Presentation 2.1 in ebook+. To test your knowledge, try Quiz 2.1 and Quiz 2.2 (solving systems of equations) in ebook+. For more on solving systems of equations, refer to Appendix G. To put your knowledge to practice, try Laboratory Exercise 2.1 in ebook+

2.5 THE NODE VOLTAGE METHOD WITH DEPENDENT SOURCES

We can apply the NVM to circuits containing dependent sources, and this just requires us to express the dependent sources in terms of the node voltages. As an example, we will solve the circuit in Figure 2.10, which contains a voltage-controlled voltage source, but the same method applies in the case of other types of dependent sources or even multiple dependent sources.

The essential nodes in the example circuit have been identified and labeled in Figure 2.11.

The two node voltage equations are the following in units of V, A, and Ω.

$$N1 \quad \frac{V_1 - 4}{2} - 1 + \frac{V_1 - V_2}{6} = 0 \tag{2.36}$$

and

$$N2 \quad \frac{V_2 - V_1}{6} + \frac{V_2}{12} + \frac{V_2 - 3v_x}{4} = 0. \tag{2.37}$$

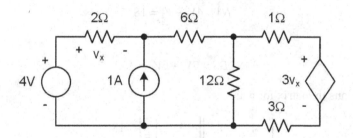

FIGURE 2.10 An electrical circuit containing a dependent source.

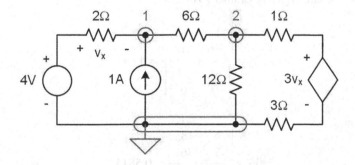

FIGURE 2.11 Essential nodes labeled in an electrical circuit containing a dependent source.

Here, we have two equations in three unknowns, so an additional equation is needed. This is the equation relating the dependent source to the node voltages, which may be found using KVL:

$$DS \quad v_x = 4 - V_1 \tag{2.38}$$

Substituting the dependent source equation into the node two equation, we obtain the following system:

$$N1 \quad \frac{V_1 - 4}{2} - 1 + \frac{V_1 - V_2}{6} = 0 \tag{2.39}$$

and

$$N2DS \quad \frac{V_2 - V_1}{6} + \frac{V_2}{12} + \frac{V_2 - 3(4 - V_1)}{4} = 0. \tag{2.40}$$

Multiplying each equation by its LCD, we have

$$N1 \quad 3V_1 - 12 - 6 + V_1 - V_2 = 0, \tag{2.41}$$

and

$$N2DS \quad 2V_2 - 2V_1 + V_2 + 3V_2 - 36 + 9V_1 = 0. \tag{2.42}$$

Collecting like terms,

$$N1 \quad 4V_1 - V_2 = 18 \tag{(2.43)}$$

and

$$N2DS \quad 7V_1 + 6V_2 = 36. \tag{2.44}$$

The equations in matrix form are

$$\begin{bmatrix} 4 & -1 \\ 7 & 6 \end{bmatrix} \begin{bmatrix} V_1 \\ V_2 \end{bmatrix} = \begin{bmatrix} 18 \\ 36 \end{bmatrix}. \tag{2.45}$$

The solution, obtained using Cramer's rule, is

$$V_1 = \frac{\begin{vmatrix} 18 & -1 \\ 36 & 6 \end{vmatrix}}{\begin{vmatrix} 4 & -1 \\ 7 & 6 \end{vmatrix}} = 4.64 \text{ V} \tag{2.46}$$

and

$$V_2 = \frac{\begin{vmatrix} 4 & 18 \\ 7 & 36 \end{vmatrix}}{\begin{vmatrix} 4 & -1 \\ 7 & 6 \end{vmatrix}} = 0.581 \text{ V}. \tag{2.47}$$

All other circuit quantities may be readily determined once the node voltages are known; for example, the controlling variable v_x is

$$v_x = 4\,\text{V} - V_1 = 4\,\text{V} - 4.64\text{V} = -0.64\,\text{V}. \tag{2.48}$$

To review, see Presentation 2.2 in ebook+. To test your knowledge, try Quiz 2.3 in ebook+.

2.6 THE NODE VOLTAGE METHOD WITH A KNOWN NODE VOLTAGE

When applying the NVM, special cases are introduced by voltage sources (independent or dependent), which are connected directly between essential nodes. When a voltage source exists between the reference node and one of the other essential nodes, this gives rise to a **known node voltage**. Such a case will be illustrated using the circuit in Figure 2.12.

Suppose we label the essential nodes as shown in Figure 2.13.

With this choice of reference node, V_1 is fixed by the 20 V source and represents a known node voltage. The node two equation is written in the usual fashion, by applying KCL. Therefore, the node equations in V, mA, and kΩ are

$$N1 \quad V_1 = 20 \text{ (known node voltage)} \tag{2.49}$$

and

$$N2 \quad \frac{V_2 - V_1}{5} + \frac{V_2}{4} + 10 = 0. \tag{2.50}$$

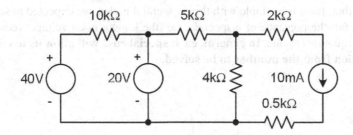

FIGURE 2.12 An electrical circuit containing a known node voltage.

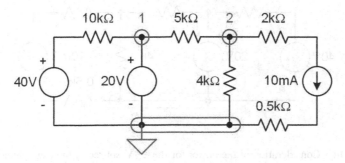

FIGURE 2.13 An electrical circuit containing a known node voltage with the nodes labeled.

Substitution of the node one equation into the node two equation results in

$$N1N2 \qquad \frac{V_2 - 20}{5} + \frac{V_2}{4} + 10 = 0. \qquad (2.51)$$

Multiplying by the LCD,

$$N1N2 \qquad 4V_2 - 80 + 5V_2 + 200 = 0 \qquad (2.52)$$

or

$$N1N2 \qquad 9V_2 = -120. \qquad (2.53)$$

Solving,

$$N1N2 \qquad V_2 = -13.33\,\text{V}. \qquad (2.54)$$

Having solved for the node voltages we can find all other circuit quantities. For example, we can determine the power for the 20 V source. Referring to the currents defined in Figure 2.14,

$$P_{20\text{V}} = (i_x)(20\,\text{V}) = (i_a + i_b)(20\,\text{V})$$

$$= \left(\frac{40\,\text{V} - V_1}{10\,\text{k}\Omega} + \frac{V_2 - V_1}{5\,\text{k}\Omega} \right)(20\,\text{V}) = -93.3\,\text{mW}. \qquad (2.55)$$

The negative value indicates that this source is developing power.

Notice that, for this example with three essential nodes, we expected to solve two equations, but the presence of a special case (the known node voltage) reduced the system of equations to one. **In general, each special case will allow us to eliminate one equation from the number to be solved.**

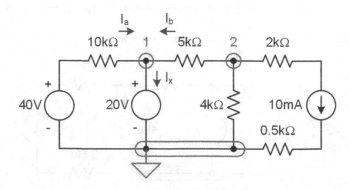

FIGURE 2.14 Consideration of the power for the 20 V source in the example circuit with a known node voltage.

Also noteworthy is our choice of V, mA, and kΩ units to solve this problem. We could have used units of V, A, and Ω, but this would have given rise to an equivalent yet cumbersome node two equation:

$$N2 \quad \frac{V_2 - V_1}{5000} + \frac{V_2}{4000} + 0.01 = 0. \quad (2.56)$$

Therefore, we will generally use units of V, mA, and kΩ whenever the resistances are all of the order of kΩ.

In the NVM, a special case can arise from any voltage source, independent or dependent. To show this, we will consider the circuit in Figure 2.15. Here, there are four essential nodes, leading to three equations. The dependent source (current-controlled voltage source) gives rise to a known node voltage and provides an opportunity to reduce to two equations, although this is less straightforward than in the case of an independent source so we may opt to solve the system of three equations instead. In Figure 2.16, the essential nodes have been identified and labeled.

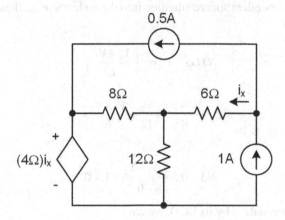

FIGURE 2.15   A circuit containing a dependent source giving rise to a known node voltage.

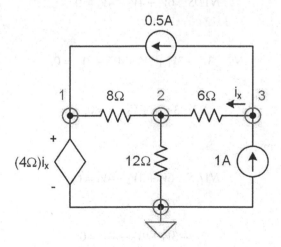

FIGURE 2.16   A circuit containing a dependent source arising from a known node voltage.

The three node voltage equations in units of V, A, and Ω are

$$N1 \quad V_1 = 4i_x \text{ (known node voltage)}, \tag{2.57}$$

$$N2 \quad \frac{V_2 - V_1}{8} + \frac{V_2}{12} + \frac{V_2 - V_3}{6} = 0, \tag{2.58}$$

and

$$N3 \quad 0.5 + \frac{V_3 - V_2}{6} - 1 = 0. \tag{2.59}$$

The equation governing the dependent source is

$$DS \quad i_x = \frac{V_3 - V_2}{6}. \tag{2.60}$$

Substituting the dependent source equation into the node one equation, we obtain the following system:

$$N1DS \quad V_1 = 4\left(\frac{V_3 - V_2}{6}\right). \tag{2.61}$$

$$N2 \quad \frac{V_2 - V_1}{8} + \frac{V_2}{12} + \frac{V_2 - V_3}{6} = 0, \tag{2.62}$$

and

$$N3 \quad 0.5 + \frac{V_3 - V_2}{6} - 1 = 0. \tag{2.63}$$

Multiplying each equation by its LCD, we have

$$N1DS \quad 6V_1 + 4V_2 - 4V_3 = 0 \tag{2.64}$$

and

$$N2 \quad 3V_2 - 3V_1 + 2V_2 + 4V_2 - 4V_3 = 0. \tag{2.65}$$

$$N3 \quad 3 + V_3 - V_2 - 6 = 0. \tag{2.66}$$

Collecting like terms,

$$N1DS \quad 6V_1 + 4V_2 - 4V_3 = 0 \tag{2.67}$$

and

$$N2 \quad -3V_1 + 9V_2 - 4V_3 = 0. \tag{2.68}$$

$$N3 - V_2 + V_3 = 3. \tag{2.69}$$

The equations in matrix form are

$$\begin{bmatrix} 6 & 4 & -4 \\ -3 & 9 & -4 \\ 0 & -1 & 1 \end{bmatrix} \begin{bmatrix} V_1 \\ V_2 \\ V_3 \end{bmatrix} = \begin{bmatrix} 0 \\ 0 \\ 3 \end{bmatrix}. \tag{2.70}$$

Solving,

$$V_1 = \frac{\begin{vmatrix} 0 & 4 & -4 \\ 0 & 9 & -4 \\ 3 & -1 & 1 \end{vmatrix}}{\begin{vmatrix} 6 & 4 & -4 \\ -3 & 9 & -4 \\ 0 & -1 & 1 \end{vmatrix}} = 2.00\,\text{V}, \tag{2.71}$$

$$V_2 = \frac{\begin{vmatrix} 6 & 0 & -4 \\ -3 & 0 & -4 \\ 0 & 3 & 1 \end{vmatrix}}{\begin{vmatrix} 6 & 4 & -4 \\ -3 & 9 & -4 \\ 0 & -1 & 1 \end{vmatrix}} = 3.60\,\text{V}, \tag{2.72}$$

and

$$V_3 = \frac{\begin{vmatrix} 6 & 4 & 0 \\ -3 & 9 & 0 \\ 0 & -1 & 3 \end{vmatrix}}{\begin{vmatrix} 6 & 4 & -4 \\ -3 & 9 & -4 \\ 0 & -1 & 1 \end{vmatrix}} = 6.60\,\text{V}. \tag{2.73}$$

Once the three node voltages are known, we can find all remaining circuit quantities. As examples,

$$i_x = \frac{V_3 - V_2}{6\,\Omega} = \frac{6.60\,\text{V} - 3.60\,\text{V}}{6\,\Omega} = 0.500\,\text{A}, \tag{2.74}$$

and the power for the dependent source is

$$P_{4ix} = (V_1)\left(0.5\,\text{A} + \frac{V_2 - V_1}{8\,\Omega}\right) = (2.0\,\text{V})\left(\frac{3.6\,\text{V} - 2.0\,\text{V}}{8\,\Omega}\right) = 0.400\,\text{W}. \tag{2.75}$$

2.7 THE NODE VOLTAGE METHOD WITH A SUPERNODE

Another special case that arises in the NVM is the **supernode**. This refers to a voltage source directly between two essential nodes, neither of which is the reference node. This means that our choice of reference node determines whether a voltage source gives rise to a known node voltage or a supernode. However, sometimes a circuit will contain a supernode no matter where we choose to place the reference node, and it is therefore necessary to understand how to solve problems containing supernodes. To see this, we can consider the circuit in Figure 2.17. This network contains two voltage sources, which are connected directly between essential nodes, and it therefore gives rise to two special cases. If we choose the bottom node as the reference, then the 12 V source gives rise to a known node voltage while the 6 V source gives rise to a supernode. There is no placement of the reference node which will avoid a supernode, and in fact placing the reference on the right-hand side of the circuit will result in two supernodes!

For our NVM analysis, we will choose the reference node and node numbers as shown in Figure 2.18. With this choice, there is a known node voltage at node one while nodes two and three form a supernode.

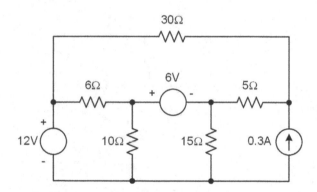

FIGURE 2.17   An electrical circuit with five essential nodes. This circuit contains at least one supernode for any choice of reference node.

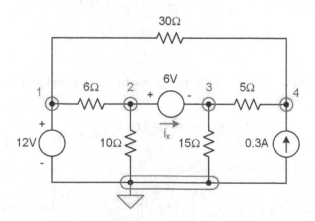

FIGURE 2.18   An electrical circuit containing a supernode.

When writing the node equations here, it is not possible to directly determine the branch current between nodes two and three by the use of Ohm's law because there is no resistor in series with the voltage source. If we invoke this branch current and define it as i_x, the node two and node three equations are, in units of V, A, and Ω:

$$N2 \quad \frac{V_2 - V_1}{6} + \frac{V_2}{10} + i_x = 0, \tag{2.76}$$

and

$$N3 \quad -i_x + \frac{V_3}{15} + \frac{V_3 - V_4}{5} = 0. \tag{2.77}$$

These equations are not as useful as written because they invoke a new unknown (the current i_x). However, if we add these two equations together we obtain a single equation for the supernode, which does not require us to invoke the current i_x:

$$N23 \quad \frac{V_2 - V_1}{6} + \frac{V_2}{10} + \frac{V_3}{15} + \frac{V_3 - V_4}{5} = 0. \tag{2.78}$$

This is the supernode equation, and it can be written directly by adding the branch currents leaving the supernode. Usually, we will do so directly, without the extra step of first writing the two node equations and adding them. Thus, the node equations for this example would be directly written, in units of V, A, and Ω, as

$$N1 \quad V_1 = 12 \text{ (known node voltage)}, \tag{2.79}$$

$$N23 \quad \frac{V_2 - V_1}{6} + \frac{V_2}{10} + \frac{V_3}{15} + \frac{V_3 - V_4}{5} = 0 \text{ (supernode)}, \tag{2.80}$$

and

$$N4 \quad \frac{V_4 - V_1}{30} + \frac{V_4 - V_3}{5} - 0.3 = 0. \tag{2.81}$$

Because we wrote a single equation for the supernode we have three equations with four unknowns. One more equation is needed, and this is the equation of the voltage source involved in the supernode, developed using KVL:

$$VS \quad V_2 = V_3 + 6. \tag{2.82}$$

Substituting the voltage source equation into the supernode equation, we obtain

$$N1 \quad V_1 = 12, \tag{2.83}$$

$$N23VS \quad \frac{V_3 + 6 - V_1}{6} + \frac{V_3 + 6}{10} + \frac{V_3}{15} + \frac{V_3 - V_4}{5} = 0, \tag{2.84}$$

and

$$N4 \quad \frac{V_4 - V_1}{30} + \frac{V_4 - V_3}{5} - 0.3 = 0. \tag{2.85}$$

We can substitute the node one equation into each of the others to obtain a system of two equations:

$$N23VS \quad \frac{V_3 + 6 - 12}{6} + \frac{V_3 + 6}{10} + \frac{V_3}{15} + \frac{V_3 - V_4}{5} = 0 \tag{2.86}$$

and

$$N4 \quad \frac{V_4 - 12}{30} + \frac{V_4 - V_3}{5} - 0.3 = 0 \tag{2.87}$$

Multiplying each equation by its LCD, we have

$$N123DS \quad 5V_3 - 30 + 3V_3 + 18 + 2V_3 + 6V_3 - 6V_4 = 0 \tag{2.88}$$

and

$$N14 \quad V_4 - 12 + 6V_4 - 6V_3 - 9 = 0. \tag{2.89}$$

Collecting like terms,

$$N123DS \quad 16V_3 - 6V_4 = 12 \tag{2.90}$$

and

$$N14 \quad -6V_3 + 7V_4 = 21. \tag{2.91}$$

The equations in matrix form are

$$\begin{bmatrix} 16 & -6 \\ -6 & 7 \end{bmatrix} \begin{bmatrix} V_3 \\ V_4 \end{bmatrix} = \begin{bmatrix} 12 \\ 21 \end{bmatrix}. \tag{2.92}$$

The solution is

$$V_3 = \frac{\begin{vmatrix} 12 & -6 \\ 21 & 7 \end{vmatrix}}{\begin{vmatrix} 16 & -6 \\ -6 & 7 \end{vmatrix}} = 2.76\,\text{V} \tag{2.93}$$

and

$$V_4 = \frac{\begin{vmatrix} 16 & 12 \\ -6 & 21 \end{vmatrix}}{\begin{vmatrix} 16 & -6 \\ -6 & 7 \end{vmatrix}} = 5.37\,\text{V} \tag{2.94}$$

Therefore,

$$V_2 = V_3 + 6V = 8.76V \qquad (2.95)$$

To review, see Presentation 2.3 in ebook+. To test your knowledge, try Quiz 2.4 in ebook+.

2.8 SUMMARY

In this chapter, we considered the NVM, which is a general-purpose tool for solving electrical circuits. By the NVM, we determine the voltages at the essential nodes ("node voltages") with respect to a reference node, and all remaining circuit quantities may be readily found once these node voltages are known.

The basic NVM involves six steps. First, we identify all n_e essential nodes (nodes connected to more than two elements). Second, we choose one of these to be the reference node. Third, we number the remaining $(n_e - 1)$ essential nodes. Fourth, we write $(n_e - 1)$ node equations based on Kirchhoff's current law. In each node equation, the sum of the branch currents leaving the node is set to zero. Fifth, we solve this set of equations using algebraic methods. Sixth, we use the node voltages to solve for other circuit quantities (voltages, currents, and power values).

If one or more dependent sources appear in the circuit, the controlling variable for each dependent source must be expressed in terms of the node voltages.

Special cases arise in the NVM when a voltage source (independent or dependent) is placed between essential nodes. If a voltage source is placed between the reference node and one of the other essential nodes, then a known node voltage results. If a voltage source appears between two essential nodes, neither of which is the reference node, a supernode results. Therefore, the choice of reference node can affect the character of individual special cases.

To evaluate your mastery of Chapters 1 and 2, solve Example Exam 2.1 in ebook+ or Example Exam 2.2 in ebook+.

PROBLEMS

Problem 2.1. Solve for V_1, I_x, and the power of the $3\,\Omega$ resistor in Figure P2.1.

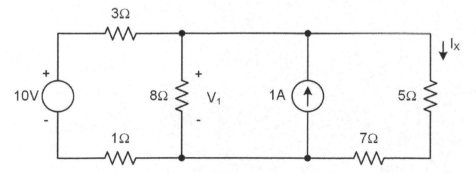

FIGURE P2.1 Circuit with seven elements and two essential nodes.

Problem 2.2. Determine I_1, I_2, and V_S of the circuit in Figure P2.2.

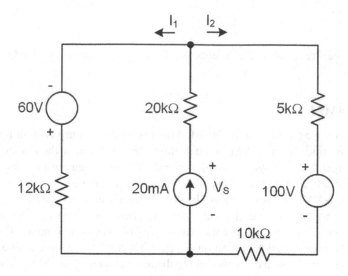

FIGURE P2.2 Circuit with three sources and two essential nodes.

Problem 2.3. Find V_A and V_B for the circuit in Figure P2.3.

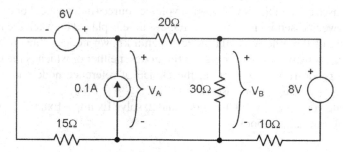

FIGURE P2.3 Circuit with seven elements and three essential nodes.

Problem 2.4. Determine the values of V_X and V_Y for the network in Figure P2.4.

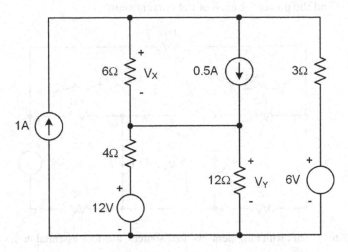

FIGURE P2.4 Circuit with eight elements and three essential nodes.

Problem 2.5. Find V_1, V_2, and V_3, and the power for the current source in Figure P2.5.

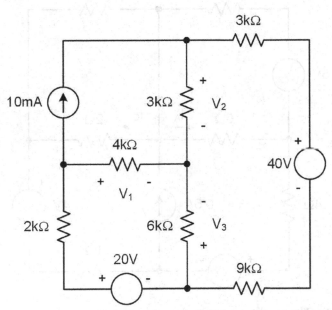

FIGURE P2.5 Circuit with nine elements, three sources, and four essential nodes.

Problem 2.6. Calculate the values of the voltages V_A, V_B, and V_C in Figure P2.6. Find the power for each of the current sources.

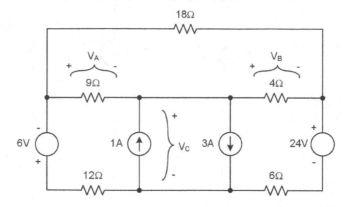

FIGURE P2.6   Circuit with nine elements, four sources, and four essential nodes.

Problem 2.7. Determine I_A, I_B, and I_C, and the power for the current source in the circuit in Figure P2.7.

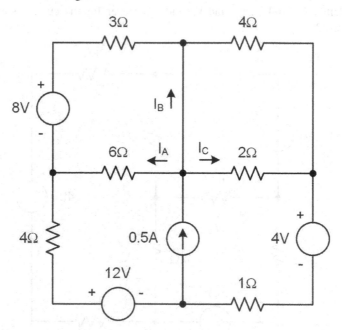

FIGURE P2.7   Circuit with ten elements, four sources, and four essential nodes.

Problem 2.8. Solve for I_A, I_B, I_C, and the power of the current source in Figure P2.8.

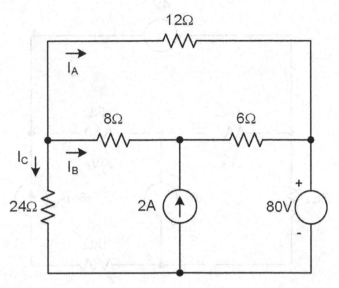

FIGURE P2.8 Circuit with six elements, four essential nodes, and one special case.

Problem 2.9. Determine the values of V_A and V_B of the circuit in Figure P2.9.

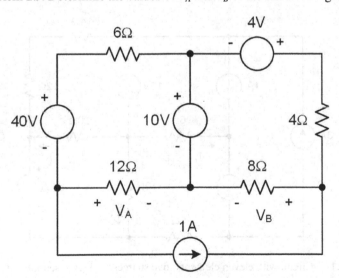

FIGURE P2.9 Circuit with eight elements, four essential nodes, and one special case.

Problem 2.10. Calculate V_A, V_B, and power of the current source in Figure P2.10.

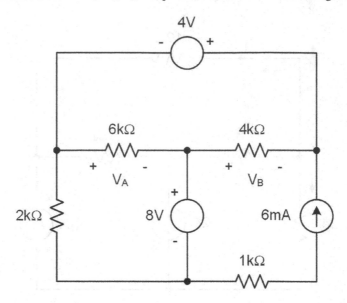

FIGURE P2.10   Circuit with seven elements, four essential nodes, and two special cases.

Problem 2.11. Find the currents I_1, I_2, and I_3, and the power for each of the current sources for the circuit in Figure P2.11.

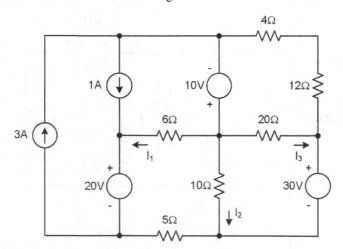

FIGURE P2.11   Circuit with eleven elements, five sources, and six essential nodes.

Problem 2.12. Calculate V_A, V_B, and V_C for the network in Figure P2.12.

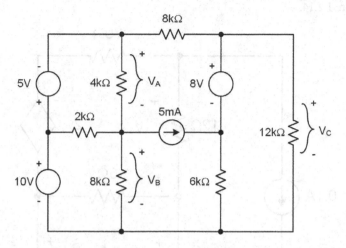

FIGURE P2.12 Circuit with ten elements, four sources, and six essential nodes.

Problem 2.13. Calculate i_x and the power for the dependent source of the circuit in Figure P2.13.

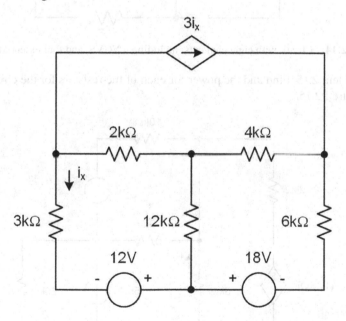

FIGURE P2.13 Circuit with eight elements, including a CCCS, and four essential nodes.

Problem 2.14. Determine v_x, v_y, and the power for each of the sources in Figure P2.14.

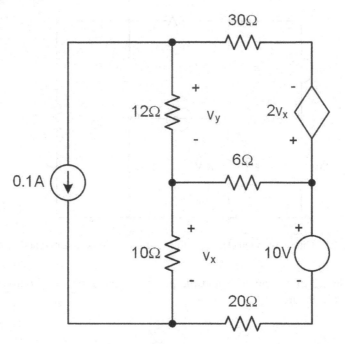

FIGURE P2.14   Circuit with eight elements, including a VCVS, and four essential nodes.

Problem 2.15. Find and the power for each of the resistors for the circuit in Figure P2.15.

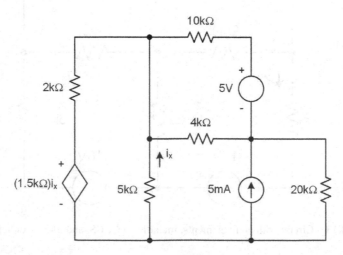

FIGURE P2.15   Circuit with eight elements, including a CCVS, and three essential nodes.

Problem 2.16. Determine v_x, v_y, and v_z for the circuit in Figure P2.16.

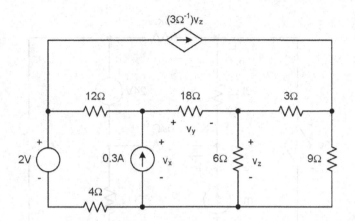

FIGURE P2.16   Circuit with nine elements, including a VCCS, and five essential nodes.

Problem 2.17. Find i_x for the network in Figure P2.17.

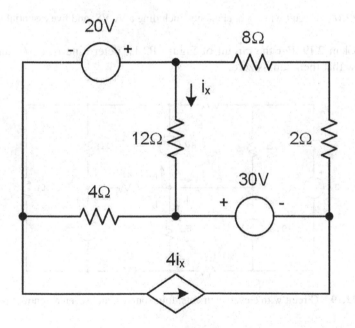

FIGURE P2.17   Circuit with seven elements, including a CCCS, and four essential nodes.

Problem 2.18. Determine the currents I_X, I_Y, and I_Z; find the voltage v_a and the power for the dependent source in Figure P2.18.

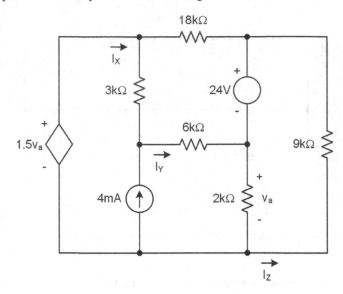

FIGURE P2.18   Circuit with eight elements, including a VCVS, and five essential nodes.

Problem 2.19. For the circuit of Figure P2.19, determine I_1, I_2, I_3, and I_4; show that their sum is zero.

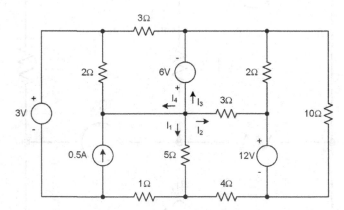

FIGURE P2.19   Circuit with eleven elements, four sources, and seven essential nodes.

Problem 2.20. Find i_x, v_x, and the power for each of the sources for the circuit in Figure P2.20.

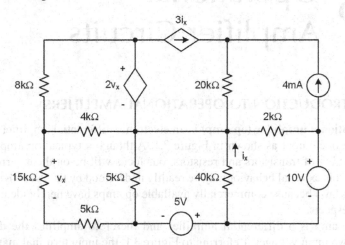

FIGURE P2.20 Circuit with thirteen elements, including two dependent sources, and seven essential nodes.

3 Operational Amplifier Circuits

3.1 INTRODUCTION TO OPERATIONAL AMPLIFIERS

The **operational amplifier (op amp)** is an electronic differential amplifier with two inputs and one output as shown in Figure 3.1. Although a typical op amp contains a large number of transistors and resistors, our focus will be on the external circuit behavior. This external behavior may be readily understood by the use of Kirchhoff's and Ohm's laws because commercially available op amps have nearly ideal behavior in many respects.

The op amp is a differential amplifier and therefore amplifies the difference between two input voltages. Referring to Figure 3.1, the input terminal marked with a plus sign is the non-inverting input with the voltage v_p applied, and the input terminal marked with a minus sign is the inverting input with the voltage v_n applied. The input currents i_p and i_n flow into the non-inverting and inverting terminals, respectively. For a real op amp, these currents are so small that they may usually be neglected. The output terminal emanates from the point of the triangular symbol, and the voltage at this terminal is V_{out}. The other two terminals shown on the op amp are the power terminals; the positive supply is $+V_{CC}$ and the negative supply is $-V_{EE}$. The power supplies must be connected to an op amp for it to function; therefore, these connections are always present even if they are omitted from some circuit diagrams for simplicity.

The voltage transfer characteristic for a typical op amp is shown in Figure 3.2. In the linear region of operation, the output voltage is proportional to the difference in the input voltages:

$$V_{out} = A(v_p - v_n),$$ (3.1)

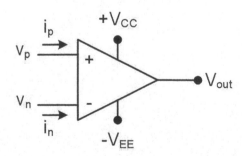

FIGURE 3.1   Operational amplifier symbol with power connections.

   DOI: 10.1201/9781003408529-3

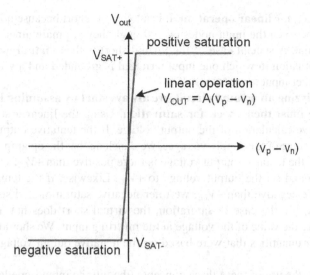

FIGURE 3.2   Operational amplifier voltage transfer characteristic.

where A is the open-loop gain of the op amp. Typically, the open-loop gain is 100,000 or more, so for any typical value of the output voltage the difference in the input voltages $(v_p - v_n) = V_{out} / A$ is exceedingly small. Therefore, for linear operation of the op amp, we will usually assume that the two input terminals are at the same voltage. However, the output voltages are limited by the supply voltages: even for an ideal op amp, V_{out} may not be more positive than $+V_{CC}$ or more negative than $-V_{EE}$. This gives rise to positive saturation at $+V_{CC}$ and negative saturation at $-V_{EE}$. It is important to realize that the open-loop gain does not apply in either case of saturation, so it is no longer true that the input voltages are approximately equal.

3.2   ANALYSIS OF OPERATIONAL AMPLIFIER CIRCUITS

For the analysis of op amp circuits, we usually assume that the op amps exhibit ideal behavior and apply three concepts: the assumption of **infinite input impedance (I^3)**, the **virtual short (VS)**, and a **check for saturation**.

   The assumption of infinite input impedance causes us to consider that the currents flowing in the inputs of the op amp (both the non-inverting and inverting terminals) are zero. Because the real input currents are typically on the order of nA, it is reasonable to neglect them as long as the resistances in the circuit are $\leq 1\,\text{M}\Omega$, resulting in circuit branch currents $\geq 1\,\mu\text{A}$. The concept of infinite input impedance applies to linear or saturated operations.

   The concept of the virtual short applies to linear operation only, for which the open-loop gain is operational. The large open-loop gain A of the op amp ensures that the difference in the input voltages, $(v_p - v_n) = V_{out} / A$, is very small (on the order of μV). **Therefore, we will assume that there is a "virtual short" between the inputs**

and that $v_p = v_n$, **for linear operation**. It is a "virtual" short because no current can flow directly between the input nodes even though they are maintained at the same electric potential. A special case of the virtual short is the **"virtual ground"**; this refers to the situation in which one input terminal is grounded and a virtual ground exists at the other input terminal.

When analyzing an op amp circuit, we always start by assuming linear operation, but we must then check for saturation. Using the linear assumption, we make a tentative calculation of the output voltage. If the tentative output voltage is outside of the range of the supply voltages, we conclude that the op amp is saturated. Specifically, if the tentative output voltage is more positive than $+V_{CC}$, we infer positive saturation and set the output voltage to $+V_{CC}$. Likewise, if the tentative output voltage is more negative than $-V_{EE}$, we infer negative saturation and set the output voltage to $-V_{EE}$. In the case of saturation, the virtual short does not apply, so we must reconsider the value of the voltage at the inverting input. We should also reconsider any other quantities that were based on the tentative output voltage, including currents.

To illustrate the use of these three concepts (the infinite input impedance, virtual short, and saturation check), we will consider the op amp circuit in Figure 3.3.

We start by assuming linear operation so we can find the tentative output voltage. The non-inverting input is connected to the ground so

$$v_p = 0. \tag{3.2}$$

Making use of the virtual short,

$$v_n = v_p = 0. \tag{3.3}$$

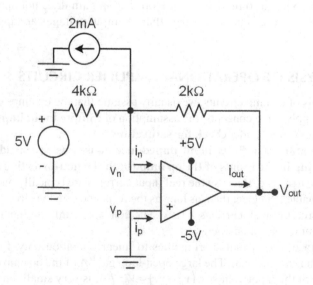

FIGURE 3.3   An op amp circuit for analysis.

We can apply Kirchhoff's current law (KCL) at the inverting input node (units of V, mA, and kΩ) with the assumption of $i_n = 0$ (infinite input impedance):

$$2 + \frac{0-5}{4} + \frac{0-V'_{out}}{2} = 0,\tag{3.4}$$

where V'_{out} is the tentative output voltage. Multiplying by the least common denominator,

$$8 - 5 - 2V'_{out} = 0.\tag{3.5}$$

Solving, we obtain

$$V'_{out} = 1.50\,\text{V}.\tag{3.6}$$

The tentative output voltage is less positive than $+V_{CC} = +5\,\text{V}$ so the op amp is not saturated, and we conclude that the tentative output voltage is the actual output voltage. Having solved for V_{out} and having verified the value of v_n, we can determine all other circuit quantities. For example,

$$i_{out} = \frac{V_{out} - v_n}{2\,\text{k}\Omega} = \frac{1.5\,\text{V} - 0}{2\,\text{k}\Omega} = 0.75\,\text{mA}.\tag{3.7}$$

It is important to note that, although the currents flowing in the inputs of the op amp are zero (infinite input impedance), the output current is generally not zero.

Another important point involves the orientation of the non-inverting and inverting input terminals. It is tempting to think that the op amp will behave in a similar fashion, though with a change in the sign of the output voltage, if we interchange the non-inverting and inverting terminals in this circuit. This is not true, however, because although we have made the assumption of the virtual short, there is actually an infinitesimally small difference between the input voltages, and this is amplified to create the output voltage. The resistor connected between the output and the inverting input terminal provides negative feedback, which stabilizes the output voltage. If we instead connect a feedback resistor between the output and the non-inverting input, positive feedback results, and the output will be driven to saturation. As such **the feedback resistor must always be connected between the output and the inverting terminal to provide negative feedback and stabilize the output.**

3.3 ANALYSIS OF A SATURATED OPERATIONAL AMPLIFIER CIRCUIT

The three concepts introduced in the previous section may be used to analyze any op amp circuit, including one in which the op amp is saturated. As in the previous example, we will start by assuming linear operation in order to find a tentative output voltage. If this tentative output voltage is outside the range of the supply voltages, we will adjust it accordingly and revisit the voltage at the inverting input. To illustrate this process, we will use the circuit in Figure 3.4.

We start by assuming linear operation so we can find the tentative output voltage. The voltage at the non-inverting input is determined by the independent voltage source:

$$v_p = 2\,\text{V}. \tag{3.8}$$

Making use of the virtual short, the tentative voltage at the inverting input v_n' is

$$v_n' = v_p = 2\,\text{V}. \tag{3.9}$$

We can apply KCL at the inverting input node (units of V, mA, and kΩ) with the assumption of $i_n = 0$ (infinite input impedance):

$$-1 + \frac{2-4}{1} + \frac{2-8}{6} + \frac{2-V_{\text{out}}'}{3} = 0. \tag{3.10}$$

Multiplying by the least common denominator,

$$-6 - 12 - 6 + 4 - 2V_{\text{out}}' = 0. \tag{3.11}$$

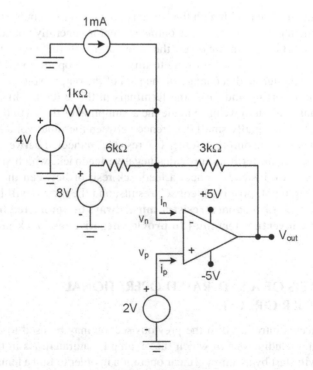

FIGURE 3.4   An op amp circuit with four independent sources.

Solving, we obtain the tentative output voltage:

$$V'_{out} = -10\,\text{V}. \tag{3.12}$$

The tentative output voltage is more negative than $-V_{EE} = -5\,\text{V}$ so the op amp is saturated, and we need to set the output voltage to the saturated voltage with the same sign as the tentative output voltage:

$$V_{out} = -5\,\text{V}. \tag{3.13}$$

Because the op amp is saturated, the virtual short does not apply and we must reconsider the value of the voltage at the inverting input. This can be done using the node voltage method at this node. In units of V, mA, and kΩ,

$$-1+\frac{v_n-4}{1}+\frac{v_n-8}{6}+\frac{v_n-V_{out}}{3}=0. \tag{3.14}$$

Multiplying by the least common denominator,

$$-6+6v_n-24+v_n-8+2v_n-2V_{out}=0. \tag{3.15}$$

Solving,

$$v_n = \frac{38+2V_{out}}{9} = 3.11\,\text{V}. \tag{3.16}$$

Having found the true values of V_{out} and v_n, we must use these true values (not tentative values) to calculate the other voltages and currents in the circuit.

To review, see Presentation 3.1 in ebook+. To test your knowledge, try Quiz 3.1 in ebook+. To put your knowledge to practice, try Laboratory Exercise 3.1 in cbook+

3.4 ANALYSIS OF A CIRCUIT INVOLVING MULTIPLE OP AMPS

We may analyze circuits containing multiple op amps using the three concepts of the infinite input impedance, virtual short, and saturation check. However, we must start the analysis upstream and work our way downstream. To show this, we will consider the circuit in Figure 3.5, which contains three op amps. Op amps OA1 and OA2 provide the inputs to OA3; therefore, we must analyze OA1 and OA2 before OA3.

We will start by analyzing OA1 with the assumption of linear operation. The non-inverting input is grounded, so assuming a virtual short exists between the inputs we find the tentative voltage at the inverting input:

$$V'_1 = 0. \tag{3.17}$$

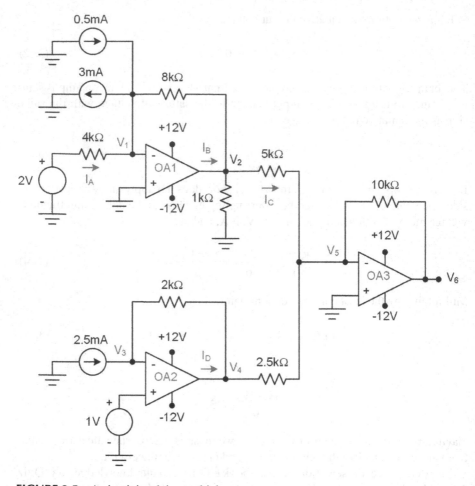

FIGURE 3.5   A circuit involving multiple op amps.

We can apply KCL at the inverting input node (units of V, mA, and kΩ) with the assumption that zero current flows into the inverting input terminal (infinite input impedance):

$$-0.5 + 3 + \frac{0-2}{4} + \frac{0-V_2'}{8} = 0. \tag{3.18}$$

Multiplying by the least common denominator,

$$-4 + 24 - 4 - V_2' = 0. \tag{3.19}$$

Solving, we obtain the tentative output voltage:

$$V_2' = 16\,\text{V}. \tag{3.20}$$

The tentative output voltage is more positive than $+V_{CC} = +12\,\text{V}$ so the op amp is saturated, and we need to set the output voltage to $+V_{CC}$:

$$V_2 = 12\,\text{V}. \tag{3.21}$$

Because the op amp is saturated, the virtual short does not apply and we must reconsider the value of the voltage at the inverting input. This can be done using the node voltage method. In units of V, mA, and kΩ,

$$-0.5 + 3 + \frac{V_1 - 2}{4} + \frac{V_1 - (12)}{8} = 0. \tag{3.22}$$

Multiplying by the least common denominator,

$$-4 + 24 + 2V_1 - 4 + V_1 - 12 = 0. \tag{3.23}$$

Collecting like terms,

$$4 + 3V_1 = 0. \tag{3.24}$$

Solving,

$$V_1 = -1.333\,\text{V}. \tag{3.25}$$

Thus, we have solved for the voltages surrounding OA1. We are not ready to find the current I_B because it depends on V_5, which is not yet known, and we will save the calculation of currents for last.

For OA2, we begin by assuming linear operation. Making use of the virtual short, we consider the tentative voltage at the inverting input to be the same as the voltage at the non-inverting input:

$$V_3' = 1\,\text{V}. \tag{3.26}$$

To find the tentative output voltage for OA2, we can apply KCL at the inverting input node (units of V, mA, and kΩ) with the assumption that zero current flows into the inverting input terminal (infinite input impedance):

$$-2.5 + \frac{1 - V_4'}{2} = 0. \tag{3.27}$$

Multiplying by the least common denominator,

$$-5 + 1 - V_4' = 0. \tag{3.28}$$

Solving, we obtain the tentative output voltage:

$$V_4' = -4\,\text{V}. \tag{3.29}$$

The tentative output voltage is within the range of the supply voltages, so OA2 is not saturated and the actual voltages are equal to the tentative voltages:

$$V_3 = 1\,\text{V} \tag{3.30}$$

and

$$V_4 = -4\,\text{V}. \tag{3.31}$$

Next, we consider OA3, starting with the assumption of linear operation. The virtual short gives rise to a tentative voltage at the inverting input:

$$V_5' = 0. \tag{3.32}$$

Making use of the infinite input impedance, the KCL equation for the inverting input node is (units of V, mA, and kΩ)

$$\frac{0-V_2}{5} + \frac{0-V_4}{2.5} + \frac{0-V_6'}{10} = 0. \tag{3.33}$$

Multiplying by the least common denominator,

$$-2V_2 - 4V_4 - V_6' = 0. \tag{3.34}$$

The tentative output voltage is

$$V_6' = -2V_2 - 4V_4 = -2(12\,\text{V}) - 4(-4\,\text{V}) = -16\,\text{V}. \tag{3.35}$$

The tentative output voltage is more negative than $-V_{EE} = -12\,\text{V}$ so OA3 is saturated and the actual output voltage is equal to $-V_{EE}$:

$$V_6 = -12\,\text{V}. \tag{3.36}$$

This causes us to reconsider the voltage at the inverting input, because the virtual short does not apply in the case of saturation. Using the node voltage method with units of V, mA, and kΩ,

$$\frac{V_5 - V_2}{5} + \frac{V_5 - V_4}{2.5} + \frac{V_5 - V_6}{10} = 0. \tag{3.37}$$

Multiplying by the least common denominator,

$$2V_5 - 2V_2 + 4V_5 - 4V_4 + V_5 - V_6 = 0. \tag{3.38}$$

Solving,

$$V_5 = \frac{2V_2 + 4V_4 + V_6}{7} = \frac{2(12\,\text{V}) + 4(-4\,\text{V}) + (-12\,\text{V})}{7} = -0.571\,\text{V}. \tag{3.39}$$

Finally, now that all of the node voltages are known, we can calculate the current values.

$$I_A = \frac{2\,\text{V} - V_1}{4\,\text{k}\Omega} = \frac{2\,\text{V} - (-1.333\,\text{V})}{4\,\text{k}\Omega} = 0.833\,\text{mA}, \tag{3.40}$$

$$I_B = \frac{V_2}{1\,\text{k}\Omega} + \frac{V_2 - V_1}{8\,\text{k}\Omega} + \frac{V_2 - V_5}{5\,\text{k}\Omega}$$

$$= \frac{12\,\text{V}}{1\,\text{k}\Omega} + \frac{12\,\text{V} - (-1.333\,\text{V})}{8\,\text{k}\Omega} + \frac{12\,\text{V} - (-0.571\,\text{V})}{5\,\text{k}\Omega} = 16.18\,\text{mA}, \tag{3.41}$$

$$I_C = \frac{V_2 - V_5}{5\,\text{k}\Omega} = \frac{12\,\text{V} - (-0.571\,\text{V})}{4\,\text{k}\Omega} = 3.14\,\text{mA}, \tag{3.42}$$

and

$$I_D = \frac{V_4 - V_3}{2\,\text{k}\Omega} + \frac{V_4 - V_5}{2.5\,\text{k}\Omega} = \frac{-4\,\text{V} - 1\,\text{V}}{2\,\text{k}\Omega} + \frac{-4\,\text{V} - (-0.571\,\text{V})}{2.5\,\text{k}\Omega} = -4.33\,\text{mA}. \tag{3.43}$$

This same general process may be used to analyze any circuit involving multiple op amps with cascading involved.

3.5 INVERTING AMPLIFIER

The three general principles applied to op amp analysis allow us to design op amp circuits as well. As an example, we will consider the inverting amplifier, which is a general-purpose building block for op amp design. The inverting amplifier has the layout shown in Figure 3.6. The non-inverting terminal is grounded. An input voltage is connected to the inverting input through an input resistor R_1 and there is a feedback resistor R_F.

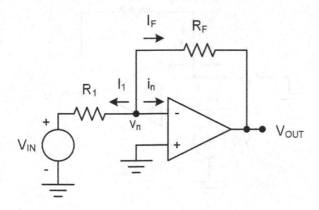

FIGURE 3.6 Inverting amplifier.

If we assume linear operation, the virtual short fixes the voltage at the inverting input:

$$v_n = v_p = 0. \tag{3.44}$$

If we apply KCL at the inverting input, making use of the infinite input impedance ($i_n = 0$), then

$$I_1 + I_F = \frac{0 - V_{IN}}{R_1} + \frac{0 - V_{OUT}}{R_F} = 0. \tag{3.45}$$

Solving for V_{OUT},

$$V_{OUT} = -\left(\frac{R_F}{R_1}\right) V_{IN}. \tag{3.46}$$

The voltage gain V_{OUT} / V_{IN} of this amplifier is $-(R_F / R_1)$, in which the minus sign indicates an inverting amplifier and can be interpreted as a phase shift of 180° for sinusoidal signals. This gain can be set by the ratio of two resistors. Moreover, variable gain may be achieved by making one of the resistors adjustable.

3.6   NON-INVERTING AMPLIFIER

Another general-purpose building block for op amp design is the non-inverting amplifier shown in Figure 3.7. An input voltage is connected directly to the non-inverting input. There is a resistor R_1 connected between the inverting input and ground, and there is a feedback resistor R_F.

If we assume linear operation, the virtual short fixes the voltage at the inverting input:

$$v_n = v_p = V_{IN}. \tag{3.47}$$

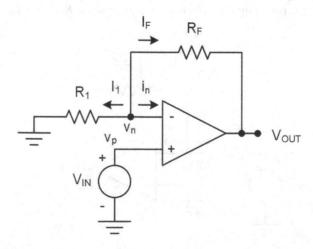

FIGURE 3.7   Non-inverting amplifier.

If we apply KCL at the inverting input, making use of the infinite input impedance $(i_n = 0)$, then

$$I_1 + I_F = \frac{V_{IN}}{R_1} + \frac{V_{IN} - V_{OUT}}{R_F} = 0. \qquad (3.48)$$

Solving for V_{OUT},

$$V_{OUT} = \left(1 + \frac{R_F}{R_1}\right) V_{IN}. \qquad (3.49)$$

The voltage gain V_{OUT} / V_{IN} of this amplifier is $(1 + R_F / R_1)$. The positive gain indicates non-inverting operation. For sinusoidal signals, this means the output voltage will be in phase with the input voltage. Also, it should be noted that the minimum absolute value of gain is unity, occurring in the limit as $R_F / R_1 \to 0$. In this limit, with R_F replaced by a wire and R_1 replaced by an open circuit, we obtain a practical circuit which is a unity-gain buffer. This circuit, shown in Figure 3.8, allows us to reproduce a voltage from a source node to a target node without drawing current from the source node. For this reason, it is called a "buffer amplifier" or a "unity-gain buffer."

3.7 SUMMING AMPLIFIER

A summing amplifier allows us to scale and add two or more voltage signals. A two-input summing amplifier is illustrated in Figure 3.9. Each input voltage is applied to the inverting input terminal through a resistor. The non-inverting input is grounded and there is a feedback resistor R_F.

If we assume linear operation, the virtual short fixes the voltage at the inverting input:

$$v_n = v_p = 0. \qquad (3.50)$$

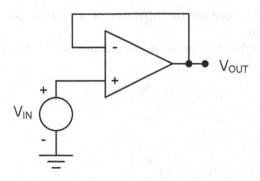

FIGURE 3.8 Buffer amplifier.

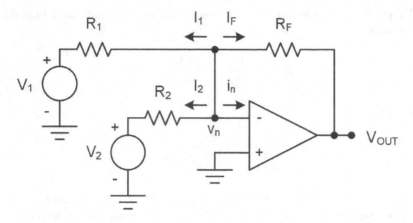

FIGURE 3.9 Summing amplifier.

If we apply KCL at the inverting input, making use of the infinite input impedance $(i_n = 0)$, then

$$I_1 + I_2 + I_F = \frac{0 - V_1}{R_1} + \frac{0 - V_2}{R_1} + \frac{0 - V_{OUT}}{R_F} = 0. \qquad (3.51)$$

Solving for V_{OUT},

$$V_{OUT} = -\left[\left(\frac{R_F}{R_1}\right)V_1 + \left(\frac{R_F}{R_2}\right)V_2\right]. \qquad (3.52)$$

The input signals have voltage gain values which may be designed independently, because although R_F is common to both ratios, R_1 and R_2 may be chosen independently. Both terms are negative, indicating inverting behavior. The summing amplifier may be scaled up to three or more inputs by simply connecting additional input resistors to the inverting input.

3.8   DIFFERENCE AMPLIFIER

A difference amplifier produces the difference in two scaled voltage signals and is shown in Figure 3.10.

Applying KCL at the inverting input, and making use of the infinite input impedance $(i_n = 0)$,

$$I_A + I_B = \frac{v_n - V_1}{R_A} + \frac{v_n - V_{OUT}}{R_B} = 0. \qquad (3.53)$$

Solving,

$$V_{OUT} = \left(1 + \frac{R_B}{R_A}\right)v_n - \left(\frac{R_B}{R_A}\right)V_1. \qquad (3.54)$$

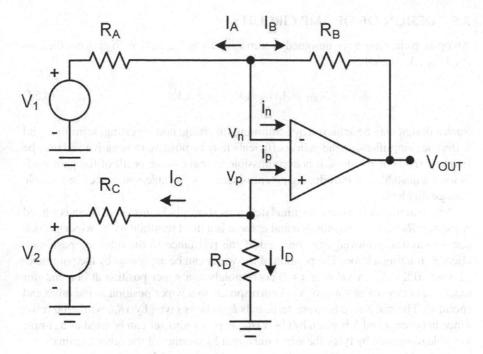

FIGURE 3.10   Difference amplifier.

Similarly, at the non-inverting input ($i_p = 0$):

$$I_C + I_D = \frac{v_p - V_2}{R_C} + \frac{v_p - V_{OUT}}{R_D} = 0 \tag{3.55}$$

Solving for v_p,

$$v_p = V_2 \left(\frac{R_D}{R_C + R_D} \right). \tag{3.56}$$

Making use of the virtual short ($v_n = v_p$), and substituting this value of into the equation for V_{OUT}, we obtain

$$V_{OUT} = \left(1 + \frac{R_B}{R_A} \right) \left(\frac{R_D}{R_C + R_D} \right) V_2 - \left(\frac{R_B}{R_A} \right) V_1. \tag{3.57}$$

Therefore, the output is the difference in the scaled input voltages, and the two scale factors may be set as desired by choice of the four resistors.

To review, see Presentation 3.2 in ebook+. To test your knowledge, try Quiz 3.2 in ebook+.

3.9   DESIGN OF OP AMP CIRCUITS

An op amp circuit may be designed to sum n inputs $V_1, V_2, \ldots, V_n$ with gain coefficients $A_1, A_2, \ldots, A_n$, such that

$$V_{\text{OUT}} = A_1 V_1 + A_2 V_2 + \cdots + A_n V_n. \tag{3.58}$$

Such a design may be achieved by combining inverting, non-inverting, summing, and difference amplifiers, so the gain coefficients may be positive or negative and may be fixed by ratios of resistors. It is even possible to make some or all of the gain coefficients adjustable, by introducing potentiometers as variable resistances or variable voltage dividers.

A potentiometer is a three-terminal device as shown in Figure 3.11. There is a fixed resistance R_P between terminals a and c; there is a third terminal b (the wiper), which can be physically moved from one end of the resistance to the other by pushing a slider or rotating a knob. The position of the wiper can be expressed by the parameter x, where $0 \le x \le 1$. A value of $x = 0$ corresponds to a wiper position at one end (for example, at c) while a value of $x = 1$ corresponds to a wiper position at the other end (point a). The resistance between terminals b and c is given by xR_P, while the resistance between a and b is given by $(1-x)R_P$. A potentiometer can be used as a simple variable resistance by tying the wiper (terminal b) to either of the other terminals.

A variable-gain inverting amplifier can be constructed by using a potentiometer as a variable feedback resistor as shown in Figure 3.12. For this circuit, the voltage gain is

$$A_V = \frac{V_{\text{OUT}}}{V_{\text{IN}}} = -\frac{xR_P}{R_1}, \tag{3.59}$$

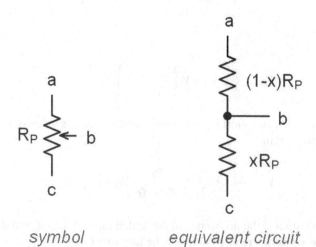

symbol   equivalent circuit

FIGURE 3.11   Potentiometer.

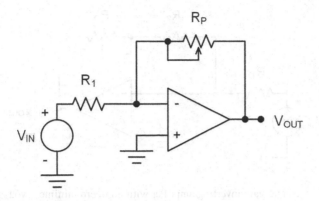

FIGURE 3.12   Variable-gain inverting amplifier.

where $0 \le x \le 1$, so the gain may be adjusted all the way to zero. As a matter of good practice, op amp circuits use resistances between $1\,\text{k}\Omega$ and $1\,\text{M}\Omega$, because smaller resistances result in excessive current loading and larger resistances result in designs which are noisy and contain voltage offsets. Thus, if a $10\,\text{k}\Omega$ potentiometer is used for the feedback resistor, the maximum absolute value of voltage gain will be 10, corresponding to $R_1 = 1\,\text{k}\Omega$.

If a non-zero minimum gain (minimum absolute value) is desired in a variable-gain inverting amplifier, this can be achieved by placing a fixed resistor in series with the variable resistance as shown in Figure 3.13. The voltage gain is

$$A_V = \frac{V_{\text{OUT}}}{V_{\text{IN}}} = -\frac{xR_P + R_2}{R_1}, \tag{3.60}$$

where $0 \le x \le 1$. The maximum value of $|A_V|$ is $(R_2 + R_P)/R_1$, whereas the minimum value is R_2 / R_1. Therefore, the ratio of maximum gain to minimum gain is $(R_2 + R_P)/R_2$.

A variable-gain non-inverting amplifier may be constructed by using a potentiometer as a variable voltage divider as shown in Figure 3.14. For this circuit, the voltage gain is

$$A_V = \frac{V_{\text{OUT}}}{V_{\text{IN}}} = x\left(1 + \frac{R_F}{R_1}\right), \tag{3.61}$$

where $0 \le x \le 1$, so the gain may be adjusted to zero.

If a non-zero minimum gain is desired in a variable-gain non-inverting amplifier, this can be achieved by placing a fixed resistor in series with the variable resistance as shown in Figure 3.15. The voltage gain is

$$A_V = \frac{V_{\text{OUT}}}{V_{\text{IN}}} = \left(\frac{R_2 + xR_P}{R_2 + R_P}\right)\left(1 + \frac{R_F}{R_1}\right), \tag{3.62}$$

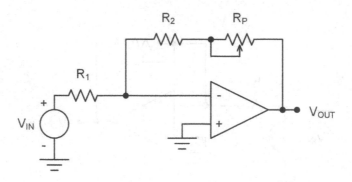

FIGURE 3.13   Variable-gain inverting amplifier with non-zero minimum voltage gain.

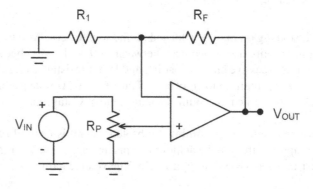

FIGURE 3.14   Variable-gain non-inverting amplifier.

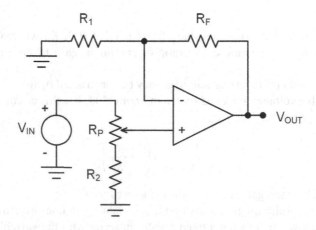

FIGURE 3.15   Variable-gain non-inverting amplifier with non-zero minimum voltage gain.

where $0 \le x \le 1$. The maximum value of A_V is $(1 + R_F / R_1)$ whereas the minimum value is $\left[R_2 / (R_2 + R_P) \right](1 + R_F / R_1)$. Therefore, the ratio of maximum gain to minimum gain is $(R_2 + R_P) / R_2$.

A summing amplifier with independent gain adjustment of two voltage signals may be implemented as shown in Figure 3.16. The output voltage of this adjustable inverting summer is given by

$$V_{OUT} \approx -\left\{ \left(\frac{R_{S1} + xR_{P1}}{R_{S1} + R_{P1}} \right) \left(\frac{R_F}{R_1} \right) V_1 + \left(\frac{R_{S2} + xR_{P2}}{R_{S2} + R_{P2}} \right) \left(\frac{R_F}{R_2} \right) V_2 \right\}. \qquad (3.63)$$

This expression is approximate because it does not take into account the loading of the variable voltage dividers by the input resistors (R_1, R_2). However, we may neglect this loading with good accuracy if R_1 and R_2 are made sufficiently large $(R_1 \gg R_{S1},$ $R_1 \gg R_{P1}, R_2 \gg R_{S2},$ and $R_2 \gg R_{P2})$.

As an example design, we can consider an op amp circuit to produce an output V_{OUT} with three inputs V_1, V_2, and V_3 as follows:

$$V_{OUT} = -A_1V_1 - A_2V_2 + A_3V_3, \qquad (3.64)$$

with A_1 fixed $(A_1 = 2)$, with A_2 variable from zero to a maximum value $(0 \le A_2 \le 10)$, and with A_3 adjustable by a factor of ten $(2 \le A_3 \le 20)$. To combine the signals, we will need a three-input summing amplifier. The summing amplifier is inherently inverting, so this will provide the minus signs for the V_1 and V_2 terms. For the V_1 channel, we can use a simple input resistor to the summing amplifier. For the V_2 channel, we can use a potentiometer as a variable voltage divider and feed this to an input resistor to the summing amplifier. For the V_3 channel, we can use a variable-gain inverting amplifier with non-zero minimum gain. The cascade of two inverting amplifiers will produce a positive coefficient for V_3. Combining these ideas, we arrive at a design layout as shown in Figure 3.17.

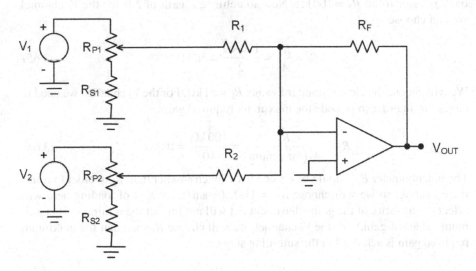

FIGURE 3.16 Variable-gain summing amplifier with independent gain adjustment of two voltage signals.

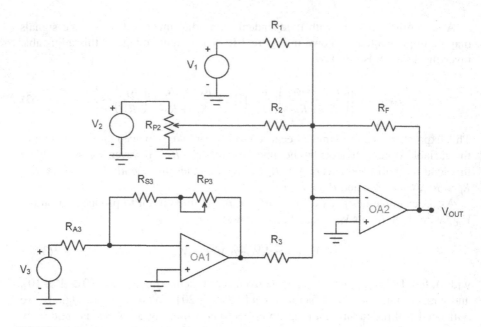

FIGURE 3.17 Three-input amplifier system with independent gain adjustment for two voltage signals.

Next, we need to choose the resistor values, keeping all resistors in the practical range of $1\,k\Omega-1\,M\Omega$. We will assume that we are using 5% resistors with standard values, and we will use potentiometers from the set of most common values ($1\,k\Omega$, $5\,k\Omega$, $10\,k\Omega$, $20\,k\Omega$, $50\,k\Omega$, and $100\,k\Omega$). We will first choose the feedback resistor to be $R_F = 100\,k\Omega$. Now to achieve a gain of 2.0 for the V_1 channel we can choose

$$R_1 = \frac{R_F}{A_1} = \frac{100\,k\Omega}{2} = 50\,k\Omega. \tag{3.65}$$

We will choose the closest standard value: $R_1 = 51\,k\Omega$. For the V_2 channel, we need to choose R_2 in order to provide the maximum required gain:

$$R_2 = \frac{R_F}{A_2\,(\text{maximum})} = \frac{100\,k\Omega}{10} = 10\,k\Omega. \tag{3.66}$$

The potentiometer R_{P2} should be chosen to be much smaller than R_2 in order to minimize loading, so we will choose $R_{P2} = 1\,k\Omega$. (A small amount of loading here will affect the linearity of the gain adjustment but will not impact the minimum or maximum values of gain.) For the V_3 channel, we will choose R_3 such that the maximum required gain is achieved in the summing stage:

$$R_3 = \frac{R_F}{A_3\,(\text{maximum})} = \frac{100\,k\Omega}{20} = 5\,k\Omega. \tag{3.67}$$

We will choose the closest standard value: $R_3 = 5.1\,\text{k}\Omega$. The inverting stage must therefore have gain adjustment in the range of $0.1-1.0$. Considering the minimum gain requirement,

$$0.1 = \frac{R_{S3}}{R_{A3}}, \tag{3.68}$$

and considering the maximum gain requirement,

$$1.0 = \frac{R_{S3} + R_{P3}}{R_{A3}}. \tag{3.69}$$

This requires $R_{P3} / R_{A3} = 0.9$. If we choose a $10\,\text{k}\Omega$ potentiometer ($R_{P3} = 10\,\text{k}\Omega$), then $R_{A3} \approx 11\,\text{k}\Omega$. (This is a standard value.) This requires $R_{S3} \approx 1.1\,\text{k}\Omega$ (also a standard value). The completed design is shown in Figure 3.18.

To review, see Presentation 3.3 in ebook+. To test your knowledge, try Quiz 3.3 in ebook+. To put your knowledge to practice, try Laboratory Exercise 3.2 in ebook+.

3.10 NON-IDEAL CHARACTERISTICS OF OP AMPS

So far we have assumed ideal op amp behavior. This simplifies the analysis and design of op amp circuits, and it is also appropriate for most purposes given that practical op amps approximate ideal ones in many ways. However, it is important to be aware of non-ideal op amp behavior when designing for high precision or high frequencies.

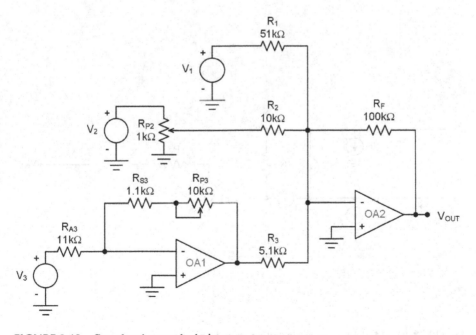

FIGURE 3.18 Completed example design.

We would not expect an op amp to respond to arbitrarily high frequencies. A real op amp has a **gain-bandwidth (GBW) product**, which limits its response at higher frequencies. If the op amp is configured in a circuit with a voltage gain of A_V, the expected upper cutoff frequency is

$$f_c = \frac{\text{GBW}}{|A_V|}. \tag{3.70}$$

For example, if we use an op amp with GBW $= 10\,\text{MHz}$ to build a circuit with a voltage gain of $|A_V| = 100$, we expect the upper cutoff frequency to be $f_c = 100\,\text{kHz}$.

For switching applications, in which the op amp output is supposed to take on a rectangular waveform, another limitation comes into play, and this is the **slew rate** s. The rise and fall time for a rectangular wave output are equal to the peak-to-peak output amplitude divided by the slew rate:

$$t_R, t_F = \frac{V_{PP}}{s}. \tag{3.71}$$

For example, with a peak-to-peak output amplitude of $V_{PP} = 10\,\text{V}$ and slew rate $s = 0.5\,\text{V/\mu s}$, the rise time will be $t_R = 20\,\mu\text{s}$.

There are several non-ideal op amp characteristics, which affect DC operation as well. These include the input bias current, input offset current, input offset voltage, open-loop gain, input resistance, and output resistance. These are shown in the op amp model of Figure 3.19. The **input bias current** I_B is the average current, which flows into each input terminal of the op amp with zero voltage difference applied.

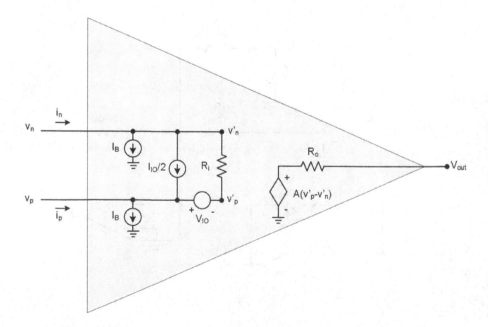

FIGURE 3.19 Model for op amp including some important non-ideal characteristics.

If the input current to the inverting terminal is I_{B-} and the input current to the non-inverting terminal is I_{B+}, the input bias current is

$$I_B = (I_{B+} + I_{B-})/2. \tag{3.72}$$

The **input offset current** I_{IO} is the difference between the two input currents

$$I_{IO} = I_{B+} - I_{B-}. \tag{3.73}$$

When a voltage difference is applied between the input terminals, there is another current proportional to this voltage difference because of the presence of a finite **input resistance** R_i. The output voltage of the op amp is not exactly zero when there is a zero voltage difference between the input terminals. The input voltage that must be applied to make the output voltage zero is called the **input offset voltage** V_{IO}. The **open-loop gain** of a real op amp is not infinite but takes on a very large number. Finally, there is a finite **output resistance** R_o. Table 3.1 summarizes some ideal characteristics and compares them to typical values for an LM741 op amp.

To see how these non-ideal characteristics could affect circuit performance, we can consider a prototypical inverting amplifier with an expected voltage gain of -2.0 as shown in Figure 3.20. If we assume ideal behavior then $i_n = i_p = 0$, $v_n = v_p = 0$, $V_{out} = -2.000\,\text{V}$, and $I_{out} = V_{out}/R_F + V_{out}/R_L = -3.000\,\text{mA}$. For comparison, we can also consider the circuit of Figure 3.21, which includes a more realistic model for the op amp including the effects of the finite input bias current, input offset current, input resistance, output resistance, and open-loop gain.

The node equations for the inverting input node (node Nn) and the output node (node Nout) are

$$Nn \quad \frac{v_n - V_{in}}{R_1} + I_B + \frac{I_{IO}}{2} + \frac{v_n + V_{IO}}{R_i} + \frac{v_n - V_{out}}{R_F} = 0 \tag{3.74}$$

TABLE 3.1
Ideal and Typical Values of Some Op Amp Characteristics

Parameter	Ideal value	Typical value for LM741
Gain-bandwidth (GBW) product	∞	3 MHz
Slew rate (s)	∞	0.5 MV/s
Input bias current (I_B)	0	30 nA
Input offset current (I_{IO})	0	3 nA
Input offset voltage (V_{IO})	0	0.8 mV
Input resistance (R_i)	∞	6 MΩ
Open-loop gain (A)	∞	2×10^5
Output resistance (R_o)	0	50 Ω

The LM741 op amp is used as an example for typical characteristics.

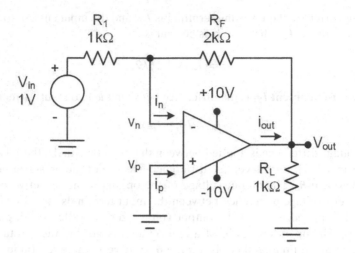

FIGURE 3.20 Inverting op amp circuit built with an ideal op amp.

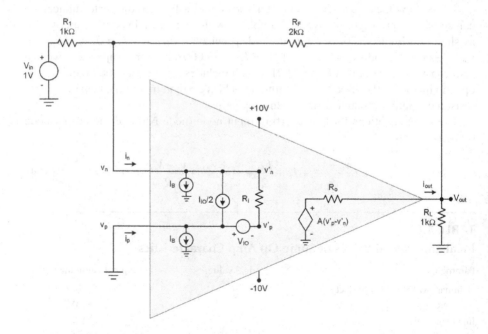

FIGURE 3.21 Inverting op amp circuit constructed with a non-ideal op amp.

and

$$Nout \quad \frac{V_{out} - v_n}{R_F} + \frac{V_{out}}{R_L} + \frac{V_{out} - A(-V_{IO} - v_n)}{R_o} = 0. \quad (3.75)$$

Collecting like terms,

$$Nn \quad v_n\left[\frac{1}{R_1}+\frac{1}{R_i}+\frac{1}{R_F}\right]+V_{\text{out}}\left[-\frac{1}{R_F}\right]=\left[\frac{V_{\text{in}}}{R_1}-I_B-\frac{I_{\text{IO}}}{2}-\frac{V_{\text{IO}}}{R_i}\right] \tag{3.76}$$

and

$$Nout \quad v_n\left[-\frac{1}{R_F}+\frac{A}{R_o}\right]+V_{\text{out}}\left[\frac{1}{R_F}+\frac{1}{R_L}+\frac{1}{R_o}\right]=\left[-\frac{AV_{\text{IO}}}{R_o}\right]. \tag{3.77}$$

In matrix form,

$$\begin{bmatrix} \dfrac{1}{R_1}+\dfrac{1}{R_i}+\dfrac{1}{R_F} & -\dfrac{1}{R_F} \\[2mm] -\dfrac{1}{R_F}+\dfrac{A}{R_o} & \dfrac{1}{R_F}+\dfrac{1}{R_L}+\dfrac{1}{R_o} \end{bmatrix}\begin{bmatrix} v_n \\[2mm] V_{\text{out}} \end{bmatrix}=\begin{bmatrix} \dfrac{V_{\text{in}}}{R_1}-I_B-\dfrac{I_{\text{IO}}}{2}-\dfrac{V_{\text{IO}}}{R_i} \\[2mm] -\dfrac{AV_{\text{IO}}}{R_o} \end{bmatrix}$$

$$\tag{3.78}$$

Solving,

$$v_n=\frac{\begin{vmatrix} \dfrac{V_{\text{in}}}{R_1}-I_B-\dfrac{I_{\text{IO}}}{2}-\dfrac{V_{\text{IO}}}{R_i} & -\dfrac{1}{R_F} \\[3mm] -\dfrac{AV_{\text{IO}}}{R_o} & \dfrac{1}{R_F}+\dfrac{1}{R_L}+\dfrac{1}{R_o} \end{vmatrix}}{\begin{vmatrix} \dfrac{1}{R_1}+\dfrac{1}{R_i}+\dfrac{1}{R_F} & -\dfrac{1}{R_F} \\[3mm] -\dfrac{1}{R_F}+\dfrac{A}{R_o} & \dfrac{1}{R_F}+\dfrac{1}{R_L}+\dfrac{1}{R_o} \end{vmatrix}} \tag{3.79}$$

$$=\frac{\left(\dfrac{V_{\text{in}}}{R_1}-I_B-\dfrac{I_{\text{IO}}}{2}-\dfrac{V_{\text{IO}}}{R_i}\right)\left(\dfrac{1}{R_F}+\dfrac{1}{R_L}+\dfrac{1}{R_o}\right)-\left(-\dfrac{1}{R_F}\right)\left(-\dfrac{AV_{\text{IO}}}{R_o}\right)}{\left(\dfrac{1}{R_1}+\dfrac{1}{R_i}+\dfrac{1}{R_F}\right)\left(\dfrac{1}{R_F}+\dfrac{1}{R_L}+\dfrac{1}{R_o}\right)-\left(-\dfrac{1}{R_F}\right)\left(-\dfrac{1}{R_F}+\dfrac{A}{R_o}\right)}, \tag{3.80}$$

and

$$V_{\text{out}}=\frac{\begin{vmatrix} \dfrac{1}{R_1}+\dfrac{1}{R_i}+\dfrac{1}{R_F} & \dfrac{V_{\text{in}}}{R_1}-I_B-\dfrac{I_{\text{IO}}}{2}-\dfrac{V_{\text{IO}}}{R_i} \\[3mm] -\dfrac{1}{R_F}+\dfrac{A}{R_o} & -\dfrac{AV_{\text{IO}}}{R_o} \end{vmatrix}}{\begin{vmatrix} \dfrac{1}{R_1}+\dfrac{1}{R_i}+\dfrac{1}{R_F} & -\dfrac{1}{R_F} \\[3mm] -\dfrac{1}{R_F}+\dfrac{A}{R_o} & \dfrac{1}{R_F}+\dfrac{1}{R_L}+\dfrac{1}{R_o} \end{vmatrix}}, \tag{3.81}$$

and

$$= \frac{\left(\dfrac{1}{R_1} + \dfrac{1}{R_i} + \dfrac{1}{R_F}\right)\left(-\dfrac{AV_{IO}}{R_o}\right) - \left(\dfrac{V_{in}}{R_1} - I_B - \dfrac{I_{IO}}{2} - \dfrac{V_{IO}}{R_i}\right)\left(-\dfrac{1}{R_F} + \dfrac{A}{R_o}\right)}{\left(\dfrac{1}{R_1} + \dfrac{1}{R_i} + \dfrac{1}{R_F}\right)\left(\dfrac{1}{R_F} + \dfrac{1}{R_L} + \dfrac{1}{R_o}\right) - \left(-\dfrac{1}{R_F}\right)\left(-\dfrac{1}{R_F} + \dfrac{A}{R_o}\right)}. \quad (3.82)$$

For example, using $V_{in} = 1\,\text{V}$, $R_1 = 1\,\text{k}\Omega$, $R_F = 2\,\text{k}\Omega$, $R_L = 1\,\text{k}\Omega$, and with the typical parameters given in Table 3.1, we obtain $v_n = -0.000789\,\text{V}$ and $V_{out} = -2.0023\,\text{V}$. These values are both within a millivolt of the values obtained using the ideal op amp model. This is consistent with the fact that the ideal op amp model will almost always provide adequate accuracy, unless we desire high voltage precision and use precision resistors with tight tolerances.

3.11 SUMMARY

An operational amplifier (op amp) is a high-gain electronic differential amplifier having two inputs and a single output. One of the inputs is called the **inverting input** and is marked by a minus sign; the other input is the **non-inverting input** and is marked by a plus sign. The voltages at the inverting and non-inverting inputs are called v_n and v_p, respectively. For **linear operation**, the output voltage is equal to the difference of the input voltages multiplied by the open-loop gain for the amplifier: $V_{OUT} = A(v_p - v_n)$, where A **is the open-loop gain and typically has a very high value** of $10^5 - 10^6$. The output voltage of the op amp is limited to be in the range of the power supply voltages. For an ideal op amp, the output will exhibit **positive saturation** limited by the positive supply $[V_{OUT}(\text{maximum}) = +V_{CC}]$ and **negative saturation** limited by the negative supply $[V_{OUT}(\text{minimum}) = -V_{EE}]$. To achieve useful values of voltage gain which are much less than the open-loop gain, **negative feedback** is employed by connecting a resistor between the output and the inverting input of the op amp.

Analysis of an arbitrary op amp circuit can be done by standard circuit analysis techniques with the use of three concepts specific to op amps. These are the **virtual short**, **infinite input impedance**, and the **check for saturation**. The virtual short applies to linear operation only; this requires us to start our analysis by first assuming linear operation so that $v_p = v_n$ because of the high open-loop gain; by so doing we can find a tentative value of the voltage at the inverting terminal (v_n'). The infinite input impedance leads us to assume that the currents flowing into the two input terminals are both zero. This applies to linear and saturation operation. After starting by assuming linear operation, we calculate the tentative output voltage (V_{OUT}'). If this tentative output voltage is in the range of the supply voltages, then the actual output voltage will be equal to the tentative value. If not, then the output will saturate at the supply voltage having the same sign as the tentative output voltage. If the output voltage is saturated, we must reconsider the value of v_n and any other circuit quantities which are affected by V_{OUT}. When analyzing a circuit that involves multiple op amps, some of which feed downstream to others, we must first analyze the upstream op amps.

Design of op amp circuits can be done to provide scaled sums or differences of multiple inputs. The scale factors (gain coefficients) are set by ratios of resistors and may be adjustable if potentiometers are used to achieve variable resistances or variable voltage dividers. Important building blocks for design of op amp circuits include the inverting amplifier, the non-inverting amplifier, the summing amplifier, and the difference amplifier.

To review, see Presentation 3.1 in ebook+. To test your knowledge, try Quiz 3.1 in ebook+. To put your knowledge to practice, try Laboratory Exercise 3.1 in ebook+.

To put your knowledge to practice, try Laboratory Exercise 3.3 in ebook+.

PROBLEMS

Problem 3.1. Find the currents I_A, I_B, and I_C; find the voltages V_1 and V_2 with respect to ground for the circuit in Figure P3.1. Assume that the op amp is ideal.

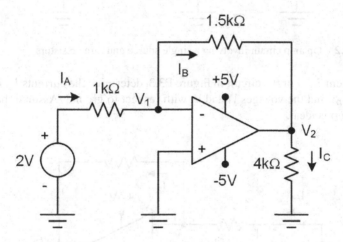

FIGURE P3.1   Op amp circuit.

Problem 3.2. For the op amp circuit in Figure P3.2, determine the currents I_A, I_B, and I_C; find the voltages V_1 and V_2 with respect to ground. Assume that the op amp is ideal.

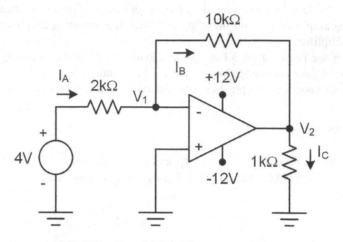

FIGURE P3.2 Op amp circuit involving a single source and three resistors.

Problem 3.3. For the circuit in Figure P3.3, determine the currents I_A, I_B, I_C, and I_D; find the voltages V_1 and V_2 with respect to ground. Assume that the op amp is ideal.

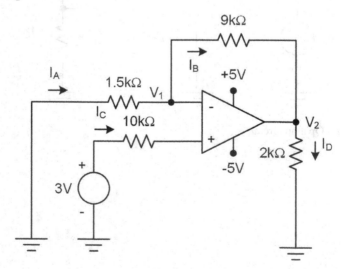

FIGURE P3.3 Op amp circuit involving a single source and four resistors.

Problem 3.4. Determine the currents I_A, I_B, I_C, and I_D and find the voltages V_1 and V_2 with respect to ground for the circuit in Figure P3.4. Assume that the op amp is ideal.

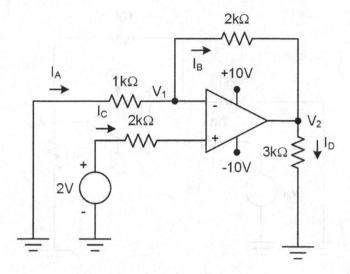

FIGURE P3.4 Op amp circuit with a fixed voltage source connected to the noninverting input and four resistors.

Problem 3.5. For the op amp circuit in Figure P3.5, determine the relationship between V_{out} and V_{in} for the case of linear operation, and using this find the range of V_{in} for linear operation of the op amp. Assume that the op amp is ideal.

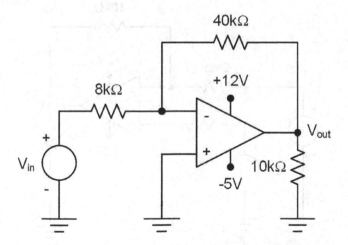

FIGURE P3.5 Op amp circuit with a variable input voltage connected to the inverting input.

Problem 3.6. Determine the relationship between V_{out} and V_{in} for the circuit of Figure P3.6 for the case of linear operation, and using this find the range of V_{in} for linear operation of the op amp. Assume that the op amp is ideal.

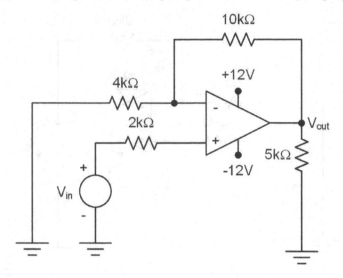

FIGURE P3.6   Op amp circuit with a variable input voltage connected to the noninverting input.

Problem 3.7. For the op amp circuit of Figure P3.7, find the relationship between V_{out} and V_{in} for linear operation, and using this determine the range of V_{in} for linear operation of the op amp. Assume that the op amp is ideal.

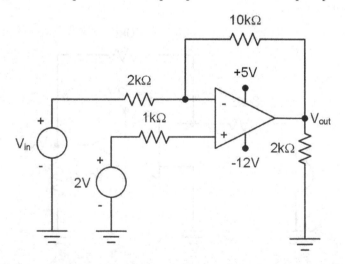

FIGURE P3.7   Op amp circuit involving a fixed voltage source, a variable input voltage, and four resistors.

Problem 3.8. For the circuit in Figure P3.8, determine the currents I_A, I_B, I_C, I_D, I_E, and I_F; find the voltages V_1, V_2, V_3, and V_4 with respect to ground. Assume that the op amps are ideal.

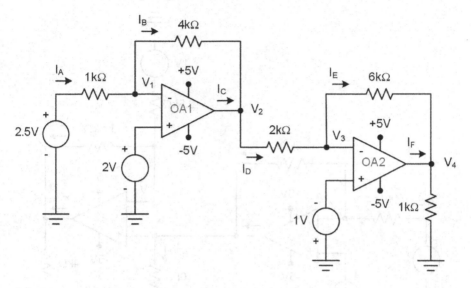

FIGURE P3.8   A circuit involving two op amps.

Problem 3.9. Determine the currents I_A, I_B, I_C, I_D, I_E, and I_F and find the voltages V_1, V_2, V_3, and V_4 with respect to ground for the circuit in Figure P3.9. Assume that the op amps are ideal.

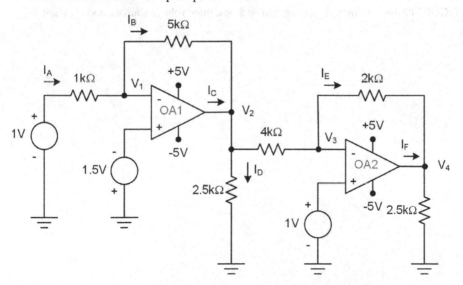

FIGURE P3.9   A circuit involving two op amps and three voltage sources.

Problem 3.10. For the circuit in Figure P3.10, determine the currents I_A, I_B, I_C, I_D, I_E, and I_F; find the voltages V_1, V_2, V_3, and V_4 with respect to ground. Assume that the op amps are ideal.

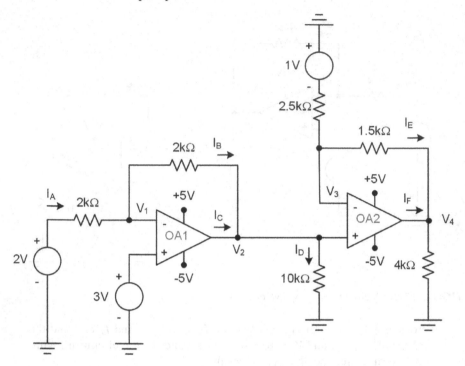

FIGURE P3.10   A circuit involving two op amps, three voltage sources, and six resistors.

Problem 3.11. Determine the currents I_A, I_B, I_C, I_D, I_E, and I_F and find the voltages V_1, V_2, V_3, and V_4 with respect to ground for the circuit in Figure P3.11. Assume that the op amps are ideal.

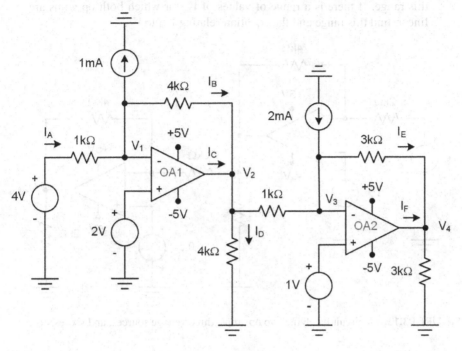

FIGURE P3.11   A circuit involving two op amps, three voltage sources, two current sources, and six resistors.

Problem 3.12. Considering the circuit in Figure P3.12, is there a range of values of V_{in} for which OA1 will be linear? If so determine this range. Is there a range of values of V_{in} for which OA2 will be linear? If so, determine this range. If there is a range of values of V_{in} for which both op amps are linear, find this range and the equation relating V_{out} to V_{in}.

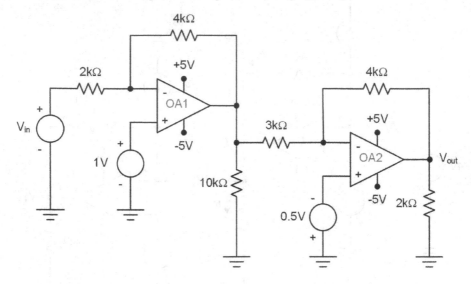

FIGURE P3.12   A circuit involving two op amps, three voltage sources, and six resistors.

Problem 3.13. Considering the circuit in Figure P3.13, is there a range of values of V_{in} for which OA1 will be linear? If so determine this range. Is there a range of values of V_{in} for which OA2 will be linear? If so, determine this range. If there is a range of values of V_{in} for which both op amps are linear, find this range and the equation relating V_{out} to V_{in}.

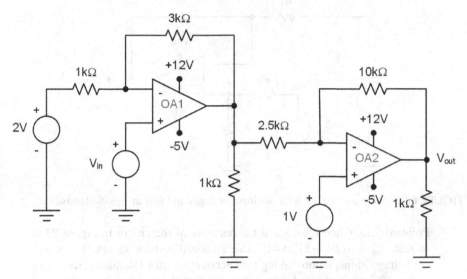

FIGURE P3.13   A circuit involving two op amps, two fixed voltage sources, a variable input voltage, and six resistors.

Problem 3.14. For the circuit in Figure P3.14, choose the values of the resistors so that $V_{out} = -4V_A - 10V_B$. Use standard resistor values. The nominal voltage gains, before taking into account resistor tolerances, should be within 10% of the desired values. Assume that the op amps are ideal.

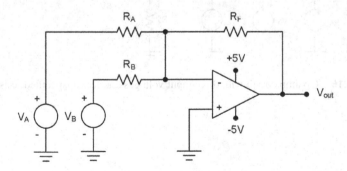

FIGURE P3.14   Op amp circuit with two input voltages and three unspecified resistors.

Problem 3.15. Choose values of the resistors in the circuit in Figure P3.15 so that $V_{out} = 20(V_B - V_A)$. Use standard resistor values. The nominal voltage gains, before taking into account resistor tolerances, should be within 10% of the desired values. Assume that the op amps are ideal (Figure P3.15).

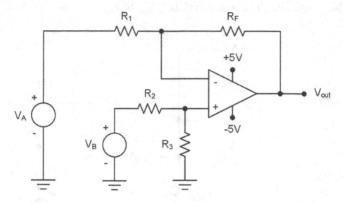

FIGURE P3.15   Op amp circuit with two input voltages and four unspecified resistors.

Problem 3.16. Choose values of the resistors of the circuit in Figure P3.16 so that $V_{out} = -12V_A - 8V_B + 4V_C$. Use standard resistor values. The nominal voltage gains, before taking into account resistor tolerances, should be within 10% of the desired values. Assume that the op amps are ideal.

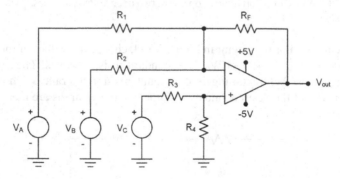

FIGURE P3.16   Op amp circuit with three input voltages and five unspecified resistors.

Problem 3.17. Choose values of the unspecified resistors of the circuit in Figure P3.17 so that the absolute value of voltage gain for V_1 is adjustable from 0 to 20 and the absolute value of voltage gain for V_2 is adjustable from 0 to 40. Use standard 5% resistor values. The nominal voltage gains, before taking into account resistor tolerances, should be within 10% of the desired values. Assume that the op amps are ideal.

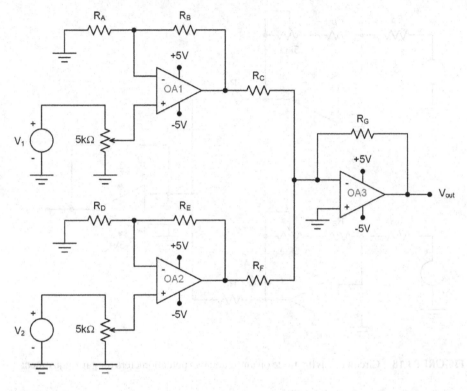

FIGURE P3.17 Circuit involving three op amps and two potentiometers for gain adjustment.

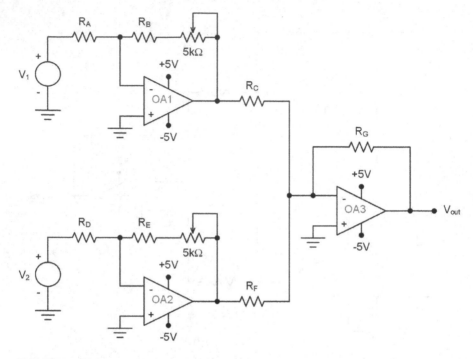

Problem 3.18. Choose values of the unspecified resistors of the circuit in Figure P3.18 so that the absolute value of voltage gain for V_1 is adjustable from 5 to 20 and the absolue value of voltage gain for V_2 is adjustable from 10 to 40. Use standard 5% resistor values. The nominal voltage gains, before taking into account resistor tolerances, should be within 10% of the desired values. Assume that the op amps are ideal.

FIGURE P3.18   Circuit involving three op amps and two potentiometers for gain adjustment.

Problem 3.19. Choose values of the potentiometers and resistors of the circuit shown in Figure P3.19 so that the absolute value of the voltage gain for V_A is adjustable from 0 to 15 and the absolute value of the voltage gain for V_B is adjustable from 10 to 100. You may use potentiometers with any of the following values: $5\,k\Omega$, $10\,k\Omega$, and $20\,k\Omega$. For the fixed resistors, use standard 5% values. The nominal voltage gains, before taking into account resistor and potentiometer tolerances, should be within 10% of the desired values. Assume that the op amps are ideal.

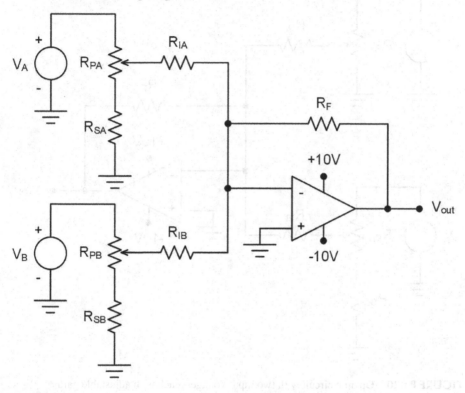

FIGURE P3.19   Op amp circuit with two potentiometers for gain adjustment.

Problem 3.20. Design an op amp circuit based on the diagram of figure P3.20 so that $V_{\text{out}} = -aV_A - bV_B$, where V_A and V_B are the input voltages and the gain coefficients are adjustable according to the following: $0 \le a \le 50$ and $10 \le b \le 100$. Potentiometers of values $5\,\text{k}\Omega$, $10\,\text{k}\Omega$, or $20\,\text{k}\Omega$ may be used. Use standard 5% fixed resistor values. The nominal voltage gains, before taking into account resistor tolerances, should be within 10% of the desired values. Assume that the op amps are ideal.

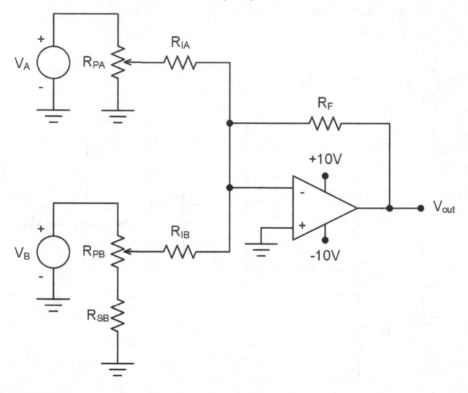

FIGURE P3.20   Op amp circuit with two input voltages, each with adjustable gain

Problem 3.21. For the op amp circuit in Figure P3.21, determine the currents I_A, I_B, I_C, I_D, I_E, and I_F; find the voltages V_1, V_2, V_3, V_4, V_5, and V_6 with respect to ground. Assume that the op amps are ideal.

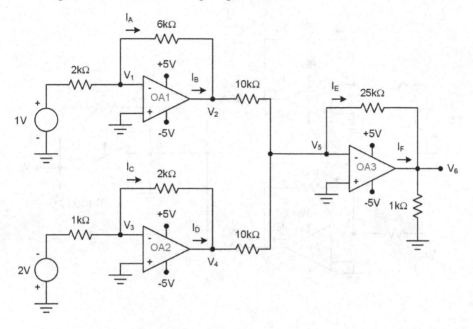

FIGURE P3.21 Circuit involving three op amps and two voltage sources.

Problem 3.22. For the circuit in Figure P3.22, determine the currents I_A, I_B, I_C, I_D, I_E, and I_F; find the voltages V_1, V_2, V_3, V_4, V_5, and V_6 with respect to ground. Assume that the op amps are ideal.

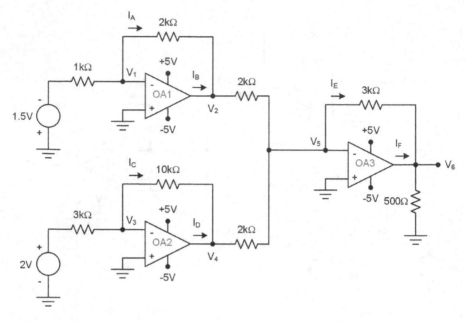

FIGURE P3.22   Circuit involving three op amps, two voltage sources, and eight resistors.

Problem 3.23. Determine the currents I_A, I_B, I_C, I_D, I_E, and I_F and find the voltages V_1, V_2, V_3, V_4, V_5, and V_6 with respect to ground of the circuit in Figure P3.23. Assume that the op amps are ideal.

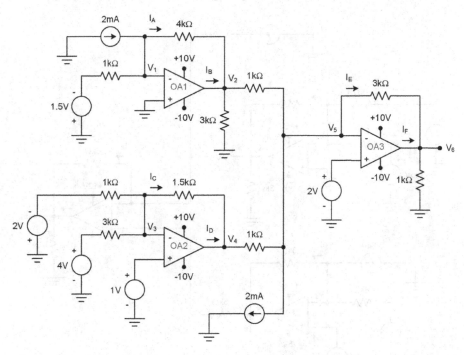

FIGURE P3.23 Circuit involving three op amps, five voltage sources, two current sources, and ten resistors.

Problem 3.24. For the circuit in Figure P3.24, determine the currents I_A, I_B, I_C, I_D, I_E, and I_F; find the voltages V_1, V_2, V_3, V_4, V_5, and V_6 with respect to ground. Assume that the op amps are ideal.

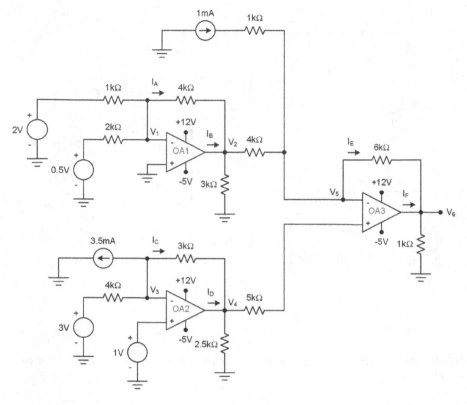

FIGURE P3.24   Circuit involving three op amps, four voltage sources, two current sources, and twelve resistors.

Problem 3.25. For the circuit in Figure P3.25, determine the currents I_A, I_B, I_C, I_D, I_E, I_F, I_G, and I_H; find the voltages V_1, V_2, V_3, V_4, V_5, V_6, V_7, and V_8 with respect to ground. Assume that the op amps are ideal.

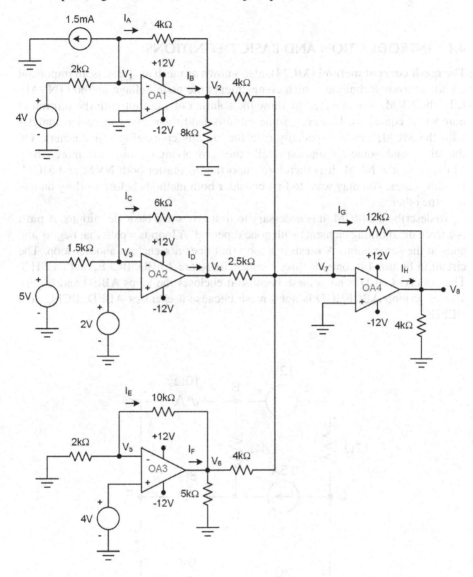

FIGURE P3.25 Circuit involving four op amps.

4 Mesh Analysis

4.1 INTRODUCTION AND BASIC DEFINITIONS

The **mesh current method (MCM)**, also known as mesh analysis, is an important circuit analysis technique, which complements the node voltage method (NVM). Like the NVM, it is intended to allow the solution of a circuit with the minimum number of equations. However, some circuits lend themselves to easier analysis with the MCM; this is especially true for circuits containing transformers. On the other hand, some circuits especially those involving op amps are more easily analyzed by the NVM. It is therefore important to master both NVM and MCM. In many cases, you may want to first consider both methods before settling on one over the other.

To describe the MCM, it is necessary to first give some basic definitions. A **path** is a trace of adjoining elements with none repeated. A **loop** is a path that begins and ends at the same point. A **mesh** is a loop that does not enclose another loop. The circuit in Figure 4.1 contains three meshes: these are ABED, BCFE, and DEFHG. The loop ABCFED is not a mesh because it encloses the loops ABED and BCFE. The outer loop ABCFHGD is not a mesh because it encloses ABED, BCFE, and DEFHG.

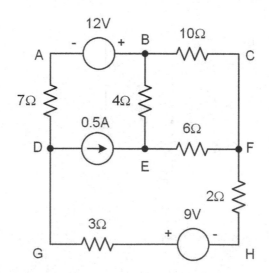

FIGURE 4.1   A circuit with three meshes.

DOI: 10.1201/9781003408529-4

4.2 APPLICABILITY OF THE MESH CURRENT METHOD

The MCM is applicable only to **planar circuits**. A planar circuit is one which may be drawn on a flat piece of paper without the use of crossovers. In the case of a non-planar circuit, we are unable to define mesh currents without ambiguities or contradictions, and therefore the NVM must be used for non-planar circuits.

The circuit shown in Figure 4.2 is non-planar because it cannot be drawn without crossovers. Therefore, the MCM is not applicable to this circuit, and we would need to use the NVM instead.

The circuit shown in Figure 4.3 initially appears to be non-planar, because it includes two crossovers. However, it may be redrawn without the use of crossovers as shown in Figure 4.4, and therefore, we could analyze this circuit using the MCM. (This circuit includes five meshes.)

4.3 THE BASIC MESH CURRENT METHOD (MCM)

To illustrate the basic MCM, we will use the circuit in Figure 4.5 as an example. There are five steps in the MCM. First, we identify the meshes; the example circuit has two.

The second step is to label and number the mesh currents as shown in Figure 4.6. As a matter of convention, we define all mesh currents to flow clockwise. Although it is possible to define counterclockwise mesh currents or even mix clockwise and counterclockwise currents, it is strongly recommended that you adhere to the clockwise convention in order to avoid sign errors. The numbering is arbitrary and may be chosen for convenience; the basic quantities in the circuit are invariant with respect to our chosen numbering of the mesh currents.

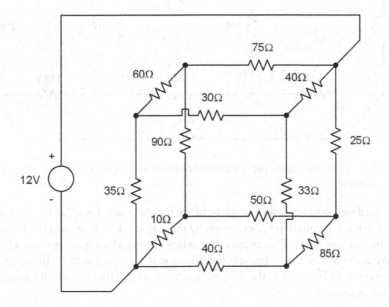

FIGURE 4.2 A non-planar circuit.

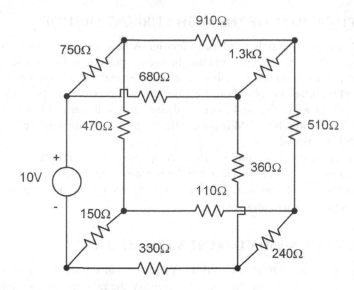

FIGURE 4.3   A planar circuit that uses unnecessary crossovers.

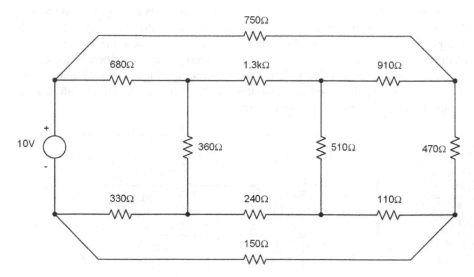

FIGURE 4.4   The circuit of the previous figure redrawn without the use of crossovers, showing that it is planar.

The third step is to write an equation for each mesh using Kirchhoff's voltage law (KVL). Each mesh equation is developed by moving clockwise around the mesh and adding the voltage drops. This means that, when moving clockwise, we add a term if we arrive at the plus sign first but subtract the term if we arrive at the minus sign first. With reference to Figure 4.7, the two mesh equations are therefore (with numerical quantities in units of V)

$$M1 \quad -5 + V_A + V_B + V_C = 0 \tag{4.1}$$

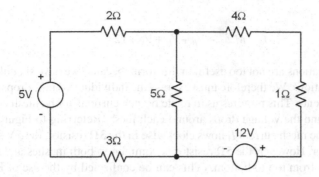

FIGURE 4.5 Example circuit with two meshes.

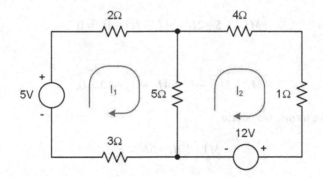

FIGURE 4.6 Example circuit with the mesh currents labeled.

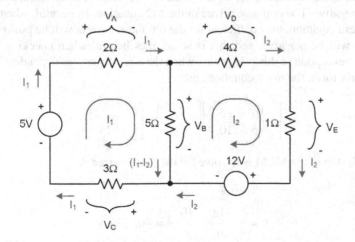

FIGURE 4.7 Example circuit with branch currents and voltage drops labeled.

and

$$M2 \quad -V_B + V_D + V_E + 12 = 0. \tag{4.2}$$

The mesh equations are not too useful in this form, because we have six unknowns but only two equations. We therefore must express the individual voltage drops in terms of the mesh currents. This requires us to relate branch currents to the mesh currents and use these to find the voltage drops around each mesh. Referring to Figure 4.7, it can be seen that the mesh current I_1 flows clockwise in the $3\,\Omega$ resistor, the $5\,\text{V}$ source, and the $2\,\Omega$ resistor. However, the $5\,\Omega$ resistor is common to both meshes and the current $(I_1 - I_2)$ flows from top to bottom. (This can be confirmed by the use of Kirchhoff's current law (KCL) at the bottom node.) Using Ohm's law, with units of V, A, and Ω, we can find the voltage drops in mesh one to be $V_A = 2I_1$, $V_B = 5(I_1 - I_2)$, and $V_C = 3I_1$. Similarly, the voltage drops in mesh two are $V_D = 4I_2$ and $V_E = 1I_2$.

The two mesh equations may therefore be written with units of V, A, and Ω as

$$M1 \quad -5 + 2I_1 + 5(I_1 - I_2) + 3I_1 = 0 \tag{4.3}$$

and

$$M2 \quad 5(I_2 - I_1) + 4I_2 + 1I_2 + 12 = 0. \tag{4.4}$$

Collecting like terms, we obtain

$$M1 \quad 10I_1 - 5I_2 = 5 \tag{4.5}$$

and

$$M2 \quad -5I_1 + 10I_2 = -12. \tag{4.6}$$

It should be noted that in the M1 equation the I_1 coefficient is positive but the I_2 coefficient is negative. The opposite is true in the M2 equation. In general, when writing the nth mesh equation, the coefficient for the nth mesh current will be positive while all others will be negative, and this is sometimes helpful when checking for sign errors. The exception to this rule occurs when the circuit contains dependent sources.

In matrix form, the mesh equations are

$$\begin{bmatrix} 10 & -5 \\ -5 & 10 \end{bmatrix} \begin{bmatrix} I_1 \\ I_2 \end{bmatrix} = \begin{bmatrix} 5 \\ -12 \end{bmatrix}. \tag{4.7}$$

The fourth step in the MCM is to solve for the mesh currents:

$$I_1 = \frac{\begin{vmatrix} 5 & -5 \\ -12 & 10 \end{vmatrix}}{\begin{vmatrix} 10 & -5 \\ -5 & 10 \end{vmatrix}} = -0.1333\,\text{A} \tag{4.8}$$

and

$$I_2 = \frac{\begin{vmatrix} 10 & 5 \\ -5 & -12 \end{vmatrix}}{\begin{vmatrix} 10 & -5 \\ -5 & 10 \end{vmatrix}} = -1.267A \qquad (4.9)$$

The fifth step in application of the MCM is to determine other circuit quantities as needed by using the mesh current values. As an example, the power in the $5\,\Omega$ resistor is

$$P_{5\Omega} = 5\Omega(I_1 - I_2)^2 = 5\Omega\left[-0.1333\,\text{A} - (-1.267\,\text{A})\right]^2 = 6.43\,\text{W}. \qquad (4.10)$$

4.4 THE MESH CURRENT METHOD WITH MORE THAN TWO MESHES

The MCM may be readily extended to more than two meshes. When solving a circuit with n_m meshes, we will have n_m mesh equations to solve for n_m mesh currents (unless there are special cases, to be described in the next two sections). To illustrate this, we will solve the circuit in Figure 4.8, which contains four meshes.

We will use the mesh numbering scheme shown in Figure 4.9. (We could use a different choice of mesh numbers if we prefer. This would change the definition of the mesh currents and therefore the values of I_1, I_2, I_3, and I_4, but it would not change basic circuit quantities such as I_X.)

The mesh equations with units of V, A, and Ω are

$$M1 \quad 15 + 5I_1 + 7(I_1 - I_3) + 4(I_1 - I_2) = 0, \qquad (4.11)$$

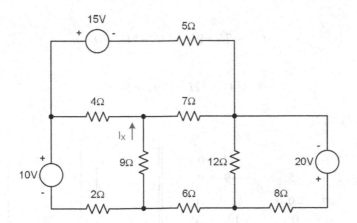

FIGURE 4.8 Example circuit with four meshes.

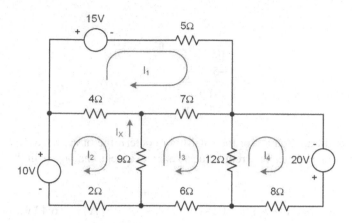

FIGURE 4.9 Example circuit with four meshes labeled.

$$M2 \quad -10+4(I_2-I_1)+9(I_2-I_3)+2I_2 = 0, \tag{4.12}$$

$$M3 \quad 9(I_3-I_2)+7(I_3-I_1)+12(I_3-I_4)+6I_3 = 0, \tag{4.13}$$

and

$$M4 \quad 12(I_4-I_3)-20+8I_4 = 0. \tag{4.14}$$

Collecting like terms,

$$M1 \quad 16I_1-4I_2-7I_3 =-15, \tag{4.15}$$

$$M2 \quad -4I_1+15I_2-9I_3 = 10, \tag{4.16}$$

$$M3 \quad -7I_1-9I_2+34I_3-12I_4 = 0, \tag{4.17}$$

and

$$M4 \quad -12I_3+20I_4 = 20. \tag{4.18}$$

In matrix form,

$$\begin{bmatrix} 16 & -4 & -7 & 0 \\ -4 & 15 & -9 & 0 \\ -7 & -9 & 34 & -12 \\ 0 & 0 & -12 & 20 \end{bmatrix} \begin{bmatrix} I_1 \\ I_2 \\ I_3 \\ I_4 \end{bmatrix} = \begin{bmatrix} -15 \\ 10 \\ 0 \\ 20 \end{bmatrix}. \tag{4.19}$$

It should be noted that the zero coefficients must be entered into the matrix. Solving,

$$I_1 = \frac{\begin{vmatrix} -15 & -4 & -7 & 0 \\ 10 & 15 & -9 & 0 \\ 0 & -9 & 34 & -12 \\ 20 & 0 & -12 & 20 \end{vmatrix}}{\begin{vmatrix} 16 & -4 & -7 & 0 \\ -4 & 15 & -9 & 0 \\ -7 & -9 & 34 & -12 \\ 0 & 0 & -12 & 20 \end{vmatrix}} = -0.410 \text{ A}, \tag{4.20}$$

$$I_2 = \frac{\begin{vmatrix} 16 & -15 & -7 & 0 \\ -4 & 10 & -9 & 0 \\ -7 & 0 & 34 & -12 \\ 0 & 20 & -12 & 20 \end{vmatrix}}{\begin{vmatrix} 16 & -4 & -7 & 0 \\ -4 & 15 & -9 & 0 \\ -7 & -9 & 34 & -12 \\ 0 & 0 & -12 & 20 \end{vmatrix}} = 0.954 \text{ A}, \tag{4.21}$$

$$I_3 = \frac{\begin{vmatrix} 16 & -4 & -15 & 0 \\ -4 & 15 & 10 & 0 \\ -7 & -9 & 0 & -12 \\ 0 & 0 & 20 & 20 \end{vmatrix}}{\begin{vmatrix} 16 & -4 & -7 & 0 \\ -4 & 15 & -9 & 0 \\ -7 & -9 & 34 & -12 \\ 0 & 0 & -12 & 20 \end{vmatrix}} = 0.661 \text{ A}, \tag{4.22}$$

and

$$I_4 = \frac{\begin{vmatrix} 16 & -4 & -7 & -15 \\ -4 & 15 & -9 & 10 \\ -7 & -9 & 34 & 0 \\ 0 & 0 & -12 & 20 \end{vmatrix}}{\begin{vmatrix} 16 & -4 & -7 & 0 \\ -4 & 15 & -9 & 0 \\ -7 & -9 & 34 & -12 \\ 0 & 0 & -12 & 20 \end{vmatrix}} = 1.397\,\text{A}. \tag{4.23}$$

Having found the mesh currents, we can easily determine all other circuit quantities. For example,

$$I_X = I_3 - I_2 = 0.661\,\text{A} - 0.954\,\text{A} = -0.293\,\text{A}. \tag{4.24}$$

The MCM is especially convenient when we need to find the power for voltage sources. In this example,

$$P_{10\text{V}} = (-1)(10\,\text{V})I_2 = (-1)(10\,\text{V})(0.954\,\text{A}) = -9.54\,\text{W}. \tag{4.25}$$

$$P_{15\text{V}} = (15\,\text{V})I_1 = (15\,\text{V})(-0.410\,\text{A}) = -6.15\,\text{W}. \tag{4.26}$$

$$P_{20\text{V}} = (-1)(20\,\text{V})I_4 = (-1)(20\,\text{V})(1.397\,\text{A}) = -27.9\,\text{W}. \tag{4.27}$$

Multiplication by −1 was necessary in the power calculations for the 10 and 20 V sources according to the passive sign convention, because in both cases the reference direction for the electrical current was entering the negative terminal for the voltage source. All three power values are negative, indicating that all three voltage sources are developing power.

To review, see Presentation 4.1 in ebook+. To test your knowledge, try Quiz 4.1 in ebook+. To put your knowledge to practice, try Laboratory Exercise 4.1 in ebook+.

4.5   THE MESH CURRENT METHOD WITH A KNOWN MESH CURRENT

In the MCM, special cases arise whenever there is a current source (independent or dependent) in the circuit. If the current source appears on the periphery of the circuit, and only in one mesh, it gives rise to a **known mesh current** (KMC). If the current source is on the interior of the circuit, and shared by two meshes, it gives rise to a **supermesh**. (The supermesh will be described in more detail in the next section.)

An example of a circuit with a KMC is shown in Figure 4.10. Here, there are three meshes, and we would generally expect to set up and solve three mesh equations. However, the presence of a special case allows us to reduce the system to two equations. In general, if there are n_m meshes and n_s special cases, the system will reduce to $(n_m - n_s)$ equations.

We will use the mesh numbering scheme shown in Figure 4.11. Here, there is a current source in mesh one, so I_1 is a KMC: $I_1 = -1\,\text{A}$. The minus sign results from the fact that the direction for the current source (counterclockwise) is opposite to the

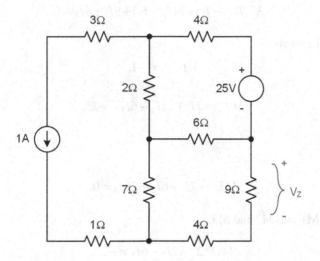

FIGURE 4.10   A circuit having a KMC.

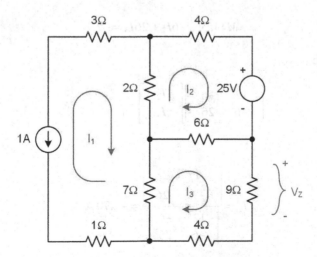

FIGURE 4.11   A circuit having a KMC with the mesh currents labeled.

reference direction for the mesh current (clockwise). The other two mesh equations can be written in the normal fashion, using KVL. Therefore, the three mesh equations with units of V, A, and Ω are

$$M1 \quad I_1 = -1, \tag{4.28}$$

$$M2 \quad 2(I_2 - I_1) + 4I_2 + 25 + 6(I_2 - I_3) = 0, \tag{4.29}$$

and

$$M3 \quad 7(I_3 - I_1) + 6(I_3 - I_2) + 9I_3 + 4I_3 = 0. \tag{4.30}$$

Collecting like terms,

$$M1 \quad I_1 = -1, \tag{4.31}$$

$$M2 \quad -2I_1 + 12I_2 - 6I_3 = -25, \tag{4.32}$$

and

$$M3 \quad -7I_1 - 6I_2 + 26I_3 = 0. \tag{4.33}$$

Substituting M1 into M2 and M3,

$$M1M2 \quad 12I_2 - 6I_3 = -27, \tag{4.34}$$

and

$$M1M3 \quad -6I_2 + 26I_3 = -7. \tag{4.35}$$

In matrix form,

$$\begin{bmatrix} 12 & -6 \\ -6 & 26 \end{bmatrix} \begin{bmatrix} I_2 \\ I_3 \end{bmatrix} = \begin{bmatrix} -27 \\ -7 \end{bmatrix}. \tag{4.36}$$

Solving,

$$I_2 = \frac{\begin{vmatrix} -27 & -6 \\ -7 & 26 \end{vmatrix}}{\begin{vmatrix} 12 & -6 \\ -6 & 26 \end{vmatrix}} = -2.70 \text{ A} \tag{4.37}$$

and

$$I_3 = \frac{\begin{vmatrix} 12 & -27 \\ -6 & -7 \end{vmatrix}}{\begin{vmatrix} 12 & -6 \\ -6 & 26 \end{vmatrix}} = -0.891\,\text{A}. \tag{4.38}$$

Other circuit quantities may be determined as needed. For example,

$$V_Z = (9\,\Omega)I_3 = (9\,\Omega)(-0.891\,\text{A}) = -8.02\,\text{V} \tag{4.39}$$

and

$$P_{25\text{V}} = (25\,\text{V})I_2 = (25\,\text{V})(-2.70\,\text{A}) = -67.5\,\text{W} \tag{4.40}$$

Whereas the MCM allows very convenient determination of power for voltage sources, power determination for a current source is somewhat cumbersome because we have to apply KVL to find the voltage across the current source. For example,

$$P_{1\text{A}} = (1\,\text{A})\left[(3\,\Omega)(I_1) + (2\,\Omega)(I_1 - I_2) + (7\,\Omega)(I_1 - I_3) + (1\,\Omega)(I_1)\right]$$
$$= -1.363\,\text{W}. \tag{4.41}$$

4.6   THE MESH CURRENT METHOD WITH A SUPERMESH

Another special case in the MCM is the **supermesh**; this is introduced by a current source (independent or dependent) which is on the interior of the circuit, and shared by two meshes. An example is shown in Figure 4.12.

   If we use the mesh numbering shown in Figure 4.13, then meshes two and three together constitute a supermesh. If we attempt to write mesh equations for each of the individual meshes, we have to invoke a new variable v_x, which represents the voltage drop across the current source in the supermesh. In units of V, A, and Ω,

$$M1 \quad -12 + 6I_1 + 1I_1 + 4 + 8(I_1 - I_3) + 3(I_1 - I_2) = 0, \tag{4.42}$$

$$M2 \quad -10 + 3(I_2 - I_1) + v_x + 4I_2 = 0, \tag{4.43}$$

and

$$M3 \quad -v_x + 8(I_3 - I_1) + 3I_3 + 2I_3 = 0. \tag{4.44}$$

As it stands, the previous set of equations may not be solved because there are four unknowns but only three equations. It is therefore more useful to write a single equation for the supermesh. This could be found by adding the M2 and M3 equations, but

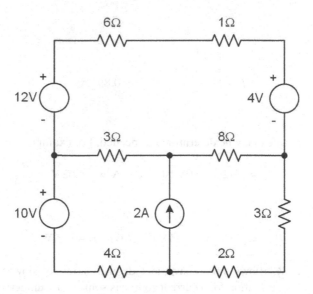

FIGURE 4.12   A circuit containing a supermesh.

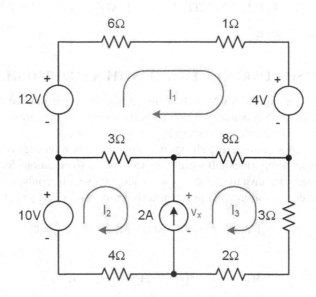

FIGURE 4.13   A circuit having a supermesh with the mesh currents labeled.

usually we will write the supermesh equation directly by use of KVL. Therefore, the mesh equations are

$$M1 \quad -12 + 6I_1 + 1I_1 + 4 + 8(I_1 - I_3) + 3(I_1 - I_2) = 0 \qquad (4.45)$$

and

$$M23 \quad -10+3(I_2-I_1)+8(I_3-I_1)+3I_3+2I_3+4I_2=0. \qquad (4.46)$$

One more equation is needed and this is the equation of the current source:

$$CS \quad I_3-I_2=2 \qquad (4.47)$$

or

$$CS \quad I_3=I_2+2. \qquad (4.48)$$

This equation may be obtained by use of KCL at the bottom node. If we substitute this current source equation into the M1 and M23 equations, we obtain a set of two equations with two unknowns:

$$M1CS \quad -12+6I_1+1I_1+4+8(I_1-I_2-2)+3(I_1-I_2)=0 \qquad (4.49)$$

and

$$M23CS \quad -10+3(I_2-I_1)+8(I_2+2-I_1)+3(I_2+2)+2(I_2+2)+4I_2=0. \qquad (4.50)$$

Collecting like terms,

$$M1CS \quad 18I_1-11I_2=24 \qquad (4.51)$$

and

$$M23CS \quad -11I_1+20I_2=-16. \qquad (4.52)$$

In matrix form,

$$\begin{bmatrix} 18 & -11 \\ -11 & 20 \end{bmatrix}\begin{bmatrix} I_1 \\ I_2 \end{bmatrix}=\begin{bmatrix} 24 \\ -16 \end{bmatrix} \qquad (4.53)$$

Solving,

$$I_1=\dfrac{\begin{vmatrix} 24 & -11 \\ -16 & 20 \end{vmatrix}}{\begin{vmatrix} 18 & -11 \\ -11 & 20 \end{vmatrix}}=1.272\,\text{A} \qquad (4.54)$$

and

$$I_2=\dfrac{\begin{vmatrix} 18 & 24 \\ -11 & -16 \end{vmatrix}}{\begin{vmatrix} 18 & -11 \\ -11 & 20 \end{vmatrix}}=-0.1004\,\text{A}. \qquad (4.55)$$

Using the current source equation,

$$I_3 = I_2 + 2 = 1.900 \, \text{A}. \tag{4.56}$$

To review, see Presentation 4.2 in ebook+.To test your knowledge, try Quiz 4.2 in ebook+.

4.7   THE DOUBLE SUPERMESH

As another example, consider the case of the double supermesh shown in Figure 4.14. The meshes are labeled in Figure 4.15, and we see that mesh two contains two current sources, each shared by an adjacent mesh. Therefore, meshes one, two, and three together constitute a **double supermesh**.

In units of V, A, and Ω, the double supermesh equation is

$$M123 \quad -9 + 3I_1 + 1.5I_2 - 6 + 1I_3 + 4I_3 + 2I_2 + 2.5I_1 = 0. \tag{4.57}$$

In addition to this, we need to write the equation for **each** of the two current sources. We will refer to the left-hand source as current source one (CS1) and the right-hand source as current source two (CS2).

$$CS1 \quad 2 = I_2 - I_1, \text{ or } I_1 = I_2 - 2 \tag{4.58}$$

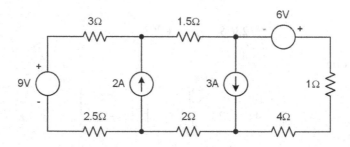

FIGURE 4.14   A circuit containing a double supermesh.

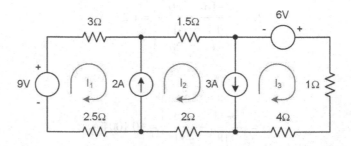

FIGURE 4.15   A circuit containing a double supermesh with the meshes labeled.

and

$$CS2 \quad 3 = I_2 - I_3, \text{ or } I_3 = I_2 - 3. \tag{4.59}$$

Collecting like terms in the double supermesh equation,

$$M123 \quad 5.5I_1 + 3.5I_2 + 5I_3 = 15. \tag{4.60}$$

Substituting the CS1 and CS2 equations into the double supermesh equation,

$$M123CS1CS2 \quad 5.5(I_2 - 2) + 3.5I_2 + 5(I_2 - 3) = 15. \tag{4.61}$$

Simplifying,

$$M123CS1CS2 \quad 14I_2 = 41. \tag{4.62}$$

Solving,

$$I_2 = 2.929 \, \text{A}, \tag{4.63}$$

$$I_1 = I_2 - 2 \, \text{A} = 0.929 \, \text{A}, \tag{4.64}$$

and

$$I_2 = I_2 - 3 \, \text{A} = -0.071 \, \text{A}. \tag{4.65}$$

Of course, it would be possible to have triple supermeshes, quadruple supermeshes, and so on, but we do not expect higher-order supermeshes to be common in practical circuits.

4.8 A SUPERMESH CONTAINING A KNOWN MESH CURRENT

In the previous section, we illustrated one combination of two special cases in the MCM: it was the combination of two supermeshes into a double supermesh. Here, we consider a second possibility of combining two special cases, which is a supermesh containing a KMC. Such a situation is shown in Figure 4.16, and with the meshes labeled in Figure 4.17.

The right-hand current source is on the periphery and within a single mesh so it gives rise to a KMC: $I_3 = -25 \, \text{mA}$. The 10 mA current source is shared by meshes two and three and therefore creates a supermesh. We will not write the supermesh equation in this case, because to do so we would need to invoke a new unknown variable to represent the voltage across the 25 mA current source. However, because I_3

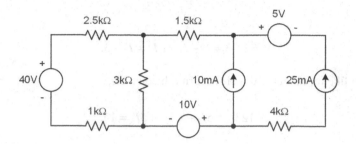

FIGURE 4.16 A circuit having a supermesh that contains a KMC.

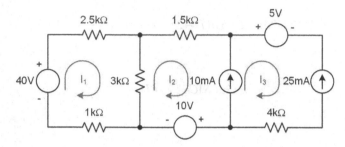

FIGURE 4.17 A circuit having a supermesh that contains a KMC with the meshes labeled.

is known we can use the equation of the 10 mA current source to find I_2. The set of three equations in units of V, mA, and kΩ is

$$M1 \quad -40 + 2.5I_1 + 3(I_1 - I_2) + 1I_1 = 0, \tag{4.66}$$

$$CS1 \quad 10 = I_3 - I_2, \text{ or } I_2 = I_3 - 10, \tag{4.67}$$

and

$$M3 \quad I_3 = -25 \, \text{mA}. \tag{4.68}$$

Substituting the M3 equation into the CS1 equation,

$$CS1M3 \quad I_2 = I_3 - 10 \, \text{mA} = -35 \, \text{mA}. \tag{4.69}$$

Substituting the value of I_2 into the M1 equation,

$$M1CS1M3 \quad -40 + 2.5I_1 + 3(I_1 - (-35)) + 1I_1 = 0. \tag{4.70}$$

Simplifying,

$$M1CS1M3 \quad 6.5I_1 = 145. \tag{4.71}$$

Solving,

$$I_1 = 22.3\,\text{mA}. \tag{4.72}$$

4.9 THE MESH CURRENT METHOD WITH A DEPENDENT SOURCE

When a dependent source appears in a circuit, the value of the controlling variable (and therefore the value of the source) must be expressed in terms of the mesh currents. If the dependent source is a current source, it will give rise to a special case – either a KMC or a supermesh. To illustrate these ideas, we can consider the circuit in Figure 4.18, which contains a current-controlled voltage source. The meshes are labeled in Figure 4.19.

This circuit has three meshes and no special cases, so we will solve a system of three mesh equations. In units of V, A, and Ω,

$$M1 \quad -4 + 2.5I_1 + 1(I_1 - I_3) + 3(I_1 - I_2) = 0, \tag{4.73}$$

$$M2 \quad 3i_x + 3(I_2 - I_1) + 2(I_2 - I_3) + 1.5I_2 = 0, \tag{4.74}$$

and

$$M3 \quad 2(I_3 - I_2) + 1(I_3 - I_1) + 0.5I_3 + 2 = 0. \tag{4.75}$$

The equation of the dependent source is

$$DS \quad i_x = I_3. \tag{4.76}$$

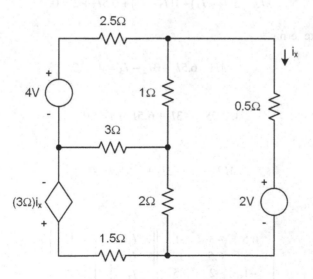

FIGURE 4.18 A circuit containing a dependent source (a current-controlled voltage source).

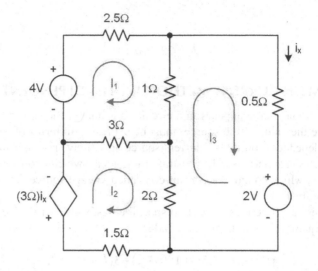

FIGURE 4.19 A circuit containing a dependent source (current-controlled voltage source) with the mesh currents labeled.

Substituting this into the M2 equation,

$$M1 \quad -4+2.5I_1+1(I_1-I_3)+3(I_1-I_2)=0, \tag{4.77}$$

$$M2DS \quad 3I_3+3(I_2-I_1)+2(I_2-I_3)+1.5I_2=0, \tag{4.78}$$

and

$$M3 \quad 2(I_3-I_2)+1(I_3-I_1)+0.5I_3+2=0. \tag{4.79}$$

Collecting like terms,

$$M1 \quad 6.5I_1-3I_2-I_3=4, \tag{4.80}$$

$$M2DS \quad -3I_1+6.5I_2+I_3=0, \tag{4.81}$$

and

$$M3 \quad -I_1-2I_2+3.5I_3=-2. \tag{4.82}$$

In matrix form,

$$\begin{bmatrix} 6.5 & -3 & -1 \\ -3 & 6.5 & 1 \\ -1 & -2 & 3.5 \end{bmatrix} \begin{bmatrix} I_1 \\ I_2 \\ I_3 \end{bmatrix} = \begin{bmatrix} 4 \\ 0 \\ -2 \end{bmatrix}. \tag{4.83}$$

Solving,

$$I_1 = \frac{\begin{vmatrix} 4 & -3 & -1 \\ 0 & 6.5 & 1 \\ -2 & -2 & 3.5 \end{vmatrix}}{\begin{vmatrix} 6.5 & -3 & -1 \\ -3 & 6.5 & 1 \\ -1 & -2 & 3.5 \end{vmatrix}} = 0.768\,\text{A}, \tag{4.84}$$

$$I_2 = \frac{\begin{vmatrix} 6.5 & 4 & -1 \\ -3 & 0 & 1 \\ -1 & -2 & 3.5 \end{vmatrix}}{\begin{vmatrix} 6.5 & -3 & -1 \\ -3 & 6.5 & 1 \\ -1 & -2 & 3.5 \end{vmatrix}} = 0.375\,\text{A}, \tag{4.85}$$

and

$$I_3 = \frac{\begin{vmatrix} 6.5 & -3 & 4 \\ -3 & 6.5 & 0 \\ -1 & -2 & -2 \end{vmatrix}}{\begin{vmatrix} 6.5 & -3 & -1 \\ -3 & 6.5 & 1 \\ -1 & -2 & 3.5 \end{vmatrix}} = -0.1376\,\text{A}. \tag{4.86}$$

As another example, consider the circuit in Figure 4.20. Here, the dependent source is a voltage-controlled current source (VCCS), and it gives rise to a special case because it is a current source.

We will use the mesh numbering shown in Figure 4.21. There are three meshes. The dependent source is a current source shared by two meshes so it gives rise to a supermesh.

The mesh equations in units of V, mA, and $k\Omega$ are

$$M12 \quad -10+2I_1+4I_2+1I_2+6(I_2-I_3)+1.5(I_1-I_3)=0 \tag{4.87}$$

and

$$M3 \quad 1.5(I_3-I_1)+6(I_3-I_2)+5I_3-20=0. \tag{4.88}$$

The equation of the dependent source is

$$DS \quad I_2-I_1 = 2v_x = 3(I_3-I_1)=0 \tag{4.89}$$

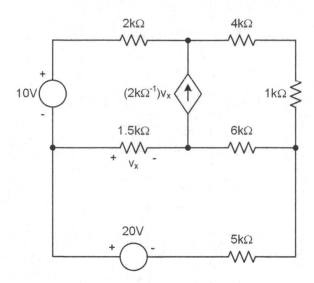

FIGURE 4.20 A circuit containing a dependent source.

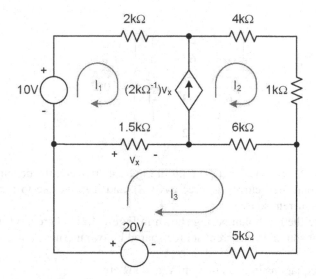

FIGURE 4.21 A circuit containing a dependent source with the mesh currents labeled.

or

$$DS \quad I_1 = 1.5I_3 - 0.5I_2. \tag{4.90}$$

Collecting like terms in the mesh equations,

$$M12 \quad 3.5I_1 + 11I_2 - 7.5I_3 = 10 \tag{4.91}$$

and

$$M3 \quad -1.5I_1 - 6I_2 + 12.5I_3 = 20. \tag{4.92}$$

Substituting the dependent source equation into the supermesh equation and the mesh three equation,

$$M12DS \quad 3.5(1.5I_3 - 0.5I_2) + 11I_2 - 7.5I_3 = 10 \tag{4.93}$$

and

$$M3DS \quad -1.5(1.5I_3 - 0.5I_2) - 6I_2 + 12.5I_3 = 20. \tag{4.94}$$

Simplifying,

$$M12DS \quad 9.25I_2 - 2.25I_3 = 10 \tag{4.95}$$

and

$$M3DS \quad -5.25I_2 + 10.25I_3 = 20. \tag{4.96}$$

In matrix form,

$$\begin{bmatrix} 9.25 & -2.25 \\ -5.25 & 10.25 \end{bmatrix} \begin{bmatrix} I_2 \\ I_3 \end{bmatrix} = \begin{bmatrix} 10 \\ 20 \end{bmatrix}. \tag{4.97}$$

Solving,

$$I_2 = \frac{\begin{vmatrix} 10 & -2.25 \\ 20 & 10.25 \end{vmatrix}}{\begin{vmatrix} 9.25 & -2.25 \\ -5.25 & 10.25 \end{vmatrix}} = 1.777 \, \text{mA} \tag{4.98}$$

and

$$I_3 = \frac{\begin{vmatrix} 9.25 & 10 \\ -5.25 & 20 \end{vmatrix}}{\begin{vmatrix} 9.25 & -2.25 \\ -5.25 & 10.25 \end{vmatrix}} = 2.86 \, \text{mA}. \tag{4.99}$$

Using the dependent source equation,

$$I_1 = 1.5I_3 - 0.5I_2 = 1.5(2.86 \, \text{mA}) - 0.5(1.777 \, \text{mA}) = 3.40 \, \text{mA}. \tag{4.100}$$

To review, see Presentation 4.3 in ebook+. To test your knowledge, try Quiz 4.3 in ebook+.

4.10 CHOOSING BETWEEN THE MCM AND THE NVM

Before solving a problem, we will generally consider both the MCM and the NVM to determine which method will require fewer equations. A solution using fewer equations is not only more efficient but also less prone to errors. To illustrate this, we will consider several example circuits, starting with the one shown in Figure 4.22.

This circuit has six meshes, as indicated in Figure 4.23. There is a single special case; the current source in mesh two gives rise to a KMC. Therefore, the number of equations that must be solved in the MCM reduces to $n_m - n_{sm} = 6 - 1 = 5$, where n_m is the number of meshes and n_{sm} is the number of special cases for the MCM.

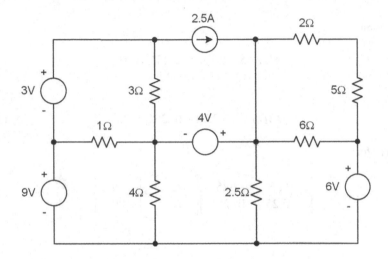

FIGURE 4.22 A circuit with six meshes and six essential nodes.

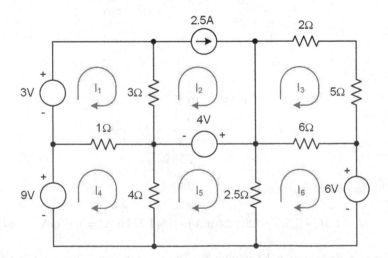

FIGURE 4.23 A circuit with six meshes and six essential nodes, with the meshes labeled.

This circuit also has six essential nodes, as shown in Figure 4.24, but there are four voltage sources resulting in four special cases for the NVM. If we make the bottom node the reference node, then we have known node voltages at node three and node five. Nodes one and three together constitute a supernode, but this involves a known node voltage so the voltage at node one can be determined immediately. Nodes two and four together constitute another supernode. Accounting for these special cases, the number of equations that must be solved when using the NVM will reduce to $n_e - n_{sn} - 1 = 6 - 4 - 1 = 1$, where n_e is the number of essential nodes and n_{sn} is the number of special cases for the NVM. Therefore, we would clearly choose the NVM, requiring one equation, over the MCM, requiring five equations, for this circuit.

Now consider the circuit of Figure 4.25.

This circuit has six meshes, as indicated in Figure 4.26. There are two special cases: meshes one and two make up a supermesh while mesh six has a KMC. The number of equations that must be solved in the MCM reduces to $n_m - n_{sm} = 6 - 2 = 4$.

This circuit has only three essential nodes, as indicated in Figure 4.27. There are no special cases for the NVM because none of the voltage sources is connected directly between essential nodes. (There is a series resistor involved in each case.) Hence, the number of equations to be solved is $n_e - n_{sn} - 1 = 3 - 0 - 1 = 2$. Once again, we would choose the NVM as the more efficient approach.

As a third example consider the circuit in Figure 4.28. This circuit is non-planar and cannot be solved using the MCM. Hence, we need to use the NVM. In this case, there are 16 essential nodes, with one supernode involving nodes two and three, so we will need to solve 14 equations (Figure 4.29).

By now you may feel that we will never use the MCM. This is not true; although we tend to use the NVM more often, we will not use it exclusively. For example,

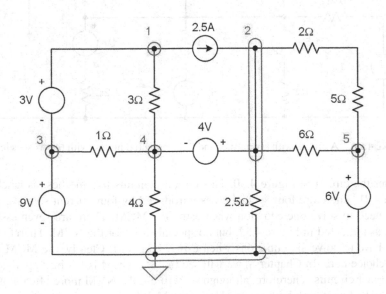

FIGURE 4.24 A circuit with six meshes and six essential nodes, with the essential nodes labeled.

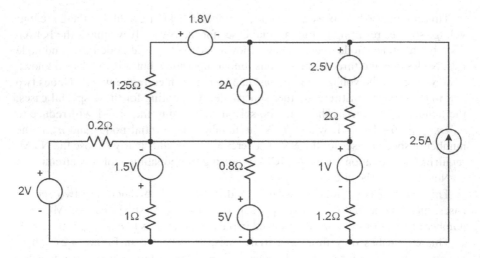

FIGURE 4.25   A circuit with six meshes and three essential nodes.

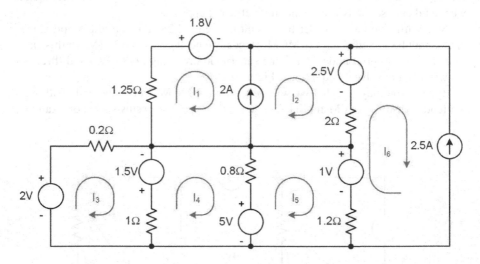

FIGURE 4.26   A circuit with six meshes and three essential nodes, with the meshes labeled.

consider the circuit in Figure 4.30. This circuit contains five meshes, as labeled in Figure 4.31. There are four special cases introduced by four current sources, so we would need to solve one equation when using the MCM. There are seven essential nodes, as indicated in Figure 4.32, but no special cases for the NVM. Therefore, we would have to solve six equations when using the NVM. Clearly, the MCM is the better choice here. In Chapter 8, we will see that the MCM is the best approach for transformer circuits. Therefore, although we will use the NVM more often, it is necessary to be fluent with both the NVM and the MCM.

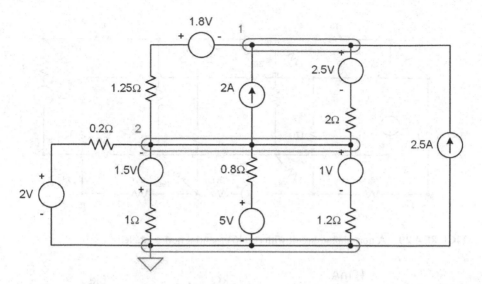

FIGURE 4.27 A circuit with six meshes and three essential nodes, with the essential nodes labeled.

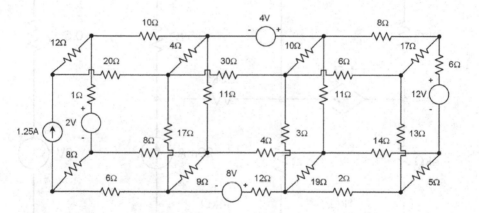

FIGURE 4.28 A non-planar circuit.

4.11 SUMMARY

The **MCM** is a general-purpose tool for solving **planar circuits**. (It is not applicable to non-planar circuits, which cannot be drawn on a flat piece of paper without the use of crossovers.) There are five steps in the basic MCM. First, we identify the meshes in the circuit. A **mesh** is a loop that does not enclose any other loops. Second, we label and number the meshes. As a convention, we define each mesh current as flowing clockwise around its mesh. Third, we write an equation for each mesh using KVL. As a convention, we travel clockwise around the mesh and add

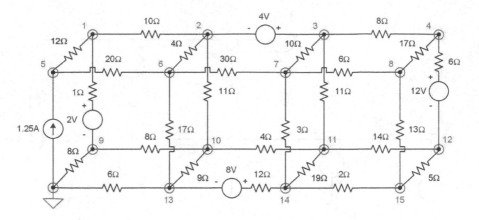

FIGURE 4.29 A non-planar circuit with the 16 essential nodes labeled.

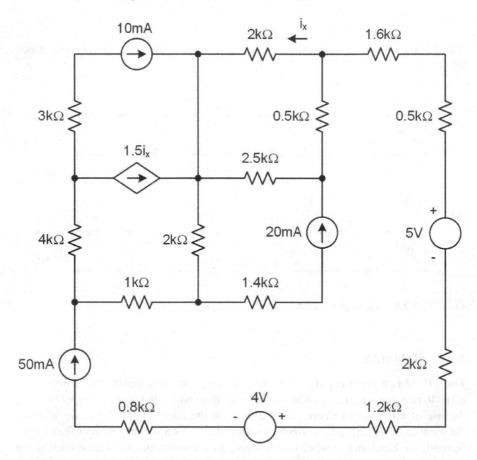

FIGURE 4.30 A circuit with five meshes and seven essential nodes.

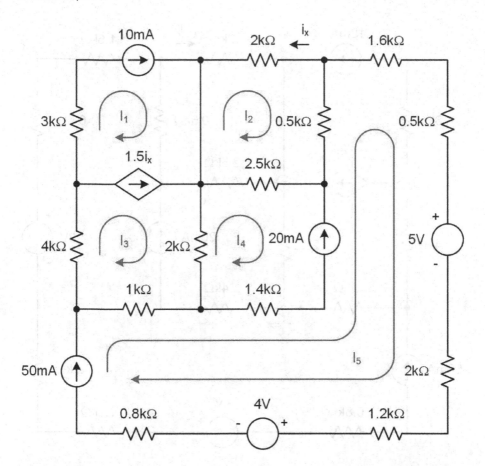

FIGURE 4.31 A circuit with five meshes and seven essential nodes labeled for implementation of the MCM.

the voltage drops. While doing this we must recognize that a branch shared by two meshes carries the difference in the individual mesh currents. Fourth, we solve the system of mesh equations using algebraic methods. Fifth, we use the mesh currents to determine all other circuit quantities as needed.

Special cases are introduced by the presence of current sources, either independent or dependent. A current source on the periphery of the circuit, which is contained in a single mesh, gives rise to a **KMC**. A current source on the interior of a circuit, shared by two meshes, gives rise to a **supermesh**. Each special case can be used to reduce the number of equations which must be solved by one. Therefore, the number of equations that must be solved is equal to $n_m - n_{sm}$, where n_m is the number of meshes and n_{sm} is the number of special cases (current sources) for the MCM.

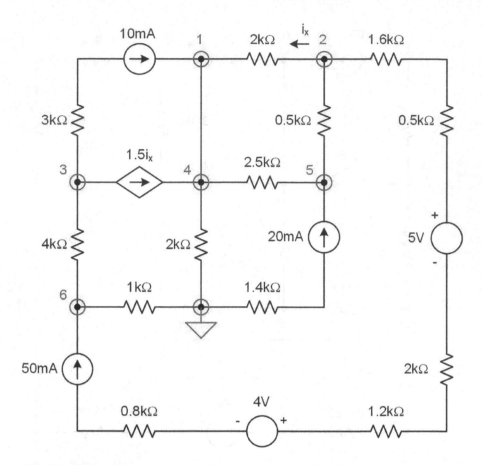

FIGURE 4.32 A circuit with five meshes and seven essential nodes labeled for the implementation of the NVM.

When using the MCM to solve a circuit containing a dependent source, we must express the value of the controlling variable, and therefore the dependent source, in terms of the mesh currents. If the dependent source is a current source, either a VCCS or a current-controlled current source, it will result in a special case as described in the previous paragraph.

Before solving a circuit, we should first evaluate the applicability and complexity (number of equations required) for the NVM and the MCM. Only the NVM is applicable to non-planar circuits, but the MCM is strongly preferred for transformer circuits, as we will see in Chapter 8. If either method is applicable, we will choose the method that results in fewer equations.

PROBLEMS

Problem 4.1. Is the circuit in Figure P4.1 planar? If it is, redraw the circuit without crossovers and determine the number of meshes. Determine the number of essential nodes.

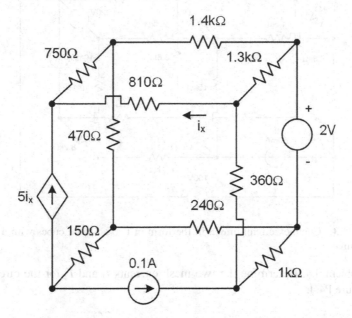

FIGURE P4.1   Complex circuit drawn in the form of a cube.

Problem 4.2. Determine if the circuit shown in Figure P4.2 is planar. If it is, determine the number of meshes. Also determine the number of essential nodes.

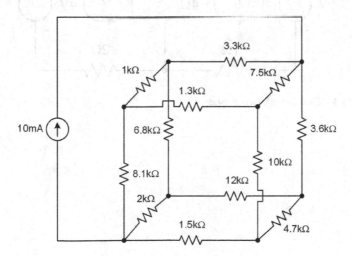

FIGURE P4.2   Complex circuit drawn using two crossovers.

Problem 4.3. Try to draw the circuit in Figure P4.3 without crossovers and thus determine if it is planar. If it is, find the number of meshes. Also find the number of essential nodes. Determine the number of essential nodes.

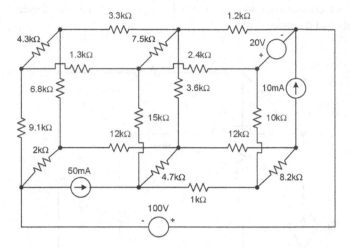

FIGURE P4.3   Complex circuit drawn in the form of two adjacent cubes with a diagonal voltage source.

Problem 4.4. Determine the two mesh currents I_1 and I_2 for the circuit in Figure P4.4.

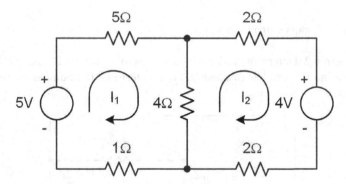

FIGURE P4.4   Circuit with two meshes.

Problem 4.5. Find the mesh currents I_1, I_2, and I_3 for the circuit in Figure P4.5. Determine the power for each of the voltage sources.

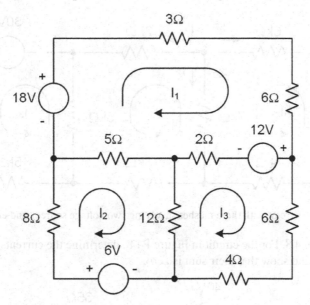

FIGURE P4.5   Circuit with three meshes.

Problem 4.6. For the circuit in Figure P4.6, determine the mesh currents I_A, I_B, and I_C. Find the power in the $8\,\Omega$ and $15\,\Omega$ resistors.

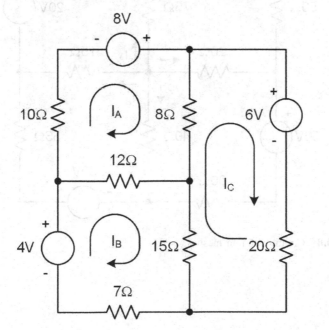

FIGURE P4.6   Circuit with three meshes involving three sources and six resistors.

Problem 4.7. Find the mesh currents I_1, I_2, and I_3 and the branch currents I_X, I_Y, and I_Z for the circuit in Figure P4.7.

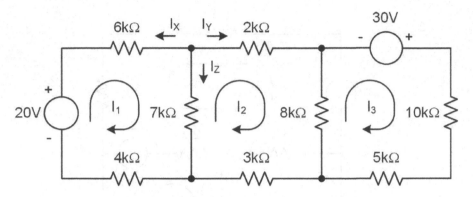

FIGURE P4.7   Circuit with three meshes involving two voltage sources and eight resistors.

Problem 4.8. For the circuit in Figure P4.8, determine the currents I_A, I_B, I_C, and I_D and show that their sum is zero.

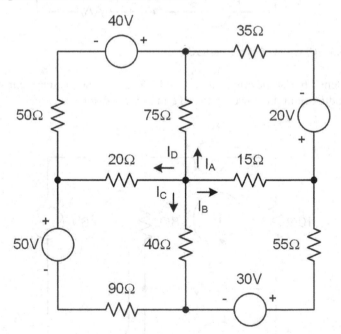

FIGURE P4.8   Circuit with four meshes.

Problem 4.9. Calculate the mesh currents I_1, I_2, I_3, and I_4 for the circuit in Figure P4.9. Find the power for each of the voltage sources.

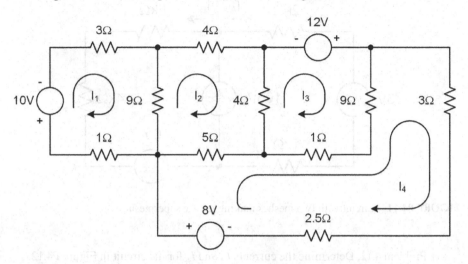

FIGURE P4.9   Circuit with four meshes, three sources, and eleven resistors.

Problem 4.10. Determine the currents I_X and I_Y for the circuit in Figure P4.10.

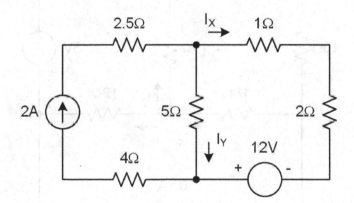

FIGURE P4.10   Circuit with two meshes and one known mesh current.

Problem 4.11. Determine the currents I_A and I_B for the circuit in Figure P4.11.

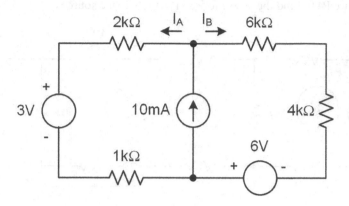

FIGURE P4.11 Circuit with two meshes making up one supermesh.

Problem 4.12. Determine the currents I_A and I_B for the circuit in Figure P4.12.

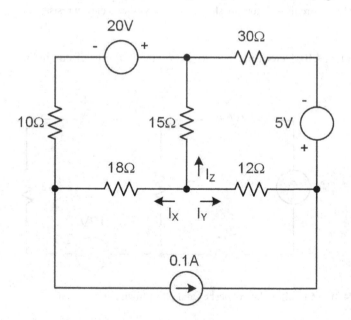

FIGURE P4.12 Circuit with three meshes and one known mesh current.

Problem 4.13. For the circuit in Figure P4.13, find the mesh currents I_1, I_2, and I_3. Determine the power for each of the sources.

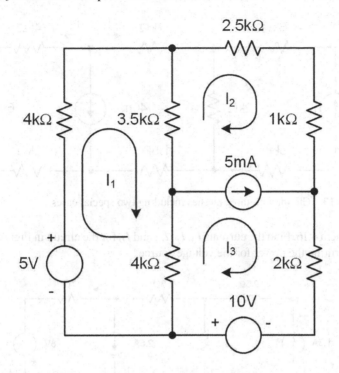

FIGURE P4.13   Circuit with three meshes including one supermesh.

Problem 4.14. For the circuit in Figure P4.14, find the currents I_A, I_B, I_C, and I_D. Calculate the power for the $8\,\Omega$ resistor.

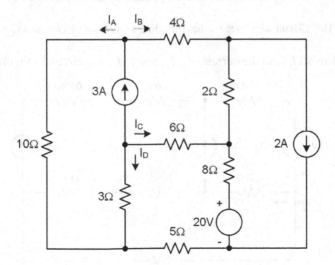

FIGURE P4.14   Circuit with four meshes including two special cases.

Problem 4.15. For the circuit in Figure P4.15, find the currents I_A and I_B. Determine the power for the voltage source.

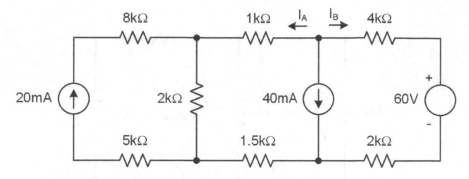

FIGURE P4.15 Circuit with three meshes including two special cases.

Problem 4.16. Find the currents I_A, I_B, I_C, and I_D for the circuit in Figure P4.16. Determine the power for the voltage source.

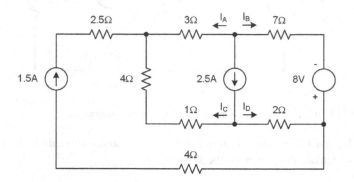

FIGURE P4.16 Circuit with three meshes including a known mesh current and a supermesh.

Problem 4.17. Find the currents I_A, I_B, and I_C for the circuit in Figure P4.17.

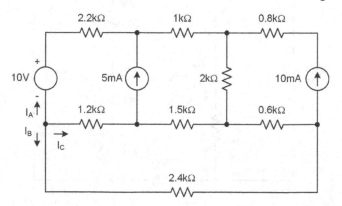

FIGURE P4.17 Circuit with four meshes including a known mesh current and a supermesh.

Problem 4.18. For the circuit in Figure P4.18, find the voltages V_1, V_2, V_3, and V_4.

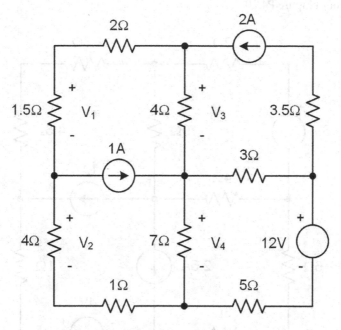

FIGURE P4.18 Circuit with four meshes including two current sources.

Problem 4.19. Determine the power for each of the voltage sources in Figure P4.19.

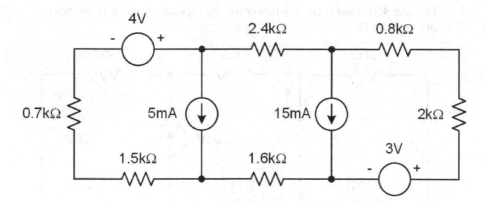

FIGURE P4.19 Circuit with three meshes which make up a double supermesh.

Problem 4.20. Determine the power for each of the voltage sources for the circuit in Figure P4.20.

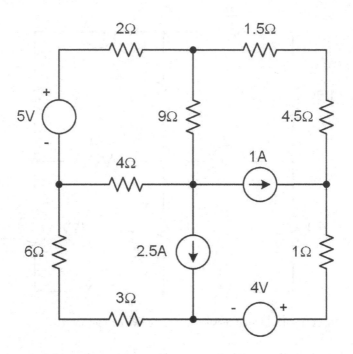

FIGURE P4.20   Circuit with four meshes including a double supermesh.

Problem 4.21. Find i_x and the power for the dependent source of the circuit in Figure P4.21.

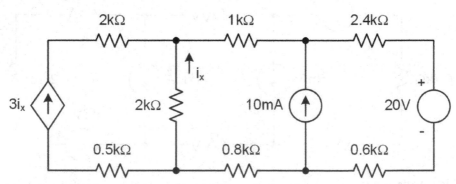

FIGURE P4.21   Circuit with three meshes.

Problem 4.22. Find v_x and the power for the dependent source in Figure P4.22.

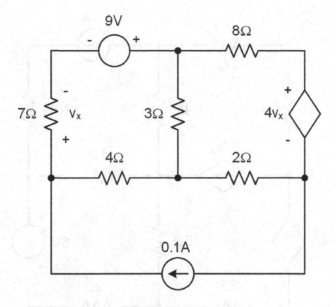

FIGURE P4.22 Circuit with three meshes.

Problem 4.23. Find v_x and i_x for the circuit in Figure P4.23.

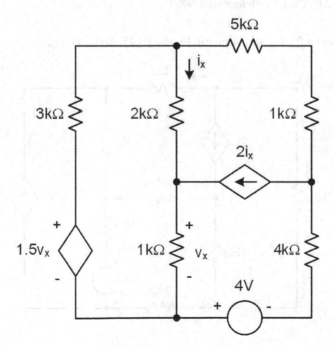

FIGURE P4.23 Circuit with three meshes.

Problem 4.24. Find v_x and i_x for the circuit in Figure P4.24.

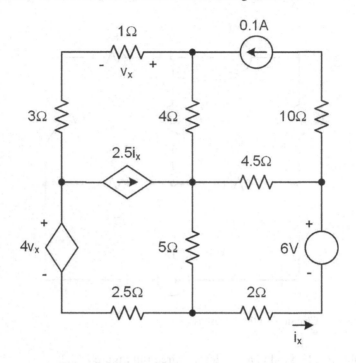

FIGURE P4.24   Circuit with four meshes.

Problem 4.25. For the circuit in Figure P4.25, find i_x and i_y.

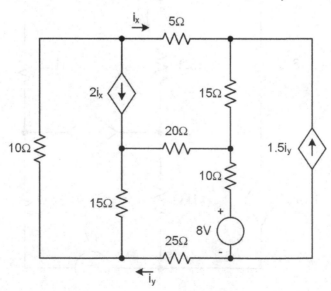

FIGURE P4.25   Circuit with four meshes.

5 Thevenin's and Norton's Theorems

5.1 THEVENIN'S THEOREM

Thevenin's theorem is a useful tool for solving certain types of problems. It states that for any two-terminal network involving resistors and sources, there is a Thevenin equivalent network involving a single voltage source V_{Th} and a single resistance R_{Th}, as shown in Figure 5.1. The Thevenin circuit is **externally equivalent** to the original circuit, in the sense that if we connect additional circuitry to the two terminals a and b, the voltages and currents in this external circuitry will be the same for the Thevenin equivalent as the original circuit, regardless of the complexity of the original circuit. They are not internally equivalent, however. This means that we cannot use the Thevenin equivalent circuit to directly find voltages and currents inside the original circuit.

Because the Thevenin circuit is externally equivalent to the original circuit, we can analyze the behavior with simple external connections to find V_{Th} and R_{Th}. One such connection involves a short circuit as shown in Figure 5.2. If we place a wire directly between the terminals a and b to create a short circuit, the same current

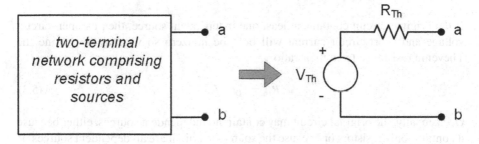

FIGURE 5.1 A network containing resistors and sources and its Thevenin equivalent.

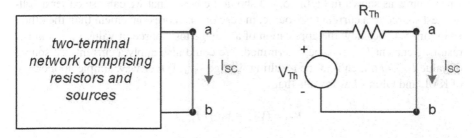

FIGURE 5.2 The short-circuit condition for a network containing resistors and sources and its Thevenin equivalent.

DOI: 10.1201/9781003408529-5

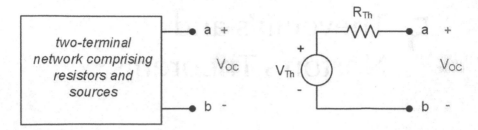

FIGURE 5.3 The open-circuit condition for a network containing resistors and sources and its Thevenin equivalent.

should flow in either case. The short-circuit current for the Thevenin equivalent is V_{Th} / R_{Th}. Therefore, the short-circuit current for the original circuit on the left should be equal to the short-circuit current for the Thevenin equivalent:

$$I_{SC} = V_{Th} / R_{Th}. \qquad (5.1)$$

Another simple external condition of importance is the open-circuit condition shown in Figure 5.3. The open-circuit voltage for the Thevenin equivalent is V_{Th}, and if the two circuits are externally equivalent, the open-circuit voltage for the original circuit on the left should be equal to the open-circuit voltage for the Thevenin equivalent:

$$V_{OC} = V_{Th}. \qquad (5.2)$$

If the original circuit contains at least one independent source, then its open-circuit voltage and short-circuit current will both be nonzero so we may determine the Thevenin resistance from their ratio:

$$R_{Th} = V_{OC} / I_{SC} \qquad (5.3)$$

Occasionally, the original circuit may contain no independent sources, either because it contains only resistors or because the sources within it are all dependent sources. If this is the case, both the open-circuit voltage and the short-circuit current will be zero and we may not use their ratio to find the Thevenin resistance. Instead, we can apply a test source as shown in Figure 5.4. It should be noted that we can use either a volt-age test source or a current test source, in case one is more convenient than the other. In Figure 5.4, we show the application of a voltage test source of value V_{Test}, and the resulting current I_{Test} is to be determined. We could also apply a current test source of value I_{Test}, and then find the resulting voltage V_{Test}. Either way, simple application of KVL and Ohm's law reveals that

$$R_{Th} = (V_{Test} - V_{Th}) / I_{Test}. \qquad (5.4)$$

In a case for which the Thevenin voltage is zero, this simplifies to

$$R_{Th} = V_{Test} / I_{Test}. \qquad (5.5)$$

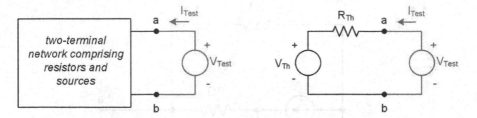

FIGURE 5.4 The application of a test source to a circuit containing resistors and sources as well as its Thevenin equivalent.

In the following sections, we will make use of these concepts to determine the Thevenin equivalent circuit for several two-terminal networks.

5.2 THEVENIN EQUIVALENT FOR A CIRCUIT WITH ONLY INDEPENDENT SOURCES

For a two-terminal circuit with only independent sources and resistors, we may use a combination of the open-circuit analysis and the short-circuit analysis to determine the Thevenin equivalent circuit. It turns out there is a third analysis, the **resistance shortcut analysis**, which may also be used to determine the Thevenin resistance. In light of this, **we may use the combination of any two of the open-circuit analysis, the short-circuit analysis, and the resistance shortcut analysis to find the Thevenin equivalent.** To illustrate this, we will find the Thevenin equivalent with respect to the terminals a and b for the circuit in Figure 5.5.

We will start by doing the open-circuit analysis by the node voltage method as shown in Figure 5.6.

In units of V, A, and Ω, the node voltage equations are

$$N1 \quad V_1 = 15, \tag{5.6}$$

$$N2 \quad -1 - 0.5 + \frac{V_2 - V_3}{2} = 0, \tag{5.7}$$

and

$$N3 \quad \frac{V_3 - V_1}{5} + \frac{V_3 - V_2}{2} + \frac{V_3}{20} = 0. \tag{5.8}$$

Substituting the N1 known node voltage into the N3 equation,

$$N2 \quad -1 - 0.5 + \frac{V_2 - V_3}{2} = 0 \tag{5.9}$$

and

$$N3 \quad \frac{V_3 - 15}{5} + \frac{V_3 - V_2}{2} + \frac{V_3}{20} = 0. \tag{5.10}$$

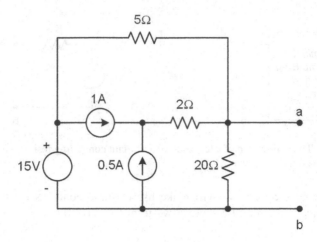

FIGURE 5.5   A two-terminal network containing four independent sources and resistors.

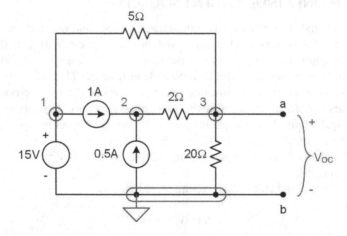

FIGURE 5.6   Open-circuit analysis of a two-terminal network containing three independent sources and resistors.

Multiplying by the least common denominators,

$$N2 \quad -2-1+V_2-V_3 = 0, \tag{5.11}$$

$$N3 \quad 4V_3-60+10V_3-10V_2+V_3 = 0. \tag{5.12}$$

Collecting like terms,

$$N2 \quad V_2-V_3 = 3, \tag{5.13}$$

$$N3 \quad -10V_2+15V_3 = 60. \tag{5.14}$$

In matrix form,

$$\begin{bmatrix} 1 & -1 \\ -10 & 15 \end{bmatrix} \begin{bmatrix} V_2 \\ V_3 \end{bmatrix} = \begin{bmatrix} 3 \\ 60 \end{bmatrix}. \tag{5.15}$$

Solving,

$$V_2 = \frac{\begin{vmatrix} 3 & -1 \\ 60 & 15 \end{vmatrix}}{\begin{vmatrix} 1 & -1 \\ -10 & 15 \end{vmatrix}} = 21.0\,\text{V} \tag{5.16}$$

and

$$V_3 = \frac{\begin{vmatrix} 1 & 3 \\ -10 & 60 \end{vmatrix}}{\begin{vmatrix} 1 & -1 \\ -10 & 15 \end{vmatrix}} = 18.0\,\text{V}. \tag{5.17}$$

The open-circuit voltage is the same as the node three voltage:

$$V_{OC} = V_3 = 18.0\,\text{V}. \tag{5.18}$$

Next, we will do the short-circuit analysis, also by the node voltage method, as shown in Figure 5.7. Notice that shorting between terminals a and b eliminates one essential node (node three is combined with the reference node).

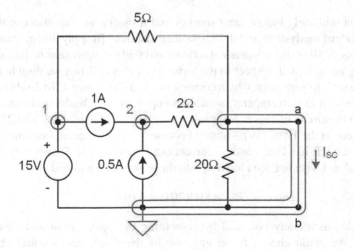

FIGURE 5.7 Short-circuit analysis of a two-terminal network containing only independent sources and resistors.

In units of V, A, and Ω, the node voltage equations are

$$N1 \quad V_1 = 15 \tag{5.19}$$

and

$$N2 \quad -1 - 0.5 + \frac{V_2}{2} = 0. \tag{5.20}$$

Simplifying,

$$N2 \quad -2 - 1 + V_2 = 0 \tag{5.21}$$

or

$$V_2 = 3\,\text{V}. \tag{5.22}$$

It is important to note that the value of V_2 under short-circuit conditions is not the same as the value of V_2 under open-circuit conditions! Having solved for the node voltages, we can use Kirchhoff's current law (KCL) to find the short-circuit current.

$$I_{SC} = \frac{V_1}{5\,\Omega} + \frac{V_2}{2\,\Omega} = 4.5\,\text{A}. \tag{5.23}$$

Now, we can determine the Thevenin equivalent circuit:

$$V_{Th} = V_{OC} = 18.0\,\text{V} \tag{5.24}$$

and

$$R_{Th} = \frac{V_{OC}}{I_{SC}} = 4.0\,\Omega. \tag{5.25}$$

In a circuit with only independent sources and resistors, we can also use the **resistance shortcut analysis** to find the Thevenin resistance. To apply the resistance shortcut, we disable all of the independent sources and find the equivalent resistance for the remaining network with respect to the terminals a and b. When we disable a voltage source we set it to zero volts, which corresponds to a short circuit. To disable a current source, we set it to zero amperes, which is an open circuit. Disabling the independent sources in the circuit of Figure 5.5 results in the circuit of Figure 5.8. The $2\,\Omega$ resistor has no effect on the Thevenin resistance because it is disconnected from the rest of the circuit on the left-hand side, and no current can flow in it. The Thevenin resistance is the equivalent resistance seen looking into the terminals a and b; it is

$$R_{Th} = 5\,\Omega \parallel 20\,\Omega = 4.0\,\Omega. \tag{5.26}$$

This result was already obtained by combining the open-circuit and short-circuit analyses. We could choose to do any two of the three analyses and obtain the Thevenin equivalent circuit. If we do all three, we can check the results for consistency and this may help to find small errors.

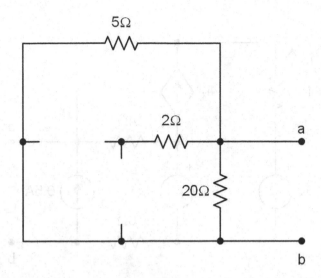

FIGURE 5.8   Resistance shortcut analysis of a two-terminal network containing only independent sources and resistors.

To review, see Presentation 5.1 in ebook+. To test your knowledge, try Quiz 5.1 in ebook+. To put your knowledge to practice, try Laboratory Exercise 5.1 in ebook+.

5.3   THEVENIN EQUIVALENT FOR A CIRCUIT WITH MIXED SOURCES

If a two-terminal network contains mixed dependent and independent sources, as well as resistors, we may still use the open-circuit and short-circuit analyses. The resistance shortcut is not applicable, but we can use a test-source analysis. We will make these points using the circuit in Figure 5.9.

We will start by doing the open-circuit analysis by the node voltage method as shown in Figure 5.10.

In units of V, A, and Ω, the node voltage equations are

$$N1 \quad V_1 = 8, \tag{5.27}$$

$$N2 \quad \frac{V_2 - 16}{12} - 4i_x + \frac{V_2 - V_3}{8} = 0, \tag{5.28}$$

and

$$N3 \quad \frac{V_3 - V_1}{4} + \frac{V_3 - V_2}{8} - 0.5 = 0. \tag{5.29}$$

The equation of the dependent source is

$$DS \quad i_x = \frac{16 - V_2}{12}. \tag{5.30}$$

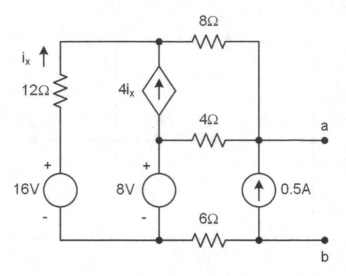

FIGURE 5.9 A two-terminal network containing mixed (independent and dependent) sources and resistors.

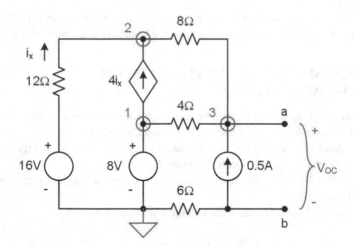

FIGURE 5.10 Open-circuit analysis of a two-terminal network containing only independent sources and resistors.

Substituting the DS equation into the N2 equation, and substituting the N1 known node voltage into the N3 equation,

$$N2DS \quad \frac{V_2 - 16}{12} - 4\left(\frac{16 - V_2}{12}\right) + \frac{V_2 - V_3}{8} = 0 \tag{5.31}$$

and

$$N1N3 \quad \frac{V_3 - 8}{4} + \frac{V_3 - V_2}{8} - 0.5 = 0. \tag{5.32}$$

Multiplying by the least common denominators,

$$N2DS \quad 2V_2 - 32 - 128 + 8V_2 + 3V_2 - 3V_3 = 0 \tag{5.33}$$

and

$$N1N3 \quad 2V_3 - 16 + V_3 - V_2 - 4 = 0. \tag{5.34}$$

Collecting like terms,

$$N2DS \quad 13V_2 - 3V_3 = 160 \tag{5.35}$$

and

$$N1N3 \quad -V_2 + 3V_3 = 20. \tag{5.36}$$

In matrix form,

$$\begin{bmatrix} 13 & -3 \\ -1 & 3 \end{bmatrix} \begin{bmatrix} V_2 \\ V_3 \end{bmatrix} = \begin{bmatrix} 160 \\ 20 \end{bmatrix}. \tag{5.37}$$

Solving,

$$V_2 = \frac{\begin{vmatrix} 160 & -3 \\ 20 & 3 \end{vmatrix}}{\begin{vmatrix} 13 & -3 \\ -1 & 3 \end{vmatrix}} = 15.00 \text{ V} \tag{5.38}$$

and

$$V_3 = \frac{\begin{vmatrix} 13 & 160 \\ -1 & 20 \end{vmatrix}}{\begin{vmatrix} 13 & -3 \\ -1 & 3 \end{vmatrix}} = 11.67 \text{ V}. \tag{5.39}$$

The open-circuit voltage is not directly equal to any of the node voltages in this case, because node B is not the reference node. (In fact, it is not even an essential node!) The open-circuit voltage is

$$V_{OC} = V_A - V_B, \tag{5.40}$$

where

$$V_A = V_3 = 11.67 \text{ V} \tag{5.41}$$

and

$$V_B = -(0.5 \text{ A})(6 \, \Omega) = -3.00 \text{ V}. \tag{5.42}$$

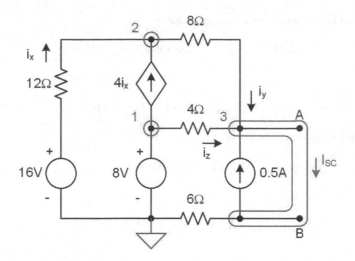

FIGURE 5.11 Short-circuit analysis of a two-terminal network containing mixed (independent and dependent) sources and resistors.

The open-circuit voltage is therefore

$$V_{OC} = V_A - V_B = 11.67\,\text{V} - (-3.00\,\text{V}) = 14.67\,\text{V}. \tag{5.43}$$

Next, we will do the short-circuit analysis, also by the node voltage method, as shown in Figure 5.11.

In units of V, A, and Ω, the node voltage equations are

$$N1 \quad V_1 = 8, \tag{5.44}$$

$$N2 \quad \frac{V_2 - 16}{12} - 4i_x + \frac{V_2 - V_3}{8} = 0, \tag{5.45}$$

and

$$N3 \quad \frac{V_3 - V_1}{4} + \frac{V_3 - V_2}{8} - 0.5 + 0.5 + \frac{V_3}{6} = 0. \tag{5.46}$$

Notice that the independent current source flows between two points on node three so the net contribution is zero when writing the node equation. (However, we will see that this current source contributes to the short-circuit current.)

The equation of the dependent source is

$$DS \quad i_x = \frac{16 - V_2}{12}. \tag{5.47}$$

Substituting the DS equation into the N2 equation, and substituting the N1 known node voltage into the N3 equation,

$$N2DS \quad \frac{V_2-16}{12}-4\left(\frac{16-V_2}{12}\right)+\frac{V_2-V_3}{8}=0 \tag{5.48}$$

and

$$N1N3 \quad \frac{V_3-8}{4}+\frac{V_3-V_2}{8}+\frac{V_3}{6}=0. \tag{5.49}$$

Multiplying by the least common denominators,

$$N2DS \quad 2V_2-32-128+8V_2+3V_2-3V_3=0 \tag{5.50}$$

and

$$N1N3 \quad 6V_3-48+3V_3-3V_2+4V_3=0. \tag{5.51}$$

Collecting like terms,

$$N2DS \quad 13V_2-3V_3=160 \tag{5.52}$$

and

$$N1N3 \quad -3V_2+13V_3=48. \tag{5.53}$$

In matrix form,

$$\begin{bmatrix} 13 & -3 \\ -3 & 13 \end{bmatrix}\begin{bmatrix} V_2 \\ V_3 \end{bmatrix}=\begin{bmatrix} 160 \\ 48 \end{bmatrix}. \tag{5.54}$$

Solving,

$$V_2=\frac{\begin{vmatrix} 160 & -3 \\ 48 & 13 \end{vmatrix}}{\begin{vmatrix} 13 & -3 \\ -3 & 13 \end{vmatrix}}=13.90\,\text{V} \tag{5.55}$$

and

$$V_3=\frac{\begin{vmatrix} 13 & 160 \\ -3 & 48 \end{vmatrix}}{\begin{vmatrix} 13 & -3 \\ -3 & 13 \end{vmatrix}}=6.90\text{V}. \tag{5.56}$$

Once again it is important to note that the node voltages under short-circuit conditions are generally different from the node voltages under open-circuit conditions. The short-circuit current may be found by using KCL with the short-circuit node voltages. Referring to Figure 5.11, we see that this is a somewhat unusual use of KCL because the short-circuit current flows between two points on the same essential node.

$$I_{SC} = i_y + i_z + 0.5\,\text{A} = \frac{V_1 - V_3}{4\,\Omega} + \frac{V_2 - V_3}{8\,\Omega} + 0.5\,\text{A} = 1.650\,\text{A}. \tag{5.57}$$

The Thevenin resistance is therefore

$$R_{\text{Th}} = \frac{V_{\text{OC}}}{I_{SC}} = \frac{14.67\,\text{V}}{1.650\,\text{A}} = 8.89\,\Omega. \tag{5.58}$$

The resistance shortcut is not applicable to circuits containing mixed independent and dependent sources. This is because dependent sources contribute to the Thevenin resistance, so disabling them alters the apparent Thevenin resistance. To see this, consider the circuit of Figure 5.9 with all sources disabled (voltage sources replaced by shorts and current sources replaced by opens). The resulting circuit appears in Figure 5.12, and the equivalent resistance with respect to the nodes a and b is

$$R_{ab} = (8\,\Omega + 12\,\Omega) \,\|\, (4\,\Omega + 6\,\Omega) = 6.59\,\Omega \neq V_{\text{Th}}. \tag{5.59}$$

Therefore, we cannot use the resistance shortcut in a network containing dependent sources.

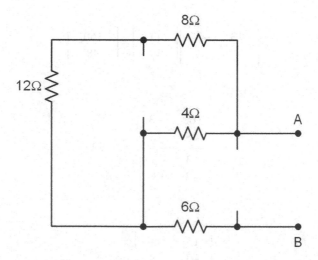

FIGURE 5.12 A two-terminal network containing mixed (independent and dependent) sources and resistors, after the sources have all been disabled to show that the resistance shortcut is not valid in the case of mixed sources.

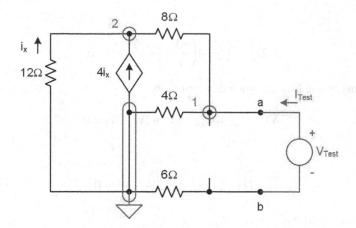

FIGURE 5.13 Test-source analysis of a two-terminal network containing mixed (independent and dependent) sources and resistors. The independent sources have been disabled but the dependent source has been left active.

On the other hand, we can use a test-source analysis for a network containing mixed sources. For this analysis, we disable only the independent sources and apply a test source to the terminals a and b. The test source may be a current source or a voltage source – and sometimes one may be more convenient than the other. Also, the test source can take a specific value (1 V, for example) or it can have an abstract value (such as V_{Test}). We will use a test source with the unspecified value of V_{Test}, as shown in Figure 5.13, but the value of the test source will cancel out in the final analysis. Notice that in this particular circuit, the number of essential nodes has been reduced to two by disabling the independent sources. This is because the node which was formerly called N1 has been joined with the reference node by disabling the 8 V source.

In units of V, A, and Ω, the node voltage equations are

$$N1 \quad \frac{V_1 - V_{Test}}{6} + \frac{V_1}{4} + \frac{V_1 - V_2}{8} = 0 \tag{5.60}$$

and

$$N2 \quad \frac{V_2}{12} - 4i_x + \frac{V_2 - V_1}{8} = 0. \tag{5.61}$$

The equation of the dependent source is

$$DS \quad i_x = \frac{-V_2}{12}. \tag{5.62}$$

Substituting the DS equations into the N2 equation, we obtain

$$N1 \quad \frac{V_1 - V_{Test}}{6} + \frac{V_1}{4} + \frac{V_1 - V_2}{8} = 0 \tag{5.63}$$

and

$$N2DS \quad \frac{V_2}{12} - 4\left(\frac{-V_2}{12}\right) + \frac{V_2 - V_1}{8} = 0. \tag{5.64}$$

Multiplying by the least common denominators,

$$N1 \quad 4V_1 - 4V_{\text{Test}} + 6V_1 + 3V_1 - 3V_2 = 0 \tag{5.65}$$

and

$$N2DS \quad 2V_2 + 8V_2 + 3V_2 - 3V_1 = 0. \tag{5.66}$$

Collecting like terms,

$$N1 \quad 13V_1 - 3V_2 = 4V_{\text{Test}} \tag{5.67}$$

and

$$N2DS \quad -3V_1 + 13V_2 = 0. \tag{5.68}$$

In matrix form,

$$\begin{bmatrix} 13 & -3 \\ -3 & 13 \end{bmatrix} \begin{bmatrix} V_1 \\ V_2 \end{bmatrix} = \begin{bmatrix} 4V_{\text{Test}} \\ 0 \end{bmatrix}. \tag{5.69}$$

Solving,

$$V_1 = \frac{\begin{vmatrix} 4V_{\text{Test}} & -3 \\ 0 & 13 \end{vmatrix}}{\begin{vmatrix} 13 & -3 \\ -3 & 13 \end{vmatrix}} = \frac{52V_{\text{Test}}}{160} = \frac{13V_{\text{Test}}}{40} \tag{5.70}$$

and

$$V_2 = \frac{\begin{vmatrix} 13 & 4V_{\text{Test}} \\ -3 & 0 \end{vmatrix}}{\begin{vmatrix} 13 & -3 \\ -3 & 13 \end{vmatrix}} = \frac{12V_{\text{Test}}}{160} = \frac{3V_{\text{Test}}}{40}. \tag{5.71}$$

Now, we can find the resulting value of test current:

$$I_{\text{Test}} = \frac{V_{\text{Test}} - V_1}{6\,\Omega} = \frac{V_{\text{Test}} - 13V_{\text{Test}}/40}{6\,\Omega}. \tag{5.72}$$

The Thevenin resistance is then

$$R_{Th} = \frac{V_{Test}}{I_{Test}} = \left[\frac{1 - 13/40}{6\,\Omega} \right]^{-1} = 8.89\,\Omega, \tag{5.73}$$

which is the same as the value found by use of the open-circuit and short-circuit analyses.

As previously mentioned, we can choose to use a specific value of the test source if it is more convenient; sometimes this simplifies the mathematics. As an example, we will solve the present example using a test source of 1 V, as shown in Figure 5.14.

In units of V, A, and Ω, the node voltage equations are

$$N1 \quad \frac{V_1 - 1}{6} + \frac{V_1}{4} + \frac{V_1 - V_2}{8} = 0 \tag{5.74}$$

and

$$N2 \quad \frac{V_2}{12} - 4i_x + \frac{V_2 - V_1}{8} = 0. \tag{5.75}$$

The equation of the dependent source is

$$DS \quad i_x = \frac{-V_2}{12}. \tag{5.76}$$

Substituting the DS equations into the N2 equation, we obtain

$$N1 \quad \frac{V_1 - 1}{6} + \frac{V_1}{4} + \frac{V_1 - V_2}{8} = 0 \tag{5.77}$$

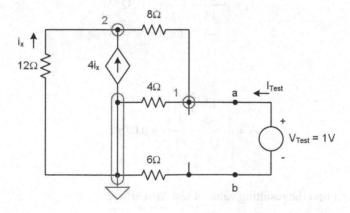

FIGURE 5.14 Use of a test source with a specific value (1 V) with a two-terminal network containing mixed (independent and dependent) sources and resistors. The independent sources have been disabled but the dependent source has been left active.

and

$$N2DS \quad \frac{V_2}{12} - 4\left(\frac{-V_2}{12}\right) + \frac{V_2 - V_1}{8} = 0. \tag{5.78}$$

Multiplying by the least common denominators,

$$N1 \quad 4V_1 - 4 + 6V_1 + 3V_1 - 3V_2 = 0 \tag{5.79}$$

and

$$N2DS \quad 2V_2 + 8V_2 + 3V_2 - 3V_1 = 0. \tag{5.80}$$

Collecting like terms,

$$N1 \quad 13V_1 - 3V_2 = 4 \tag{5.81}$$

and

$$N2DS \quad -3V_1 + 13V_2 = 0. \tag{5.82}$$

In matrix form,

$$\begin{bmatrix} 13 & -3 \\ -3 & 13 \end{bmatrix} \begin{bmatrix} V_1 \\ V_2 \end{bmatrix} = \begin{bmatrix} 4 \\ 0 \end{bmatrix}. \tag{5.83}$$

Solving,

$$V_1 = \frac{\begin{vmatrix} 4 & -3 \\ 0 & 13 \end{vmatrix}}{\begin{vmatrix} 13 & -3 \\ -3 & 13 \end{vmatrix}} = 0.325\,\text{V} \tag{5.84}$$

and

$$V_2 = \frac{\begin{vmatrix} 13 & 4 \\ -3 & 0 \end{vmatrix}}{\begin{vmatrix} 13 & -3 \\ -3 & 13 \end{vmatrix}} = 0.0750\,\text{V}. \tag{5.85}$$

Now, we can find the resulting value of test current:

$$I_{\text{Test}} = \frac{1\,\text{V} - V_1}{6\,\Omega} = \frac{1\,\text{V} - 0.325\,\text{V}}{6\,\Omega} = 0.1125\,\text{A}. \tag{5.86}$$

The Thevenin resistance is then

$$R_{Th} = \frac{V_{Test}}{I_{Test}} = \frac{1\,V}{0.1125\,A} = 8.89\,\Omega, \tag{5.87}$$

The same as before.

Finally, we could choose to use a test current source as shown in Figure 5.15.

In units of V, A, and Ω, the node voltage equations are

$$N1 \quad -I_{Test} + \frac{V_1}{4} + \frac{V_1 - V_2}{8} = 0 \tag{5.88}$$

and

$$N2 \quad \frac{V_2}{12} - 4i_x + \frac{V_2 - V_1}{8} = 0. \tag{5.89}$$

The equation of the dependent source is

$$DS \quad i_x = \frac{-V_2}{12}. \tag{5.90}$$

Substituting the DS equations into the N2 equation, we obtain

$$N1 \quad -I_{Test} + \frac{V_1}{4} + \frac{V_1 - V_2}{8} = 0 \tag{5.91}$$

and

$$N2DS \quad \frac{V_2}{12} - 4\left(\frac{-V_2}{12}\right) + \frac{V_2 - V_1}{8} = 0. \tag{5.92}$$

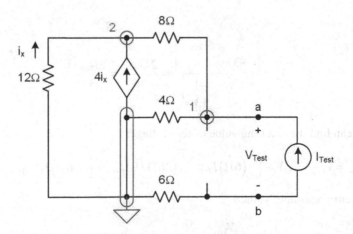

FIGURE 5.15 Use of a test current source with a two-terminal network containing mixed (independent and dependent) sources and resistors. The independent sources have been disabled but the dependent source has been left active.

Multiplying by the least common denominators,

$$N1 \quad -8I_{\text{Test}} + 2V_1 + V_1 - V_2 = 0 \tag{5.93}$$

and

$$N2DS \quad 2V_2 + 8V_2 + 3V_2 - 3V_1 = 0. \tag{5.94}$$

Collecting like terms,

$$N1 \quad 3V_1 - V_2 = 8I_{\text{Test}} \tag{5.95}$$

and

$$N2DS \quad -3V_1 + 13V_2 = 0. \tag{5.96}$$

In matrix form,

$$\begin{bmatrix} 3 & -1 \\ -3 & 13 \end{bmatrix} \begin{bmatrix} V_1 \\ V_2 \end{bmatrix} = \begin{bmatrix} 8I_{\text{Test}} \\ 0 \end{bmatrix}. \tag{5.97}$$

Solving,

$$V_1 = \frac{\begin{vmatrix} 8I_{\text{Test}} & -1 \\ 0 & 13 \end{vmatrix}}{\begin{vmatrix} 3 & -1 \\ -3 & 13 \end{vmatrix}} = \frac{104 I_{\text{Test}}}{36} = \frac{26 I_{\text{Test}}}{9} \tag{5.98}$$

and

$$V_2 = \frac{\begin{vmatrix} 3 & 8I_{\text{Test}} \\ -3 & 0 \end{vmatrix}}{\begin{vmatrix} 3 & -1 \\ -3 & 13 \end{vmatrix}} = \frac{24 I_{\text{Test}}}{36} = \frac{2 I_{\text{Test}}}{3}. \tag{5.99}$$

Now, we can find the resulting value of test voltage:

$$V_{\text{Test}} = V_a - V_b = V_1 - \left[-\left(6\,\Omega\right) I_{\text{Test}} \right] = \left(26\,\Omega/9\right) I_{\text{Test}} - \left[-\left(6\,\Omega\right) I_{\text{Test}} \right]. \tag{5.100}$$

The Thevenin resistance is then

$$R_{\text{Th}} = \frac{V_{\text{Test}}}{I_{\text{Test}}} = \frac{26}{9}\,\Omega + 6\,\Omega = 8.89\,\Omega, \tag{5.101}$$

same as before. When using a current source, we could also assume a specific value for I_{Test}, such as 1 A, if desired.

To review, see Presentation 5.2 in ebook+. To test your knowledge, try Quiz 5.2 in ebook+.

5.4 THEVENIN EQUIVALENT FOR A CIRCUIT CONTAINING ONLY DEPENDENT SOURCES

If a two-terminal network contains only dependent sources, along with resistors, the open-circuit voltage is zero (and therefore the Thevenin voltage is zero). This is known without performing the open-circuit analysis, so the open-circuit analysis provides no new useful information. The short-circuit current is also zero, and the Thevenin resistance cannot be found from the ratio V_{OC} / I_{SC}. The resistance shortcut method is not applicable because of the presence of dependent sources. **Only the test-source analysis is applicable.**

Consider the circuit in Figure 5.16, which contains only dependent sources and resistors.

First, we will show that the open-circuit voltage is zero for this circuit, using the node voltage method as shown in Figure 5.17.

In units of V, mA, and kΩ, the node voltage equations are

$$N1 \quad -4i_x + \frac{V_1 - V_2}{1} + \frac{V_1 + 2i_x - V_2}{10} = 0 \tag{5.102}$$

and

$$N2 \quad \frac{V_2 - V_1}{1} + \frac{V_2}{10} + \frac{V_2}{2} + \frac{V_2 - 2i_x - V_1}{10} = 0. \tag{5.103}$$

The equation of the current controlling the dependent sources is

$$DS \quad i_x = -\frac{V_2}{10}. \tag{5.104}$$

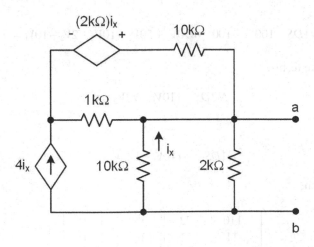

FIGURE 5.16 A two-terminal network containing only dependent sources and resistors.

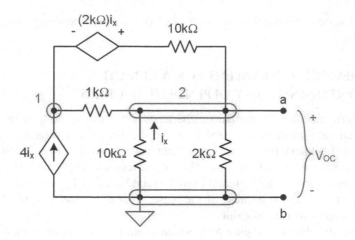

FIGURE 5.17 Open-circuit analysis of a two-terminal network containing only dependent sources and resistors.

Substituting the DS equation into the N1 and N2 equations,

$$N1DS \quad -4\left(-\frac{V_2}{10}\right)+\frac{V_1-V_2}{1}+\frac{V_1-2V_2/10-V_2}{10}=0 \qquad (5.105)$$

and

$$N2DS \quad \frac{V_2-V_1}{1}+\frac{V_2}{10}+\frac{V_2}{2}+\frac{V_2+2V_2/10-V_1}{10}=0. \qquad (5.106)$$

Multiplying by the least common denominators,

$$N2DS \quad 40V_2+100V_1-100V_2+10V_1-2V_2-10V_2=0 \qquad (5.107)$$

and

$$N1DS \quad 100V_2-100V_1+10V_2+50V_2+10V_2+2V_2-10V_1=0. \qquad (5.108)$$

Collecting like terms,

$$N2DS \quad 110V_1-72V_2=0 \qquad (5.109)$$

and

$$N1DS \quad -110V_1+172V_2=0. \qquad (5.110)$$

In matrix form,

$$\begin{bmatrix} 110 & -72 \\ -110 & 172 \end{bmatrix}\begin{bmatrix} V_2 \\ V_3 \end{bmatrix}=\begin{bmatrix} 0 \\ 0 \end{bmatrix}. \qquad (5.111)$$

Solving,

$$V_1 = \frac{\begin{vmatrix} 0 & -72 \\ 0 & 172 \end{vmatrix}}{\begin{vmatrix} 110 & -72 \\ -110 & 172 \end{vmatrix}} = 0 \tag{5.112}$$

and

$$V_2 = \frac{\begin{vmatrix} 110 & 0 \\ -110 & 0 \end{vmatrix}}{\begin{vmatrix} 110 & -72 \\ -110 & 172 \end{vmatrix}} = 0. \tag{5.113}$$

The open-circuit voltage is equal to the node two voltage and is therefore zero:

$$V_{OC} = V_2 = 0. \tag{5.114}$$

In general, we can show that both **the open-circuit voltage and the short-circuit current will be zero for any two-terminal network containing only dependent sources and resistors. Therefore, the Thevenin voltage is zero for such a network, and the Thevenin equivalent circuit comprises only a resistor R_{Th}.**

The Thevenin resistance may be determined for such a network by a test-source analysis. This may be done using a voltage test source (either generic or of a specific value) or a current test source (either generic or of a specific value). In Figure 5.18, we apply a generic voltage source to the network in Figure 5.16.

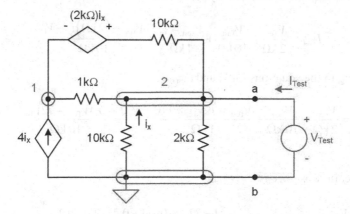

FIGURE 5.18 Use of a test voltage source with a two-terminal network containing only dependent sources and resistors.

In units of V, mA, and kΩ, the node voltage equations are

$$N1 \quad -4i_x + \frac{V_1 - V_2}{1} + \frac{V_1 + 2i_x - V_2}{10} = 0 \tag{5.115}$$

and

$$N2 \quad V_2 = V_{\text{Test}}. \tag{5.116}$$

The equation for the current controlling the dependent sources is

$$DS \quad i_x = -\frac{V_2}{10}. \tag{5.117}$$

Substituting the DS and N2 equations into the N1 equation,

$$N1N2DS \quad -4\left(-\frac{V_{\text{Test}}}{10}\right) + \frac{V_1 - V_{\text{Test}}}{1} + \frac{V_1 - 2V_{\text{Test}}/10 - V_{\text{Test}}}{10} = 0. \tag{5.118}$$

Multiplying by the least common denominator,

$$N1N2DS \quad 40V_{\text{Test}} + 100V_1 - 100V_{\text{Test}} + 10V_1 - 2V_{\text{Test}} - 10V_{\text{Test}} = 0. \tag{5.119}$$

Collecting like terms,

$$N1N2DS \quad 110V_1 - 72V_{\text{Test}} = 0 \tag{5.120}$$

or

$$V_1 = 72V_{\text{Test}}/110. \tag{5.121}$$

Next, we can find the test current by using KCL:

$$I_{\text{Test}} = \frac{V_{\text{Test}}}{2\,\text{k}\Omega} + \frac{V_{\text{Test}}}{10\,\text{k}\Omega} + \frac{V_{\text{Test}} - V_1}{1\,\text{k}\Omega} + \frac{V_{\text{Test}} - (2\,\text{k}\Omega)i_x - V_1}{10\,\text{k}\Omega}. \tag{5.122}$$

Substituting in the equations for V_1 and i_{Test},

$$I_{\text{Test}} = \frac{V_{\text{Test}}}{2\,\text{k}\Omega} + \frac{V_{\text{Test}}}{10\,\text{k}\Omega} + \frac{V_{\text{Test}} - 72V_{\text{Test}}/110}{1\,\text{k}\Omega} + \frac{V_{\text{Test}} + 0.4V_{\text{Test}} - 72V_{\text{Test}}/110}{10\,\text{k}\Omega}.$$

$$\tag{5.123}$$

The Thevenin resistance is

$$R_{\text{Th}} = \frac{V_{\text{Test}}}{I_{\text{Test}}} = \left(\frac{1}{2\,\text{k}\Omega} + \frac{1}{10\,\text{k}\Omega} + \frac{1 - 72/110}{1\,\text{k}\Omega} + \frac{1 + 0.4 - 72/110}{10\,\text{k}\Omega}\right)^{-1} = 0.980\,\text{k}\Omega. \tag{5.124}$$

It should be noted that the Thevenin equivalent circuit comprises only this resistance because the Thevenin voltage is zero and there is no need to include a voltage source. It should also be noted that the resistance shortcut may not be used in this circuit due to the presence of dependent sources, which contribute to the Thevenin resistance. (The resistance shortcut would yield an equivalent resistance of $R_{ab} = 10\,\text{k}\Omega \parallel 2\,\text{k}\Omega = 1.667\,\text{k}\Omega$, which is not valid.)

To review, see Presentation 5.3 in ebook+. To test your knowledge, try Quiz 5.3 in ebook+.

5.5 SOURCE TRANSFORMATIONS

A two-terminal network comprising a voltage source and a series resistance may be transformed to a two-terminal network comprising a current source and a parallel resistor, and these two will be externally equivalent as long as the open-circuit voltage and short-circuit current are preserved by the transformation. This has broad applicability because we have seen that any two-terminal network involving resistors and sources may be represented by its Thevenin equivalent involving a voltage source and series resistance.

Referring to the top two networks in Figure 5.19, the open-circuit voltages of the two are equal provided that $V_{\text{Th}} = I_N R_N$. The two short-circuit currents are equal if $V_{\text{Th}} / R_{\text{Th}} = I_N$. These two conditions are satisfied if

$$I_N = \frac{V_{\text{Th}}}{R_{\text{Th}}} \tag{5.125}$$

and

$$R_N = R_{\text{Th}}. \tag{5.126}$$

These equations allow us to transform from the circuit on the left (with the voltage source) to the one on the right (containing the current source). By rearranging the

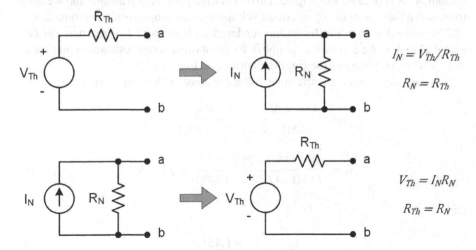

$$I_N = V_{Th}/R_{Th}$$
$$R_N = R_{Th}$$

$$V_{Th} = I_N R_N$$
$$R_{Th} = R_N$$

FIGURE 5.19 Source transformations.

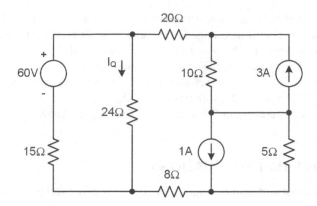

FIGURE 5.20   A circuit containing four meshes and five essential nodes.

first equation, we can transform in the other direction as illustrated on the bottom of Figure 5.19. As long as the two networks behave equivalently for the open-circuit and short-circuit conditions, they will behave equivalently when connected to a general network of sources and resistors. The two are externally equivalent; that is, all external voltages and currents will be unchanged by transforming from one to the other.

The use of source transformations can sometimes aid in the analysis of a circuit if it allows us to reduce the number of essential nodes or meshes or both. Consider the determination of I_Q for the circuit in Figure 5.20, which contains four meshes and five essential nodes. We can simplify this circuit without obscuring the current I_Q if we perform two source transformations on the right-hand side.

First, we will transform the 3 A in parallel with 10 Ω to 30 V in series with 10 Ω, resulting in the circuit of Figure 5.21. This has reduced the meshes to three and the essential nodes to four.

Next, we will transform the 1 A in parallel with 5 Ω to 5 V in series with 5 Ω, resulting in the circuit of Figure 5.22. Notice that the 5 V source has the plus sign on the bottom. This is because the original current source pointed downward, and we need to preserve the sign of the open-circuit voltage for this two-terminal combination.

The second source transformation rendered a circuit with two meshes and two essential nodes. We can further simplify by combining series resistances and series voltage sources, giving us the simplified circuit in Figure 5.23.

Now, we can easily apply the node voltage method with a single equation:

$$N1 \quad \frac{V_1 - 60\,\text{V}}{15\,\Omega} + \frac{V_1}{24\,\Omega} + \frac{V_1 - 25\,\text{V}}{43\,\Omega} = 0; \tag{5.127}$$

$$V_1 = \frac{60\,\text{V}/15\,\Omega + 25\,\text{V}/43\,\Omega}{1/15\,\Omega + 1/24\,\Omega + 1/43\,\Omega} = 34.82\,\text{V} \tag{5.128}$$

and

$$I_Q = \frac{V_1}{24\,\Omega} = 1.451\,\text{A}. \tag{5.129}$$

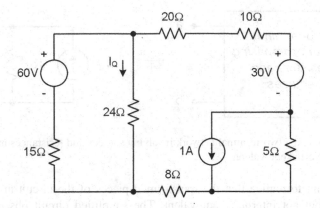

FIGURE 5.21 The previous circuit simplified by one source transformation.

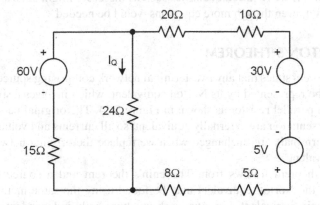

FIGURE 5.22 The previous circuit simplified by a second source transformation.

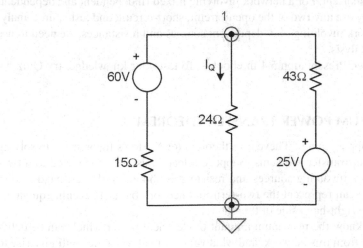

FIGURE 5.23 The previous circuit simplified by combining series resistors and series voltage sources.

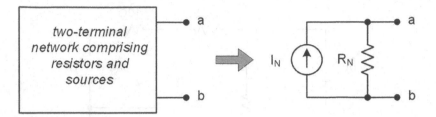

FIGURE 5.24 Any two-terminal network involving sources and resistances may be represented by its Norton equivalent.

It is important to realize that the transformed pieces of the circuit are externally equivalent, but not internally equivalent. The simplified circuit obscures certain quantities in the original circuit, such as the power in the 3 A source or the current in the 5 Ω resistor. If we required quantities such as these, we might choose to solve the original circuit, even though more equations would be needed.

5.6 NORTON'S THEOREM

Norton's theorem states that any two-terminal network comprising sources and resistances may be represented by its Norton equivalent, which involves a single current source and a parallel resistor as shown in Figure 5.24. The original network and its Norton representation are externally equivalent, so all currents and voltages external to the two terminals are unchanged when we replace the original network with its Norton equivalent.

Norton's theorem follows from Thevenin's theorem and a source transformation. Hence, the Norton equivalent may be found using the same methods used to find a Thevenin equivalent. For a network involving only independent sources and resistances, we can use any two of the open-circuit analysis, short-circuit analysis, and resistance shortcut. For a network involving mixed (independent and dependent) sources, we may use any two of the open-circuit, short-circuit, and test-source analyses. For a network involving only dependent sources and resistances, we need to use a test-source analysis.

To review, see Presentation 5.4 in ebook+. To test your knowledge, try Quiz 5.4 in ebook+.

5.7 MAXIMUM POWER TRANSFER THEOREM

An important application of Thevenin's theorem (or Norton's theorem) is in solving maximum power transfer problems. Suppose a load resistor is connected to a two-terminal network involving sources and resistors as shown on the left-hand side of Figure 5.25. We can represent the two-terminal network by its Thevenin equivalent as shown on the right-hand side of this figure.

We want to know the maximum amount of electrical power which can be delivered to the load from this network, and what value of load resistance will give rise to this maximum power transfer. This problem can be easily solved using the Thevenin representation for the original circuit.

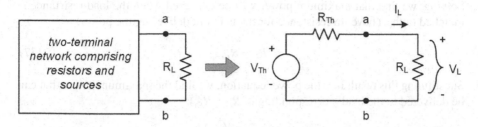

FIGURE 5.25 A load resistor connected to a two-terminal network for consideration of maximum power transfer.

The load voltage may be found using the voltage divider rule:

$$V_L = \frac{V_{Th}R_L}{R_{Th} + R_L}. \tag{5.130}$$

The load current may be determined using Ohm's law:

$$I_L = \frac{V_{Th}}{R_{Th} + R_L}. \tag{5.131}$$

The load power is

$$P_L = V_L I_L = \frac{V_{Th}^2 R_L}{\left(R_{Th} + R_L\right)^2}. \tag{5.132}$$

To find the value of load resistance that will maximize this power, we can differentiate the power expression with respect to the load resistance, set the derivative to zero, and solve. We can use the quotient rule to find the partial derivative of the power with respect to the load resistance:

$$\frac{\partial P_L}{\partial R_L} = \frac{V_{Th}^2 \left(R_{Th} + R_L\right)^2 - V_{Th}^2 R_L 2\left(R_{Th} + R_L\right)}{\left(R_{Th} + R_L\right)^4}. \tag{5.133}$$

We set this partial derivative to zero in order to maximize the power:

$$\frac{V_{Th}^2 \left(R_{Th} + R_L\right)^2 - V_{Th}^2 R_L 2\left(R_{Th} + R_L\right)}{\left(R_{Th} + R_L\right)^4} = 0. \tag{5.134}$$

We may multiply both sides by $\left(R_{Th} + R_L\right)^3$:

$$V_{Th}^2 \left(R_{Th} + R_L\right) - V_{Th}^2 R_L 2 = 0. \tag{5.135}$$

It is assumed that the Thevenin voltage is nonzero, so we may divide both sides by V_{Th}^2:

$$\left(R_{Th} + R_L\right) - 2R_L = 0. \tag{5.136}$$

Solving, we find that maximum power will be delivered when the load resistance is matched to the Thevenin resistance for the network delivering the power:

$$R_L = R_{\text{Th}}. \tag{5.137}$$

Substituting this result into the power equation, we find the maximum power that can be delivered to the load (corresponding to $R_L = R_{\text{Th}}$):

$$P_{L,\text{max}} = \frac{V_{\text{Th}}^2}{4R_{\text{Th}}}. \tag{5.138}$$

The maximum power that can be delivered to a resistive load R_L by a two-terminal network with Thevenin voltage V_{Th} and Thevenin resistance R_{Th} is equal to $V_{\text{Th}}^2/(4R_{\text{Th}})$, and this maximum power is delivered to the load when its resistance is matched to the Thevenin resistance: $R_L = R_{\text{Th}}$.

As an example, we can find the maximum power $P_{L,\text{max}}$, which may be delivered to the load resistor R_L in Figure 5.26, and the corresponding value of this load resistor. The most efficient approach to this problem is to find the Thevenin equivalent with respect to the terminals where the load resistor is connected.

In Figure 5.27, the load resistor is removed so we can perform the open-circuit analysis with respect to the terminals where this load had been applied.

Using the node voltage method with units of V, A, and Ω,

$$N1 \quad \frac{V_1 - V_2}{1} + 3i_\alpha + \frac{V_1 - V_4}{3} = 0, \tag{5.139}$$

$$N2 \quad \frac{V_2 - 8}{2} + \frac{V_2 - V_1}{1} + \frac{V_2 - V_3}{6} = 0, \tag{5.140}$$

$$N3 \quad \frac{V_3 - V_2}{6} - 3i_\alpha - 0.5 + \frac{V_3 - V_4}{4} = 0, \tag{5.141}$$

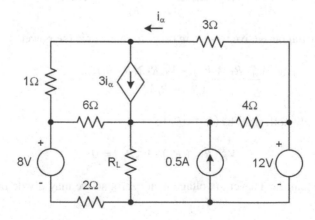

FIGURE 5.26 A two-terminal network with a load resistor R_L connected.

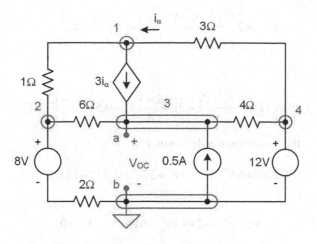

FIGURE 5.27 Open-circuit analysis of the previous circuit.

$$N4 \quad V_4 = 12. \tag{5.142}$$

The equation controlling the dependent source is

$$DS \quad i_\alpha = \frac{V_4 - V_1}{3}. \tag{5.143}$$

Substituting the DS equation into the N1 and N3 equations,

$$N1DS \quad \frac{V_1 - V_2}{1} + 3\left(\frac{V_4 - V_1}{3}\right) + \frac{V_1 - V_4}{3} = 0, \tag{5.144}$$

$$N2 \quad \frac{V_2 - 8}{2} + \frac{V_2 - V_1}{1} + \frac{V_2 - V_3}{6} = 0, \tag{5.145}$$

$$N3DS \quad \frac{V_3 - V_2}{6} - 3\left(\frac{V_4 - V_1}{3}\right) - 0.5 + \frac{V_3 - V_4}{4} = 0, \tag{5.146}$$

and

$$N4 \quad V_4 = 12. \tag{5.147}$$

Substituting the N4 equation into the N1 and N3 equations,

$$N1N4DS \quad \frac{V_1 - V_2}{1} + 3\left(\frac{12 - V_1}{3}\right) + \frac{V_1 - 12}{3} = 0, \tag{5.148}$$

$$N2 \quad \frac{V_2 - 8}{2} + \frac{V_2 - V_1}{1} + \frac{V_2 - V_3}{6} = 0, \tag{5.149}$$

and

$$N3N4DS \quad \frac{V_3 - V_2}{6} - 3\left(\frac{12 - V_1}{3}\right) - 0.5 + \frac{V_3 - 12}{4} = 0. \tag{5.150}$$

Multiplying by the least common denominators,

$$N1N4DS \quad 3V_1 - 3V_2 + 36 - 3V_1 + V_1 - 12 = 0, \tag{5.151}$$

$$N2 \quad 3V_2 - 24 + 6V_2 - 6V_1 + V_2 - V_3 = 0, \tag{5.152}$$

and

$$N3N4DS \quad 2V_3 - 2V_2 - 144 + 12V_1 - 6 + 3V_3 - 36 = 0. \tag{5.153}$$

Collecting like terms,

$$N1N4DS \quad V_1 - 3V_2 = -24, \tag{5.154}$$

$$N2 \quad -6V_1 + 10V_2 - V_3 = 24, \tag{5.155}$$

and

$$N3N4DS \quad 12V_1 - 2V_2 + 5V_3 = 186. \tag{5.156}$$

In matrix form,

$$\begin{bmatrix} 1 & -3 & 0 \\ -6 & 10 & -1 \\ 12 & -2 & 5 \end{bmatrix} \begin{bmatrix} V_1 \\ V_2 \\ V_3 \end{bmatrix} = \begin{bmatrix} -24 \\ 24 \\ 186 \end{bmatrix}. \tag{5.157}$$

Solving,

$$V_1 = \frac{\begin{vmatrix} -24 & -3 & 0 \\ 24 & 10 & -1 \\ 186 & -2 & 5 \end{vmatrix}}{\begin{vmatrix} 1 & -3 & 0 \\ -6 & 10 & -1 \\ 12 & -2 & 5 \end{vmatrix}} = 39.0\,\text{V}, \tag{5.158}$$

$$V_2 = \frac{\begin{vmatrix} 1 & -24 & 0 \\ -6 & 24 & -1 \\ 12 & 186 & 5 \end{vmatrix}}{\begin{vmatrix} 1 & -3 & 0 \\ -6 & 10 & -1 \\ 12 & -2 & 5 \end{vmatrix}} = 21.0\,\text{V}, \tag{5.159}$$

and

$$V_3 = \frac{\begin{vmatrix} 1 & -3 & -24 \\ -6 & 10 & 24 \\ 12 & -2 & 186 \end{vmatrix}}{\begin{vmatrix} 1 & -3 & 0 \\ -6 & 10 & -1 \\ 12 & -2 & 5 \end{vmatrix}} = -48.0\,\text{V}. \tag{5.160}$$

The open-circuit voltage is equal to the node three voltage:

$$V_{OC} = V_3 = -48.0\,\text{V}. \tag{5.161}$$

Next, we can consider the short-circuit analysis. Referring to Figure 5.28, we see that the short eliminates one essential node.

Using the node voltage method with units of V, A, and Ω,

$$N1 \quad \frac{V_1 - V_2}{1} + 3i_\alpha + \frac{V_1 - V_3}{3} = 0, \tag{5.162}$$

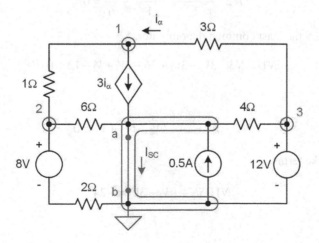

FIGURE 5.28 Short-circuit analysis of the previous circuit.

$$N2 \quad \frac{V_2-8}{2}+\frac{V_2-V_1}{1}+\frac{V_2}{6}=0, \tag{5.163}$$

and

$$N3 \quad V_3=12. \tag{5.164}$$

The equation controlling the dependent source is

$$DS \quad i_\alpha=\frac{V_3-V_1}{3}. \tag{5.165}$$

Substituting the DS equation into the N1 equations,

$$N1DS \quad \frac{V_1-V_2}{1}+3\left(\frac{V_3-V_1}{3}\right)+\frac{V_1-V_3}{3}=0, \tag{5.166}$$

$$N2 \quad \frac{V_2-8}{2}+\frac{V_2-V_1}{1}+\frac{V_2}{6}=0, \tag{5.167}$$

and

$$N3 \quad V_3=12. \tag{5.168}$$

Substituting the N3 equation into the N1 equation,

$$N1DSN3 \quad \frac{V_1-V_2}{1}+3\left(\frac{12-V_1}{3}\right)+\frac{V_1-12}{3}=0 \tag{5.169}$$

and

$$N2 \quad \frac{V_2-8}{2}+\frac{V_2-V_1}{1}+\frac{V_2}{6}=0. \tag{5.170}$$

Multiplying by the least common denominators,

$$N1DSN3 \quad 3V_1-3V_2+36-3V_1+V_1-12=0 \tag{5.171}$$

and

$$N2 \quad 3V_2-24+6V_2-6V_1+V_2=0. \tag{5.172}$$

Collecting like terms,

$$N1DSN4 \quad V_1-3V_2=-24 \tag{5.173}$$

and

$$N2 \quad -6V_1+10V_2=24. \tag{5.174}$$

In matrix form,

$$\begin{bmatrix} 1 & -3 \\ -6 & 10 \end{bmatrix} \begin{bmatrix} V_1 \\ V_2 \end{bmatrix} = \begin{bmatrix} -24 \\ 24 \end{bmatrix}. \tag{5.175}$$

Solving,

$$V_1 = \frac{\begin{vmatrix} 1 & -3 \\ -6 & 10 \end{vmatrix}}{\begin{vmatrix} 1 & -3 \\ -6 & 10 \end{vmatrix}} = 21.0\,\text{V} \tag{5.176}$$

and

$$V_2 = \frac{\begin{vmatrix} 1 & -3 \\ -6 & 10 \end{vmatrix}}{\begin{vmatrix} 1 & -3 \\ -6 & 10 \end{vmatrix}} = 15.0\,\text{V}. \tag{5.177}$$

The short-circuit current may be found using KCL:

$$I_{SC} = \frac{V_2}{6\,\Omega} + 3i_\alpha + 0.5\,\text{A} + \frac{V_3}{4\,\Omega}$$

$$= \frac{15.0\,\text{V}}{6\,\Omega} + 3\left(\frac{12.0\,\text{V} - 21.0\,\text{V}}{3\,\Omega}\right) + 0.5\,\text{A} + \frac{12.0\,\text{V}}{4\,\Omega} = -3.00\,\text{A}. \tag{5.178}$$

The Thevenin resistance is

$$R_{Th} = \frac{V_{OC}}{I_{SC}} = \frac{-48.0\,\text{V}}{-3.0\,\text{A}} - 16.0\,\Omega. \tag{5.179}$$

Notice that the usual sign conventions resulted in negative values for both the open-circuit voltage and short-circuit current, and this yielded a positive value of Thevenin resistance. Mixed signs for V_{OC} and I_{SC} (one positive and one negative) would indicate a sign error.

Knowing the Thevenin equivalent for the two-terminal network connected to the load resistor allows us to easily solve the maximum power transfer problem. The maximum power that may be delivered to the load is

$$P_{L,\max} = \frac{V_{Th}^2}{4R_{Th}} = 36.0\,\text{W}, \tag{5.180}$$

and this corresponds to a load resistor which is matched to the Thevenin resistance:

$$R_L = R_{Th} = 16.0\,\Omega. \tag{5.181}$$

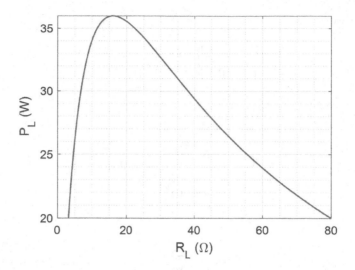

FIGURE 5.29   Load power as a function of load resistance for the example considered above.

Another advantage of the Thevenin approach is that we can easily determine the power delivered to an arbitrary load resistor even if does not correspond to the maximum power condition. In this more general case, the power delivered to the load is

$$P_L = \frac{V_{Th}^2 R_L}{\left(R_{Th} + R_L\right)^2},\tag{5.182}$$

as previously determined. Using V_{Th} and R_{Th} from the example just considered, we can plot the load power as a function of the load resistance and this is shown in Figure 5.29. As expected, the load power peaks at $V_{Th}^2 / (4R_{Th}) = 36.0\,\text{W}$ and this corresponds to a load resistance of $R_L = R_{Th} = 16.0\,\Omega$. If the load resistance is halved, the load power is reduced to $V_{Th}^2 (0.5R_{Th}) / (R_{Th} + 0.5R_{Th})^2 = (8/9)V_{Th}^2 / (4R_{Th}) = 32.0\,\text{W}$, which is reduced by a factor of $8/9$ compared to the maximum power. If the load resistance is doubled, the load power is reduced to $V_{Th}^2 (2R_{Th}) / (R_{Th} + 2R_{Th})^2 = (8/9)V_{Th}^2 / (4R_{Th}) = 32.0\,\text{W}$, which is also reduced by a factor of $8/9$ compared to the maximum power!

To review, see Presentation 5.5 in ebook+. To test your knowledge, try Quiz 5.5 in ebook+.

5.8   SUMMARY

Thevenin's theorem states that any two-terminal network made up of sources and resistors may be represented by its Thevenin equivalent, which comprises a single voltage source V_{Th} and a series resistance R_{Th}. The Thevenin model and the original circuit are externally equivalent; that is, they produce identical responses in external circuitry connected to the two terminals. **Norton's theorem** states that any two-terminal network including sources and resistors may be represented by its Norton equivalent, which comprises a current source I_N and a parallel resistance R_N. In a two-terminal circuit containing only independent sources with resistors, we may find the Thevenin (or Norton) equivalent by performing any two of the **open-circuit**, **short-circuit**, and **resistance shortcut** analyses. In a two-terminal network involving mixed (independent and dependent) sources as well as resistors, we may use any two of the **short-circuit**, **open-circuit**, and **test-source** analyses to find the Thevenin (or Norton equivalent). For a two-terminal network involving only dependent sources and resistors (no independent sources), the Thevenin voltage is zero and the Thevenin equivalent reduces to a single resistance. This resistance may be found using a **test-source analysis**. The Norton current also vanishes in this case, so the Norton equivalent reduces to a single resistance and is identical to the Thevenin equivalent in this case!

We may translate between the Thevenin and Norton equivalent circuits by using **source transformations**. More generally, a source transformation allows us to find a voltage source in series with a resistance, which is externally equivalent to a current source with a parallel resistance, or vice versa. This can be very useful in circuit analysis because it may allow us to simplify a circuit, this reducing the number of essential nodes or the number of meshes. We must be careful when simplifying a circuit this way, because it is possible to obscure some of the quantities (currents or voltages) which need to be determined. In principle, source transformations may be done with dependent sources, but extreme care must be exercised so the controlling variable for a dependent source is not obscured.

Thevenin's and Norton's theorems are extremely helpful when solving a complex circuit for a number of values of a single parameter. An example is a **maximum power transfer** problem, in which we want to find the maximum amount of power which may be delivered to a load resistor by a two-terminal circuit, as well as the corresponding value of the load resistor. If we find the Thevenin equivalent for the circuitry connected to the two terminals of the load resistor, then the maximum power that may be delivered to the load is $V_{Th}^2 / (4R_{Th})$ and this corresponds to a load which is matched to the Thevenin resistance ($R_L = R_{Th}$). In the next two chapters, we will see other important applications for Thevenin and Norton equivalents in first-order and second-order circuits.

To evaluate your mastery of Chapters 3–5, solve Example Exam 5.1 in ebook+ or Example Exam 5.2 in ebook+.

PROBLEMS

Problem 5.1. For the network in Figure P5.1, find the Thevenin equivalent with respect to the terminals a and b.

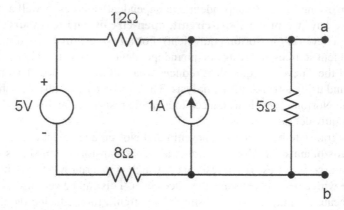

FIGURE P5.1   A two-terminal network containing only independent sources and resistors.

Problem 5.2. Find the Thevenin equivalent for the network in Figure P5.2 with respect to the terminals a and b.

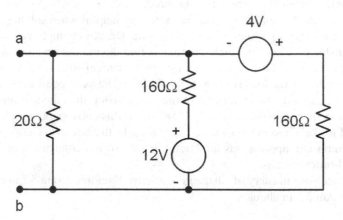

FIGURE P5.2   A two-terminal network containing two independent sources and three resistors.

Problem 5.3. For the two-terminal network in Figure P5.3, find the Norton equivalent with respect to the terminals a and b.

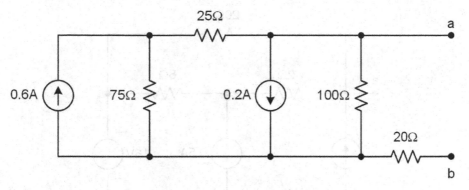

FIGURE P5.3   A two-terminal network containing two independent sources and four resistors.

Problem 5.4. For the network in Figure P5.4, determine the Thevenin equivalent with respect to the terminals a and b.

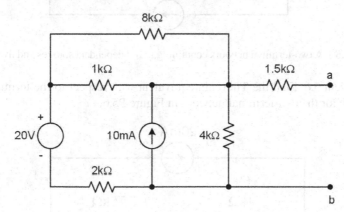

FIGURE P5.4   A two-terminal network containing two independent sources and five resistors.

Problem 5.5. For the circuit in Figure P5.5, find the Norton equivalent with respect to the terminals a and b.

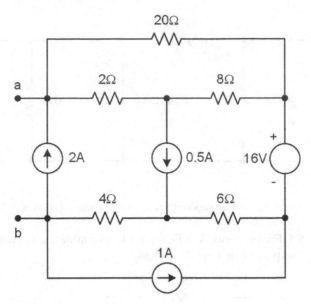

FIGURE P5.5   A two-terminal network containing four independent sources and five resistors.

Problem 5.6. Find the Thevenin equivalent with respect to the terminals a and b for the two-terminal network in Figure P5.6.

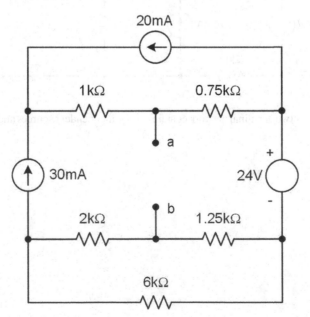

FIGURE P5.6   A two-terminal network containing only independent sources and resistors.

Problem 5.7. For the circuit in Figure P5.7, use source transformations to solve for the current I_X.

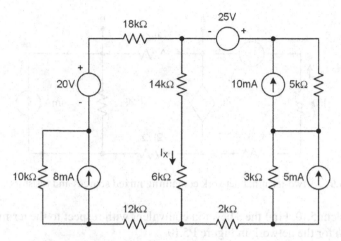

FIGURE P5.7   Circuit with five meshes.

Problem 5.8. For the circuit in Figure P5.8, use the number of source transformations to find the Thevenin equivalent with respect to the terminals where the load resistor R_L is connected. Using this Thevenin equivalent, find and plot I_L as a function of R_L, V_L as a function of R_L, and P_L as a function of R_L, for a range of load resistance $0.1R_{Th} \leq R_L \leq 10R_{Th}$.

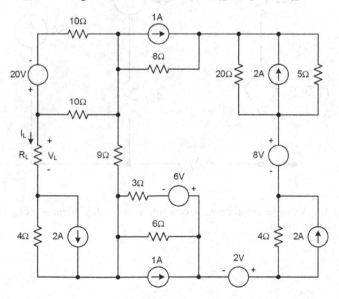

FIGURE P5.8   Complex circuit involving independent sources and resistors, including an unspecified load resistor R_L.

Problem 5.9. For the network in Figure P5.9, find the Thevenin equivalent with respect to the terminals a and b.

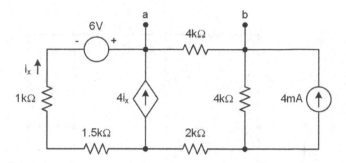

FIGURE P5.9 A two-terminal network containing mixed sources and resistors.

Problem 5.10. Find the Thevenin equivalent with respect to the terminals a and b for the network in Figure P5.10.

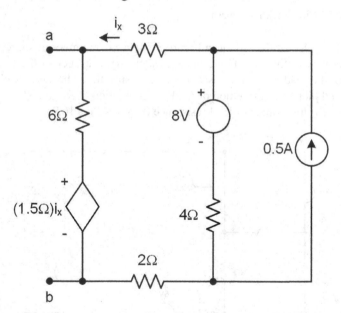

FIGURE P5.10 A two-terminal network containing two independent sources, a dependent source, and three resistors.

Problem 5.11. For the circuit in Figure P5.11, find the Thevenin equivalent with respect to the terminals a and b.

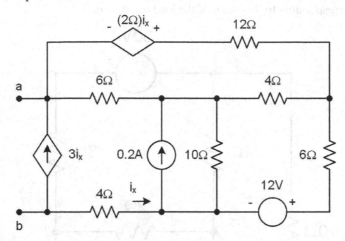

FIGURE P5.11   A two-terminal network containing two independent sources, two dependent sources and six resistors.

Problem 5.12. For the circuit in Figure P5.12, find the Thevenin equivalent with respect to the terminals a and b.

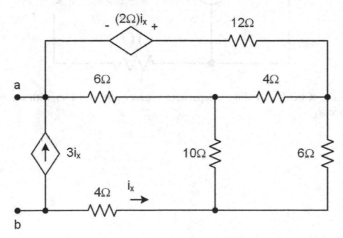

FIGURE P5.12   A two-terminal network containing only dependent sources and resistors.

Problem 5.13. Determine the value of load resistance R_L which will dissipate maximum power of the circuit in Figure P5.13. Find the values of I_L, V_L, and P_L corresponding to this value of the load resistance.

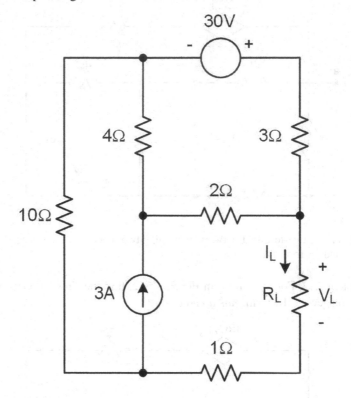

FIGURE P5.13   Circuit including independent sources and resistors with an unspecified load resistance R_L.

Problem 5.14. Find the value of the load resistance R_L which will dissipate maximum power in Figure P5.14. Find the power delivered to the load resistance when it is set to this value.

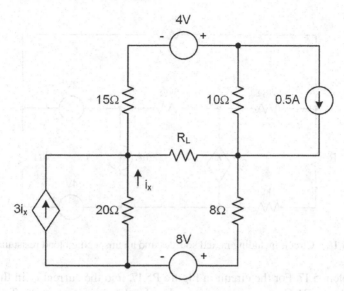

FIGURE P5.14   Circuit involving mixed sources and resistors, including an unspecified load resistance R_L.

Problem 5.15. For the circuit in Figure P5.15, find the value of the load resistance R_L which will dissipate maximum power. Find the power delivered to the load resistance when it is set to this value.

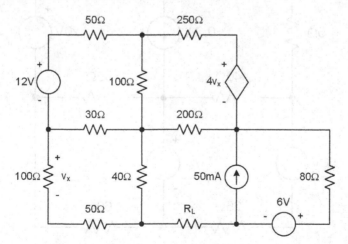

FIGURE P5.15   Circuit including three independent sources, a dependent source, and ten resistors, one of which is an unspecified load resistance R_L.

Problem 5.16. Find the value of the load resistance R_L in Figure P5.16 which will dissipate maximum power. Find the power delivered to the load resistance when it is set to this value.

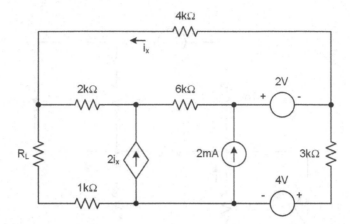

FIGURE P5.16   Circuit including mixed sources and an unspecified load resistance R_L.

Problem 5.17. For the circuit in Figure P5.17, find the current I_S in the voltage source V_S as a function of its value, by determination of the Thevenin equivalent for the rest of the circuit.

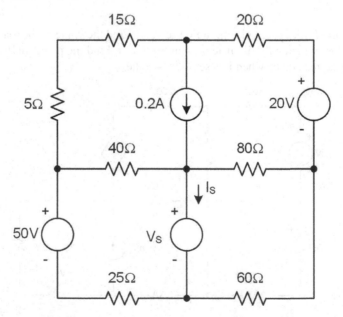

FIGURE P5.17   Circuit with an unspecified voltage source V_S.

Problem 5.18. For the circuit in Figure P5.18, calculate the value of V_x for each of the following values of R_x: 1 kΩ, 2 kΩ, 4 kΩ, 8 kΩ, and 16 kΩ.

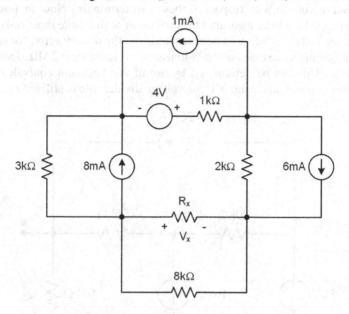

FIGURE P5.18   Circuit with an unspecified resistance R_X.

Problem 5.19. For the circuit in Figure P5.19, calculate the value of I_Q for each of the following values of R_Q: 0.3 Ω, 1 Ω, 3 Ω, 10 Ω, and 30 Ω.

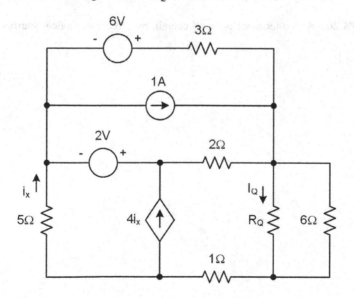

FIGURE P5.19   Circuit with an unspecified resistance R_Q.

Problem 5.20. Refer to the circuit in Figure P5.20. Find the voltage V_Z that exists between the terminals a and b in the circuit below. (This is the Thevenin voltage with respect to these two terminals.) Now suppose that the voltage V_Z is to be measured by a voltmeter with a finite input resistance. Find the voltage which will be measured, and the percent error, for each of the following three cases of the voltmeter input resistance: $2\,M\Omega$, $5\,M\Omega$, and $10\,M\Omega$. (This can be determined by use of the Thevenin equivalent with respect to terminals a and b, if the voltage divider rule is utilized.)

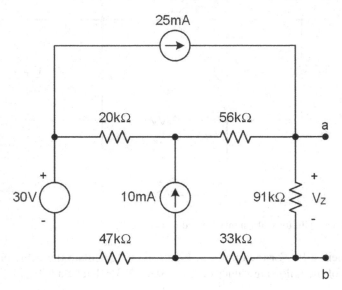

FIGURE P5.20   A two-terminal network containing three independent sources and five resistors.

6 First-Order Circuits

6.1 INTRODUCTION

Up to this point, we have considered only circuits with constant (not time-varying) voltages and currents. In this chapter, we forge ahead to first-order circuits that have time-varying circuit quantities. The voltages and currents in these circuits are solutions to first-order homogeneous linear differential equations, and end up being exponential in nature.

There are two important types of first-order circuits; these are the RC circuit and the RL circuit. The RC circuit involves a capacitor connected to a resistance (which may be part of a Thevenin equivalent), whereas an RL circuit involves an inductor connected with a resistance (which may also be part of a Thevenin equivalent). In this chapter, we will therefore need to consider two new circuit elements, the capacitor and inductor, to set up our discussions of first-order circuits.

6.2 THE CAPACITOR

The capacitor is a two-terminal element comprising two metal plates separated by an insulating material, or dielectric, as shown in Figure 6.1.

The application of a voltage across the capacitor with the polarity given in the figure places positive charges on the top plate and negative charges on the bottom plate. The positive and negative charges are equal in number, so the capacitor as a whole is charge neutral, but the separation of charge supports the voltage applied across the plates. As a consequence of Gauss' law, the charge on either plate is directly proportional to the applied voltage,

$$Q = Cv, \tag{6.1}$$

where Q is the electrical charge on the top plate in Coulombs, v is the applied voltage in Volts, and C is the capacitance in Farads (in basic units, Farad = Coulomb/Volt).

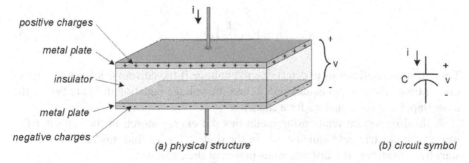

FIGURE 6.1   The capacitor: (a) physical structure and (b) circuit symbol.

DOI: 10.1201/9781003408529-6

The capacitance depends on the geometry of the structure (the area A and separation of the plates d) as well as the permittivity ε_i of the insulating material between the plates: $C = A\varepsilon_i / d$. It is important to recognize that the Farad is a very large quantity; although supercapacitors of 1 Farad or more have become available, we will commonly use capacitors measured in microfarads ($1\mu F = 10^{-6}\,F$), nanofarads ($1nF = 10^{-9}\,F$), or even picofarads ($1pF = 10^{-12}\,F$).

In an ideal capacitor, no electrical current may flow from one plate to the other through the insulating material. Nonetheless, an external current in the leads is necessary to charge and discharge the plates, and this is referred to as a **displacement current**. Combining the defining relation for electrical current with equation (6.1),

$$i = \frac{dQ}{dt} = C\frac{dv}{dt}. \tag{6.2}$$

Here, we have assumed that the capacitance is constant and not a function of applied voltage. (This is generally a good assumption but may not hold for some special-purpose capacitive devices.) Therefore, the above relationship allows us to find the displacement current if we know the time variation of the voltage. An important consequence of this equation is that **the voltage across a capacitor cannot change instantaneously in time**, because this would require infinite current.

In some cases, we may know the time dependence of the current and desire to find the resulting voltage. To analyze this situation, we can rearrange the current equation:

$$dv = \frac{i}{C}dt. \tag{6.3}$$

Integrating both sides, with x as a variable of integration for voltage and y as a variable of integration for time,

$$\int_{V_0}^{v} dx = \frac{1}{C}\int_{0}^{t} i\,dy, \tag{6.4}$$

where V_0 is the initial value of the voltage: $V_0 = v(0)$. Evaluating the left-hand integral at the limits,

$$v = V_0 + \frac{1}{C}\int_{0}^{t} i\,dy. \tag{6.5}$$

This expression allows us to determine the voltage if the current is known as a function of time. The current equation (6.2) and the voltage equation (6.5) are two of the most important relationships for a capacitor.

A third important relationship quantifies the energy stored in the capacitor (by way of the electric field which exists in the dielectric). To find this relationship, we start by considering the instantaneous power of the capacitor:

$$p = vi. \tag{6.6}$$

Using the current equation (6.2), we may write this as

$$p = vC\frac{dv}{dt}. \tag{6.7}$$

Rearranging,

$$pdt = Cvdv. \tag{6.8}$$

Integrating both sides, with x as a variable of integration for voltage and y as a variable of integration for time,

$$\int_0^t pdy = C\int_0^v xdx. \tag{6.9}$$

The expression on the left is the definition of work, so the energy stored in the capacitor is

$$w = C\left[\frac{x^2}{2}\right]_0^v = \frac{1}{2}Cv^2. \tag{6.10}$$

Notice that, because we are squaring the voltage, the energy stored in the capacitor depends only on the absolute value of the voltage and not on its algebraic sign.

6.3 PARALLEL CAPACITORS

When two or more capacitors are connected in parallel, as shown in Figure 6.2, we may find the equivalent capacitance for the combination, and thereby simplify the overall circuit.

The total current i may be found by applying KCL, with the recognition that the three parallel capacitors share the same voltage:

$$i = i_1 + i_2 + i_3 = C_1\frac{dv}{dt} + C_2\frac{dv}{dt} + C_3\frac{dv}{dt} = (C_1 + C_2 + C_3)\frac{dv}{dt}. \tag{6.11}$$

The equivalent capacitance for the parallel combination may be found as follows:

$$i = C_{eq}\frac{dv}{dt} = (C_1 + C_2 + C_3)\frac{dv}{dt}; \tag{6.12}$$

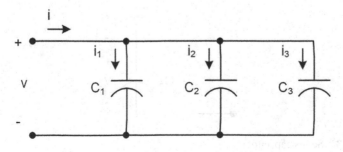

FIGURE 6.2 Parallel capacitors.

therefore,

$$C_{eq} = C_1 + C_2 + C_3. \tag{6.13}$$

Thus, the equivalent capacitance for parallel capacitors is equal to the sum of the individual capacitors. Although we considered three parallel capacitors here, this rule applies to any number of parallel capacitors. Moreover, because parallel capacitors share the same voltage, they share the same initial voltage for any integration.

6.4 SERIES CAPACITORS

Sometimes, we may have series combinations of capacitors, as shown in Figure 6.3.

The current i is common to the series capacitors, but they have different voltages. By KVL,

$$v = v_1 + v_2 + v_3$$

$$= V_{01} + \frac{1}{C_1}\int_0^t idy + V_{02} + \frac{1}{C_2}\int_0^t idy + V_{03} + \frac{1}{C_3}\int_0^t idy$$

$$= (V_{01} + V_{02} + V_{03}) + \left(\frac{1}{C_1} + \frac{1}{C_2} + \frac{1}{C_3}\right)\int_0^t idy \tag{6.14}$$

$$= V_{0eq} + \frac{1}{C_{eq}}\int_0^t idy.$$

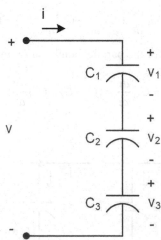

FIGURE 6.3 Series capacitors.

Therefore, the initial voltage for the combination is the sum of the individual initial voltages:

$$V_{0eq} = V_{01} + V_{02} + V_{03}, \tag{6.15}$$

and the equivalent capacitance is the reciprocal of the sum of the reciprocals of the individual capacitances:

$$C_{eq} = \left(\frac{1}{C_1} + \frac{1}{C_2} + \frac{1}{C_3} \right)^{-1}. \tag{6.16}$$

In terms of the equivalent capacitance, parallel capacitors combine like series resistors but series capacitors combine like parallel resistors.

To review, see Presentation 6.1 in ebook+.

To test your knowledge, try Quiz 6.1 in ebook+.

6.5 NATURAL RESPONSE OF AN RC CIRCUIT

Suppose that the switch in Figure 6.4 has been in position a for a "long time" before instantaneously moving to position b at $t = 0$. The circuitry on the left could represent any two-terminal network, which has been replaced by its Thevenin equivalent. Although we are considering a simple resistor on the right for now, we will later generalize to the case of a Thevenin circuit on the right as well.

When we say that the switch has been in position a for a "long time," we mean that the circuit has settled, and therefore all time derivatives have vanished to zero. Therefore, the capacitor current $C dv / dt$ will have settled to zero. If we apply KVL at $t = 0^-$, the point in time right before the switch moves

$$-V_{Th0} + i(0^-) R_{Th0} + v(0^-) = 0. \tag{6.17}$$

Because the capacitor current (and therefore the middle term) will have settled to zero,

$$v(0^-) = V_{Th0}. \tag{6.18}$$

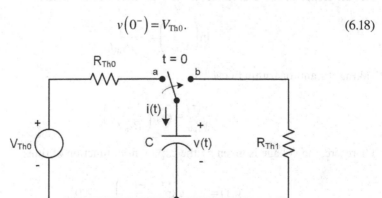

FIGURE 6.4 A switched RC circuit for the consideration of the natural response.

Therefore, the capacitor voltage will have settled to the Thevenin voltage for the circuitry connected prior to the switch movement. Because the capacitor voltage cannot change instantaneously, it will still have this value immediately after the switch movement:

$$v(0^+) = v(0^-) = V_{Th0}. \tag{6.19}$$

If we consider KCL for node b after the switch has moved, then

$$C\frac{dv}{dt} + \frac{v}{R_{Th1}} = 0. \tag{6.20}$$

This is a first-order linear homogeneous differential equation. Rearranging,

$$\frac{dv}{v} = -\frac{dt}{R_{Th1}C}. \tag{6.21}$$

We can solve this by integrating both sides. Using x as a variable of integration for voltage and y as a variable of integration for time,

$$\int_{V_{Th0}}^{v(t)} \frac{dx}{x} = -\frac{1}{R_{Th1}C}\int_0^t dy. \tag{6.22}$$

Integrating,

$$[\ln(x)]\big|_{V_{Th0}}^{v(t)} = -\frac{1}{R_{Th1}C}[y]\big|_0^t \tag{6.23}$$

Applying the limits,

$$\ln[v(t)] - \ln[V_{Th0}] = -\frac{t}{R_{Th1}C}. \tag{6.24}$$

The difference of two logarithms is the logarithm of the quotient:

$$\ln\left(\frac{v(t)}{V_{Th0}}\right) = -\frac{t}{R_{Th1}C}. \tag{6.25}$$

Taking the antilogarithm of each side,

$$\frac{v(t)}{V_{Th0}} = \exp\left(\frac{-t}{R_{Th1}C}\right). \tag{6.26}$$

Therefore, the voltage is given by the exponential function of time:

$$v(t) = V_{Th0}\exp\left(\frac{-t}{R_{Th1}C}\right); \quad t \geq 0^+. \tag{6.27}$$

The exponential function may also be expressed as

$$v(t) = V_0 \exp(-t / \tau); \quad t \geq 0^+,$$ (6.28)

where V_0 is the initial voltage across the capacitor and τ is the "time constant" for the circuit, given by $\tau = R_{Th1}C$. Figure 6.5 shows the normalized natural response for an RC circuit; the voltage is normalized by dividing by the initial value and time is normalized by dividing by the time constant. We can see that the normalized value is 0.05 (95% settled to zero) after three time constants and 0.007 (99.3% settled to zero) after five time constants. As a practical rule of thumb, we often say that a circuit is settled after 3–5 time constants, although the actual characteristic reaches its limiting value asymptotically. Therefore, if we require that the switch was in position a for a "long time" before moving at $t = 0$, a practical interpretation of this is that the switch was in its starting position for at least five time constants. When applying this requirement, we must consider the time constant which existed for $t < 0$, or $R_{Th0}C$.

It should be noted that **all voltages and currents in an RC circuit will be given by exponential functions with the same time constant as the capacitor voltage.** For example, the capacitor current may be found by using the capacitor–current relationship:

$$i(t) = C\frac{dv}{dt} = C\frac{d}{dt}\left[V_{Th0} \exp\left(\frac{-t}{R_{Th1}C}\right)\right]$$

(6.29)

$$= \frac{V_{Th0}}{R_{Th1}} \exp\left(\frac{-t}{R_{Th1}C}\right); \quad t \geq 0^+.$$

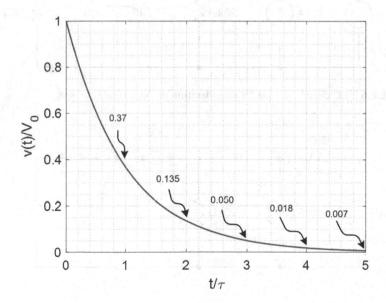

FIGURE 6.5   Normalized natural response for an RC circuit.

As an example, we will find $v(t)$ and $i(t)$ for $t > 0^+$ of the circuit in Figure 6.6. Here, the 2 kΩ resistor is connected both before and after the movement of the switch, so it affects both R_{Th0} and R_{Th1}. However, the exact value of R_{Th0} is unimportant if we are to determine $v(t)$ and $i(t)$ for $t > 0^+$, just as long as the switch was in its original position for a long time. In order to solve this problem, we will first find $v(t)$ and then differentiate to determine $i(t)$. This might be more convenient than first finding $i(t)$ and then integrating to obtain $v(t)$.

In order to find V_{Th0}, we will perform the open-circuit analysis for $t < 0$ (switch closed) with respect to the terminals where the capacitor had been connected as shown in Figure 6.7.

We can apply the node voltage method with a single equation:

$$N1 \quad \frac{V_{Th0} - 60\,V}{5\,k\Omega} - 24\,mA + \frac{V_{Th0}}{20\,k\Omega} + \frac{V_{Th0}}{2\,k\Omega} = 0; \tag{6.30}$$

$$V_{Th0} = \frac{60\,V\,/\,5\,k\Omega + 24\,mA}{1\,/\,5\,k\Omega + 1\,/\,20\,k\Omega + 1\,/\,2\,k\Omega} = 48.0\,V. \tag{6.31}$$

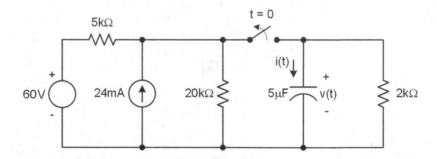

FIGURE 6.6   RC circuit example for consideration of the natural response.

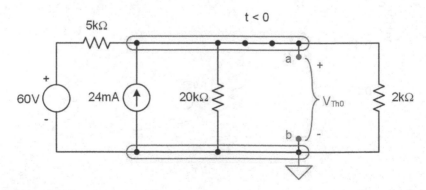

FIGURE 6.7   Open-circuit analysis of the previous circuit for $t < 0$ with respect to the terminals where the capacitor had been connected.

For $t > 0$, with the switch open, there are no sources connected to the capacitor so we are considering a natural response. By inspection, $R_{Th1} = 2\,k\Omega$. Therefore, $V_0 = V_{Th0} = 48.0\,V$, $\tau = R_{Th1}C = (2\,k\Omega)(5\,\mu F) = 10\,ms$, and

$$v(t) = 48.0\,V \exp(-t/10\,ms); \quad t \geq 0^+.$$ (6.32)

The capacitor current may be found by differentiating

$$i(t) = C\frac{dv}{dt} = (5\,\mu F)\frac{d}{dt}\left[48.0\,V \exp\left(\frac{-t}{10\,ms}\right)\right]$$ (6.33)

$$= \frac{(5\,\mu F)(48.0\,V)}{-10\,ms}\exp\left(\frac{-t}{10\,ms}\right) = -24.0\,mA \exp\left(\frac{-t}{10\,ms}\right); \quad t \geq 0^+.$$

The voltage response is shown in Figure 6.8 and the current response is shown in Figure 6.9. It can be seen that both settle in approximately five time constants, which is 50 ms in this case. As stated previously, the voltage across the capacitor cannot change instantaneously so $v(0^+) = v(0^-) = 48.0\,V$. The same is not true for the capacitor current, which is proportional to the time derivative of the voltage and may change instantaneously. Therefore, although the current will have settled to zero before the switch is moved $\left[i(0^-)\right] = 0$, the value of current right after the switch movement is non-zero $\left[i(0^+) = -24\,mA\right]$.

To review, see Presentation 6.2 in ebook+.

To test your knowledge, try Quiz 6.2 in ebook+.

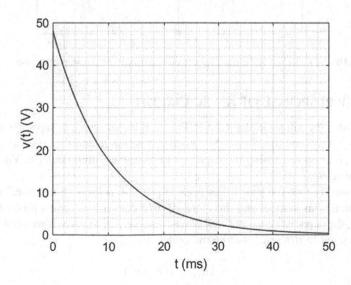

FIGURE 6.8 Natural response (voltage) for the previous example.

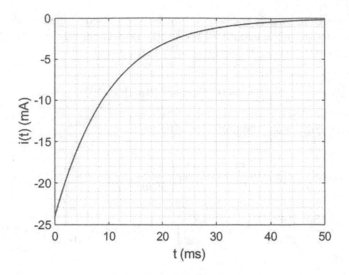

FIGURE 6.9   Natural response (current) for the previous example.

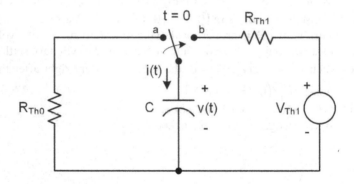

FIGURE 6.10   A switched RC circuit for the consideration of the step response.

6.6   STEP RESPONSE OF AN RC CIRCUIT

Now consider the situation depicted in Figure 6.10, in which a capacitor has been discharged for a "long time" with the switch in position a, and then the switch moves to position b at $t = 0$, causing the capacitor to charge asymptotically to V_{Th1} through the resistance R_{Th1}.

When we say that the switch has been in position a for a "long time," we mean that the circuit has settled, and therefore all time derivatives have vanished to zero. Therefore, the capacitor current Cdv/dt will have settled to zero. If we apply KVL at $t = 0^-$, the point in time right before the switch moves,

$$i\left(0^-\right)R_{Th0} + v\left(0^-\right) = 0. \tag{6.34}$$

Because the capacitor current (and therefore the left-hand term) will have settled to zero,

$$v(0^-) = 0. \tag{6.35}$$

Because the capacitor voltage cannot change instantaneously, it will still have this value immediately after the switch movement:

$$v(0^+) = v(0^-) = 0. \tag{6.36}$$

If we consider KCL for node b after the switch has moved, then

$$C\frac{dv}{dt} + \frac{v - V_{Th1}}{R_{Th1}} = 0. \tag{6.37}$$

This is a first-order linear differential equation. Rearranging,

$$\frac{dv}{V_{Th1} - v} = \frac{dt}{R_{Th1}C}. \tag{6.38}$$

We can solve by integrating both sides. Using x as a variable of integration for voltage and y as a variable of integration for time,

$$\int_0^{v(t)} \frac{dx}{V_{Th1} - x} = \frac{1}{R_{Th1}C}\int_0^t dy. \tag{6.39}$$

Integrating,

$$[\ln(V_{Th1} - x)]\Big|_0^{v(t)} = -\frac{1}{R_{Th1}C}\Big[y\Big|_0^t. \tag{6.40}$$

Applying the limits,

$$\ln[V_{Th1} - v(t)] - \ln[V_{Th1}] = -\frac{t}{R_{Th1}C}. \tag{6.41}$$

The difference of two logarithms is the logarithm of the quotient:

$$\ln\left(\frac{V_{Th1} - v(t)}{V_{Th1}}\right) = -\frac{t}{R_{Th1}C}. \tag{6.42}$$

Taking the antilogarithm of each side,

$$\frac{V_{Th1} - v(t)}{V_{Th1}} = \exp\left(\frac{-t}{R_{Th1}C}\right). \tag{6.43}$$

Therefore, the voltage is given by the exponential function of time:

$$v(t) = V_{\text{Th1}}\left[1 - \exp\left(\frac{-t}{R_{\text{Th1}}C}\right)\right]; \quad t \geq 0^{+}. \tag{6.44}$$

The exponential function may also be expressed as

$$v(t) = V_F\left[1 - \exp(-t/\tau)\right]; \quad t \geq 0^{+}, \tag{6.45}$$

where V_F is the final voltage across the capacitor and τ is the "time constant" for the circuit, given by $\tau = R_{\text{Th1}}C$. Figure 6.11 shows the normalized step response for an RC circuit. Similar to the case of the natural response, the value is 95% settled after three constants and 99.3% settled after five time constants; the difference is that the normalized step response asymptotically approaches unity rather than zero.

As an example, consider the circuit in Figure 6.12. This circuit is almost exactly the same as the one considered in the previous section; the only difference is that the

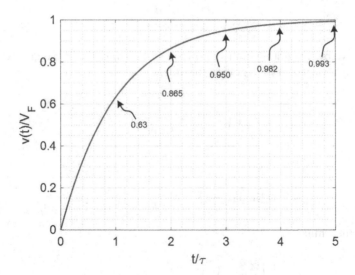

FIGURE 6.11 Normalized step response for an RC circuit.

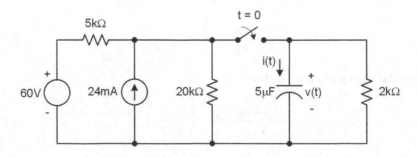

FIGURE 6.12 RC circuit example for consideration of the step response.

switch closes rather than opening at $t = 0$, and this produces a step response. There is no source connected to the capacitor for $t < 0$ so $V_{Th0} = 0$. The value of R_{Th0} is unimportant provided that the switch has been in its starting position for a long time, allowing the circuit to have settled before we move the switch.

There is no source connected to the capacitor for $t < 0$ so $V_{Th0} = 0$. In order to find V_{Th1}, we will perform the open-circuit analysis for $t > 0$ (switch closed) with respect to the terminals where the capacitor had been connected as shown in Figure 6.13.

We can apply the node voltage method with a single equation:

$$N1 \quad \frac{V_{Th1} - 60\,V}{5\,k\Omega} - 24\,mA + \frac{V_{Th1}}{20\,k\Omega} + \frac{V_{Th1}}{2\,k\Omega} = 0 \qquad (6.46)$$

and

$$V_{Th1} = \frac{60\,V / 5\,k\Omega + 24\,mA}{1/5\,k\Omega + 1/20\,k\Omega + 1/2\,k\Omega} = 48.0\,V. \qquad (6.47)$$

Next, we will perform the short-circuit analysis for $t > 0$ with respect to the terminals where the capacitor had been connected, as shown in Figure 6.14. There is only one

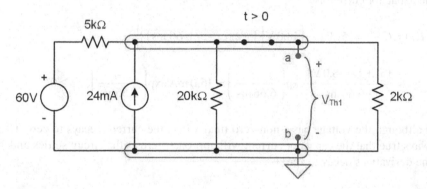

FIGURE 6.13 Open-circuit analysis of the previous circuit for $t > 0$ with respect to the terminals where the capacitor had been connected.

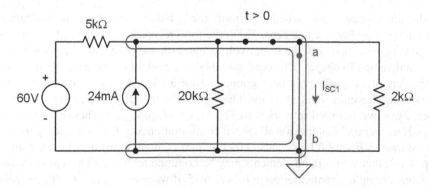

FIGURE 6.14 Short-circuit analysis of the previous circuit for $t > 0$ with respect to the terminals where the capacitor had been connected.

essential node (the reference node) so it is unnecessary to solve for any node voltages. Also, under short-circuit conditions, no current flows in either the $20\,\text{k}\Omega$ resistor or the $2\,\text{k}\Omega$ resistor, because there is zero voltage applied across both of them.

The short-circuit current is given by

$$I_{SC1} = \frac{60\,\text{V}}{5\,\text{k}\Omega} + 24\,\text{mA} = 36.0\,\text{mA}. \tag{6.48}$$

The Thevenin resistance R_{Th1} for $t > 0$ is therefore

$$R_{Th1} = \frac{V_{OC1}}{I_{SC1}} = \frac{V_{Th1}}{I_{SC1}} = \frac{48.0\,\text{V}}{36.0\,\text{mA}} = 1.333\,\text{k}\Omega. \tag{6.49}$$

The final voltage is $V_F = V_{Th1} = 48.0\,\text{V}$ and the time constant is $\tau = R_{Th1}C = (1.333\,\text{k}\Omega)(5\,\mu\text{F}) = 6.66\,\text{ms}$ so the capacitor voltage is therefore

$$v(t) = 48.0\,\text{V}\left[1 - \exp(-t/6.66\,\text{ms})\right], \quad t \geq 0^+. \tag{6.50}$$

The capacitor current is

$$i(t) = C\frac{dv}{dt} = (5\,\mu\text{F})\frac{d}{dt}\left\{48.0\,\text{V}\left[1 - \exp(-t/6.66\,\text{ms})\right]\right\}$$

$$= \frac{(5\,\mu\text{F})(-48.0\,\text{V})}{-6.66\,\text{ms}}\exp\left(\frac{-t}{6.66\,\text{ms}}\right) = 36.0\,\text{mA}\exp\left(\frac{-t}{6.66\,\text{ms}}\right), \quad t \geq 0^+. \tag{6.51}$$

So although the voltage has a non-zero final value, the current decays to zero. (It is always true that the capacitor current will approach zero as the circuit settles and all time derivatives decay to zero.)

6.7   GENERAL CASE OF NATURAL AND STEP RESPONSE IN AN RC CIRCUIT

In the general case of a switched RC circuit, the initial and final values for the capacitor voltage may both be non-zero. Then, the total response is the sum of the natural response (associated with the initial voltage) and the step response (associated with the final value of voltage). This corresponds to a circuit of the type shown in Figure 6.15. Here, the capacitor has been connected for a long time to a two-terminal network with Thevenin voltage V_{Th0} and Thevenin resistance R_{Th0}, and after $t = 0$ is connected to a two-terminal network with Thevenin voltage V_{Th1} and Thevenin resistance R_{Th1}. Keep in mind that the overall circuit configuration may not look exactly like the one shown in Figure 6.15, because other switch configurations may be used and it is possible that some circuit elements may be common to the two Thevenin circuits.

Considering the configuration in Figure 6.15, if we apply KVL at $t = 0^-$, the point in time right before the switch moves,

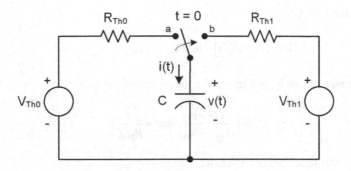

FIGURE 6.15 A switched RC circuit for the consideration of the general case of the natural and step response.

$$-V_{\text{Th0}} + i\left(0^-\right)R_{\text{Th0}} + v\left(0^-\right) = 0. \tag{6.52}$$

Because the capacitor current (and therefore the middle term) will have settled to zero,

$$v\left(0^-\right) = V_{\text{Th0}}. \tag{6.53}$$

Because the capacitor voltage cannot change instantaneously, it will still have this value immediately after the switch movement:

$$v\left(0^+\right) = v\left(0^-\right) = V_{\text{Th0}}. \tag{6.54}$$

If we consider KCL for node b after the switch has moved, then

$$C\frac{dv}{dt} + \frac{v - V_{\text{Th1}}}{R_{\text{Th1}}} = 0. \tag{6.55}$$

This is a first-order linear differential equation. Rearranging,

$$\frac{dv}{v - V_{\text{Th1}}} = -\frac{dt}{R_{\text{Th1}}C}. \tag{6.56}$$

We can solve by integrating both sides. Using x as a variable of integration for voltage and y as a variable of integration for time,

$$\int_{V_{\text{Th0}}}^{v(t)} \frac{dx}{x - V_{\text{Th1}}} = -\frac{1}{R_{\text{Th1}}C}\int_0^t dy. \tag{6.57}$$

Integrating,

$$\left[\ln\left(x - V_{\text{Th1}}\right)\right]_{V_{\text{Th0}}}^{v(t)} = -\frac{1}{R_{\text{Th1}}C}\left[y\right]_0^t. \tag{6.58}$$

Applying the limits,

$$\ln\left[v(t)-V_{\text{Th1}}\right]-\ln\left[V_{\text{Th0}}-V_{\text{Th1}}\right]=-\frac{t}{R_{\text{Th1}}C}. \tag{6.59}$$

The difference of two logarithms is the logarithm of the quotient:

$$\ln\left(\frac{v(t)-V_{\text{Th1}}}{V_{\text{Th0}}-V_{\text{Th1}}}\right)=-\frac{t}{R_{\text{Th1}}C}. \tag{6.60}$$

Taking the antilogarithm of each side,

$$\frac{v(t)-V_{\text{Th1}}}{V_{\text{Th0}}-V_{\text{Th1}}}=\exp\left(\frac{-t}{R_{\text{Th1}}C}\right). \tag{6.61}$$

Therefore, the voltage is given by the exponential function of time:

$$v(t)=V_{\text{Th1}}+\left(V_{\text{Th0}}-V_{\text{Th1}}\right)\exp\left(\frac{-t}{R_{\text{Th1}}C}\right); \quad t\geq 0^{+}. \tag{6.62}$$

This function may also be expressed as

$$v(t)=V_F+\left(V_0-V_F\right)\exp(-t/\tau); \quad t\geq 0^{+}, \tag{6.63}$$

where V_0 is the initial voltage $(V_0=V_{\text{Th0}})$, V_F is the final voltage $(V_F=V_{\text{Th1}})$, and τ is the "time constant" given by $\tau=R_{\text{Th1}}C$. The same form is applicable to any voltage or current in an RC circuit. This is extremely useful because it allows us to immediately write the solution for any such voltage or current as long as the initial value, final value, and time constant are known. For example, the current in the capacitor may be written as

$$i(t)=I_F+\left(I_0-I_F\right)\exp(-t/\tau); \quad t\geq 0^{+}, \tag{6.64}$$

where I_0 is the initial current, I_F is the final current, and τ is the "time constant." (The time constant takes on the same value for any voltage or current in an RC circuit.)

Lastly, we should note that the overall response can be considered to be the sum of the natural and step response:

$$v(t)=\underbrace{V_0\exp(-t/\tau)}_{\text{natural response}}+\underbrace{V_F\left[1-\exp(-t/\tau)\right]}_{\text{step response}}; \quad t\geq 0^{+}. \tag{6.65}$$

Here, the natural response is associated with the initial value term while the step response is associated with the final value terms.

As an example, consider the RC circuit in Figure 6.16. The switch has been in its starting position for a "long time" before moving at $t=0$, and we want to determine $v(t)$ and $i(t)$ for $t>0^{+}$. By inspection, $V_{\text{Th0}}=15\,\text{V}$ and $R_{\text{Th0}}=2\,\text{k}\Omega$.

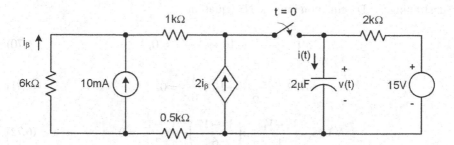

FIGURE 6.16 RC circuit example for consideration of the natural and step response.

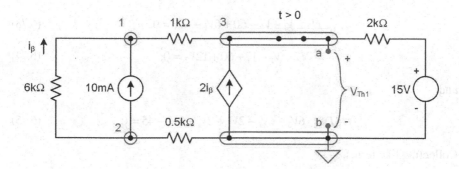

FIGURE 6.17 Open-circuit analysis of the previous circuit for $t > 0$ with respect to the terminals where the capacitor had been connected.

With the switch closed, we may use the open-circuit and short-circuit analyses to find V_{Th1} and R_{Th1}. Figure 6.17 shows the labeling of nodes for use of the node voltage method with the open-circuit analysis.

Using units of V, mA, and kΩ,

$$N1 \quad \frac{V_1 - V_2}{6} - 10 + \frac{V_1 - V_3}{1} = 0, \tag{6.66}$$

$$N2 \quad \frac{V_2 - V_1}{6} + 10 + \frac{V_2}{0.5} = 0, \tag{6.67}$$

and

$$N3 \quad \frac{V_3 - V_1}{1} - 2i_\beta + \frac{V_3 - 15}{2} = 0. \tag{6.68}$$

The equation of the current controlling the dependent source is

$$DS \quad i_\beta = \frac{V_2 - V_1}{6}. \tag{6.69}$$

Substituting the DS equation into the N3 equation,

$$N1 \quad \frac{V_1 - V_2}{6} - 10 + \frac{V_1 - V_3}{1} = 0, \tag{6.70}$$

$$N2 \quad \frac{V_2 - V_1}{6} + 10 + \frac{V_2}{0.5} = 0, \tag{6.71}$$

$$N3DS \quad \frac{V_3 - V_1}{1} - 2\left(\frac{V_2 - V_1}{6}\right) + \frac{V_3 - 15}{2} = 0. \tag{6.72}$$

Multiplying by the least common denominators,

$$N1 \quad V_1 - V_2 - 60 + 6V_1 - 6V_3 = 0, \tag{6.73}$$

$$N2 \quad V_2 - V_1 + 60 + 12V_2 = 0, \tag{6.74}$$

and

$$N3DS \quad 6V_3 - 6V_1 - 2V_2 + 2V_1 + 3V_3 - 45 = 0. \tag{6.75}$$

Collecting like terms,

$$N1 \quad 7V_1 - V_2 - 6V_3 = 60, \tag{6.76}$$

$$N2 \quad -V_1 + 13V_2 = -60, \tag{6.77}$$

and

$$N3DS \quad -4V_1 - 2V_2 + 9V_3 = 45. \tag{6.78}$$

In matrix form,

$$\begin{bmatrix} 7 & -1 & -6 \\ -1 & 13 & 0 \\ -4 & -2 & 9 \end{bmatrix} \begin{bmatrix} V_1 \\ V_2 \\ V_3 \end{bmatrix} = \begin{bmatrix} 60 \\ -60 \\ 45 \end{bmatrix}. \tag{6.79}$$

Solving,

$$V_1 = \frac{\begin{vmatrix} 60 & -1 & -6 \\ -60 & 13 & 0 \\ 45 & -2 & 9 \end{vmatrix}}{\begin{vmatrix} 7 & -1 & -6 \\ -1 & 13 & 0 \\ -4 & -2 & 9 \end{vmatrix}} = -19.07\,\text{V}, \tag{6.80}$$

$$V_2 = \dfrac{\begin{vmatrix} 7 & 60 & -6 \\ -1 & -60 & 0 \\ -4 & 45 & 9 \end{vmatrix}}{\begin{vmatrix} 7 & -1 & -6 \\ -1 & 13 & 0 \\ -4 & -2 & 9 \end{vmatrix}} = -3.15\,\text{V},$$

(6.81)

and

$$V_3 = \dfrac{\begin{vmatrix} 7 & -1 & 60 \\ -1 & 13 & -60 \\ -4 & -2 & 45 \end{vmatrix}}{\begin{vmatrix} 7 & -1 & -6 \\ -1 & 13 & 0 \\ -4 & -2 & 9 \end{vmatrix}} = 12.78\,\text{V}.$$

(6.82)

Therefore,

$$V_{\text{Th1}} = V_{\text{OC1}} = V_3 = 12.78\,\text{V}.$$

(6.83)

Next, we will perform the short-circuit analysis for $t > 0$ with respect to the terminals where the capacitor had been connected, as shown in Figure 6.18.

Using units of V, mA, and kΩ,

$$N1 \quad \frac{V_1 - V_2}{6} - 10 + \frac{V_1}{1} = 0$$

(6.84)

and

$$N2 \quad \frac{V_2 - V_1}{6} + 10 + \frac{V_2}{0.5} = 0.$$

(6.85)

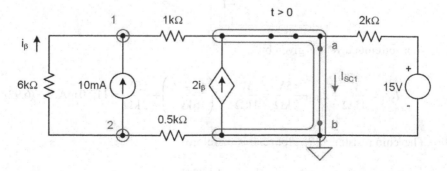

FIGURE 6.18 Short-circuit analysis of the previous circuit for $t > 0$ with respect to the terminals where the capacitor had been connected.

Multiplying by the least common denominators,

$$N1 \quad V_1 - V_2 - 60 + 6V_1 = 0 \tag{6.86}$$

and

$$N2 \quad V_2 - V_1 + 60 + 12V_2 = 0. \tag{6.87}$$

Collecting like terms,

$$N1 \quad 7V_1 - V_2 = 60 \tag{6.88}$$

and

$$N2 \quad -V_1 + 13V_2 = -60. \tag{6.89}$$

In matrix form,

$$\begin{bmatrix} 7 & -1 \\ -1 & 13 \end{bmatrix} \begin{bmatrix} V_1 \\ V_2 \end{bmatrix} = \begin{bmatrix} 60 \\ -60 \end{bmatrix}. \tag{6.90}$$

Solving,

$$V_1 = \frac{\begin{vmatrix} 60 & -1 \\ -60 & 13 \end{vmatrix}}{\begin{vmatrix} 7 & -1 \\ -1 & 13 \end{vmatrix}} = 8.00 \, \text{V} \tag{6.91}$$

and

$$V_2 = \frac{\begin{vmatrix} 7 & 60 \\ -1 & -60 \end{vmatrix}}{\begin{vmatrix} 7 & -1 \\ -1 & 13 \end{vmatrix}} = -4.00 \, \text{V}. \tag{6.92}$$

The short-circuit current is given by

$$I_{SC1} = \frac{V_1}{1\,\text{k}\Omega} + 2i_\beta + \frac{15\,\text{V}}{2\,\text{k}\Omega} = \frac{V_1}{1\,\text{k}\Omega} + 2\left(\frac{V_2 - V_1}{6\,\text{k}\Omega}\right) + \frac{15\,\text{V}}{2\,\text{k}\Omega} = 11.50\,\text{mA}. \tag{6.93}$$

The Thevenin resistance R_{Th1} for $t > 0$ is therefore

$$R_{Th1} = \frac{V_{OC1}}{I_{SC1}} = \frac{V_{Th1}}{I_{SC1}} = \frac{12.78\,\text{V}}{11.50\,\text{mA}} = 1.111\,\text{k}\Omega \tag{6.94}$$

The initial voltage is $V_0 = V_{Th0} = 15\,\text{V}$, the final voltage is $V_F = V_{Th1} = 12.78\,\text{V}$ and the time constant is $\tau = R_{Th1}C = (1.111\,\text{k}\Omega)(2\,\mu\text{F}) = 2.22\,\text{ms}$ so the capacitor voltage is therefore

$$v(t) = V_F + (V_0 - V_F)\exp(-t/\tau)$$

$$= 12.78\,\text{V} + 2.22\,\text{V}\exp(-t/2.22\,\text{ms}); \quad t \geq 0^+$$

(6.95)

The capacitor current is

$$i(t) = C\frac{dv}{dt} = (2\,\mu\text{F})\frac{d}{dt}\{12.78\,\text{V} + 2.22\,\text{V}\exp(-t/2.22\,\text{ms})\}$$

(6.96)

$$= \frac{(2\,\mu\text{F})(2.22\,\text{V})}{-2.22\,\text{ms}}\exp\left(\frac{-t}{2.22\,\text{ms}}\right) = -2.00\,\text{mA}\exp\left(\frac{-t}{2.22\,\text{ms}}\right). \quad t \geq 0^+.$$

It should be noted that the general form of the solution for the step plus natural response is always applicable, although either V_F or V_0 may be zero in the special cases of the natural response or step response, respectively.

To review, see Presentation 6.3 in ebook+.

To test your knowledge, try Quiz 6.3 in ebook+. To put your knowledge to practice, try Laboratory Exercise 6.1 in ebook+.

6.8 THE INDUCTOR

The inductor is a two-terminal element which involves a coil of wire wrapped around a core which could be air or some magnetic material shown in Figure 6.19. Flowing a current through the coil gives rise to magnetic flux, which links the coil and is in a direction determined by the "right-hand rule."

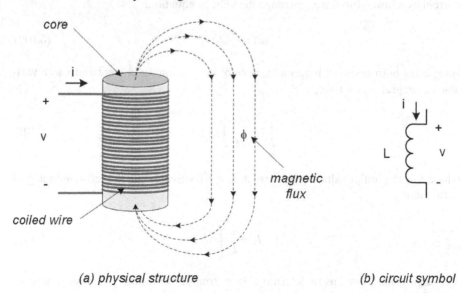

(a) physical structure   *(b) circuit symbol*

FIGURE 6.19 The inductor; (a) physical structure and (b) circuit symbol.

Quantitatively, the magnetic flux density B in a linear magnetic medium is given by

$$B = \mu H = \mu N i, \tag{6.97}$$

where μ is the permeability of the medium (Wb m^{-2} A^{-1}), H is the magnetic field intensity (A), N is the number of turns in the coil, and i is the electrical current (A). The magnetic flux ϕ is the product of the flux density B and the cross-sectional area A:

$$\phi = AB = A\mu N i. \tag{6.98}$$

By the Faraday law, the voltage developed across the coil is proportional to the time rate of change of the magnetic flux linkage:

$$v = \frac{d\lambda}{dt} = \frac{dN\phi}{dt} = A\mu N^2 \frac{di}{dt} = L \frac{di}{dt}, \tag{6.99}$$

where $\lambda = N\phi$ is the magnetic flux linkage (Wb) and L is the inductance in units of Henries (H). It is clear from this equation that the inductance depends on the device geometry as well as the magnetic medium. Like the Farad, the Henry is a very large quantity. In most practical applications, we will use inductors measured in millihenries (1 mH $= 10^{-3}$ H) or microhenries (1 μH $= 10^{-6}$ H).

The voltage equation, $v = Ldi/dt$, is one of the three most important equations for the inductor, and it allows us to find the voltage if the current is known as a function of time. It also reveals that the inductor current may not change instantaneously in time, because that would require infinite voltage. We can derive an equation for the current as follows. First, we rearrange the voltage equation:

$$di = \frac{v}{L} dt. \tag{6.100}$$

Integrating both sides, with x as a variable of integration for current and y as a variable of integration for time,

$$\int_{I_0}^{i} dx = \frac{1}{L} \int_{0}^{t} v \, dy, \tag{6.101}$$

where I_0 is the initial value of the current: $I_0 = i(0)$. Evaluating the left-hand integral at the limits,

$$i = I_0 + \frac{1}{L} \int_{0}^{t} v \, dy. \tag{6.102}$$

This expression allows us to determine the current if the voltage is known as a function of time.

A third important relationship quantifies the energy stored in the inductor (by way of the magnetic field). To find this relationship, we start by considering the instantaneous power of the inductor:

$$p = iv. \tag{6.103}$$

Using the voltage equation, we may write this as

$$p = iL\frac{di}{dt}. \tag{6.104}$$

Rearranging,

$$pdt = Lidi. \tag{6.105}$$

Integrating both sides, with x as a variable of integration for current and y as a variable of integration for time,

$$\int_0^t pdy = L\int_0^i xdx. \tag{6.106}$$

The expression on the left is the definition of work, so the energy stored in the capacitor is

$$w = L\left[\frac{x^2}{2}\right]_0^i = \frac{1}{2}Li^2. \tag{6.107}$$

Because we are squaring the current, the energy stored in the inductor depends only on the absolute value of the current and not on its algebraic sign. This is approximately true for a real inductor if it is approximately linear and free from saturation and residual magnetization effects.

6.9   SERIES INDUCTORS

When two or more inductors are connected in series, as shown in Figure 6.20, we can find the equivalent inductance for the combination and this may allow us to simplify the overall circuit.

The total voltage v may be found by applying KVL, with the recognition that the three series inductors share the same current:

$$v = v_1 + v_2 + v_3 = L_1\frac{di}{dt} + L_2\frac{di}{dt} + L_3\frac{di}{dt} = (L_1 + L_2 + L_3)\frac{di}{dt}. \tag{6.108}$$

The equivalent inductance for the series combination may then be found.

$$v = L_{eq}\frac{di}{dt} = (L_1 + L_2 + L_3)\frac{di}{dt}; \tag{6.109}$$

therefore,

$$L_{eq} = L_1 + L_2 + L_3. \tag{6.110}$$

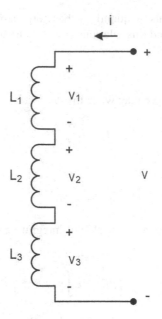

FIGURE 6.20 Series inductors.

Thus, the equivalent inductance for series inductors is equal to the sum of the individual inductances. Although we considered three series inductors here, this rule applies to any number. Moreover, because series inductors share the same current, they share the same initial current for any integration.

6.10 PARALLEL INDUCTORS

Sometimes, we may have parallel combinations of inductors, as shown in Figure 6.21.

The voltage v is common to the parallel inductors, but they have distinct currents. By KCL,

$$i = i_1 + i_2 + i_3$$

$$= I_{01} + \frac{1}{L_1} \int_0^t v \, dy + I_{02} + \frac{1}{L_2} \int_0^t v \, dy + I_{03} + \frac{1}{L_3} \int_0^t v \, dy$$

$$= (I_{01} + I_{02} + I_{03}) + \left(\frac{1}{L_1} + \frac{1}{L_2} + \frac{1}{L_3} \right) \int_0^t v \, dy \qquad (6.111)$$

$$= I_{0eq} + \frac{1}{L_{eq}} \int_0^t v \, dy.$$

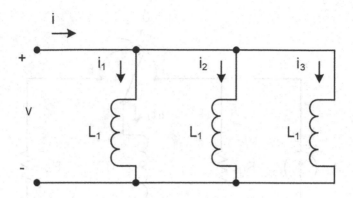

FIGURE 6.21 Parallel inductors.

Therefore, the initial current for the combination is the sum of the individual initial voltages:

$$I_{0eq} = I_{01} + I_{02} + I_{03}, \qquad (6.112)$$

and the equivalent inductance is the reciprocal of the sum of the reciprocals of the individual inductances:

$$L_{eq} = \left(\frac{1}{L_1} + \frac{1}{L_2} + \frac{1}{L_3} \right)^{-1}. \qquad (6.113)$$

In terms of the equivalent value, series and parallel inductances combine in a manner like resistors.

To review, see Presentation 6.4 in ebook+.

To test your knowledge, try Quiz 6.4 in ebook+.

6.11 NATURAL RESPONSE OF AN RL CIRCUIT

Suppose that the switch in Figure 6.22 has been in position a for a "long time" before instantaneously moving to position b at $t = 0$. Prior to the switch movement, the inductor is connected to a two-terminal network represented by its Norton equivalent comprising a current source I_{N0} and a parallel resistance R_{N0}. (This is completely general; any two-terminal circuit containing sources and resistors may be represented by its Norton equivalent.) After movement of the switch, the inductor is connected to a simple resistance R_{N1}.

The switch used here is a special type, known as a "make-before-break" switch because it makes contact with b before breaking contact with a. The configuration shown requires a make-before-break switch to avoid a momentary interruption of the inductor current, which would give rise to a large induced voltage (Ldi / dt) and cause arcing across the physical contacts. We will see that in some circuit configurations it is possible to use a simple single-pole, single-throw switch as long as a path for inductor current exists for both positions of the switch.

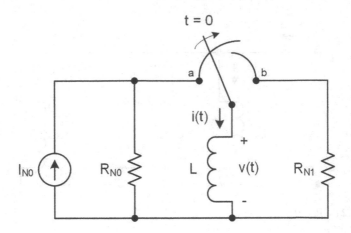

FIGURE 6.22 A switched RL circuit for the consideration of the natural response.

When we say that the switch has been in position a for a "long time," we mean that the circuit has settled, and therefore all time derivatives have vanished to zero. Therefore, the inductor voltage Ldi/dt will have settled to zero. If we apply KCL at $t = 0^-$, the point in time right before the switch moves,

$$-I_{N0} + \frac{v(0^-)}{R_{N0}} + i(0^-) = 0. \qquad (6.114)$$

Because the inductor voltage (and therefore the middle term) will have settled to zero,

$$i(0^-) = I_{N0}. \qquad (6.115)$$

Therefore, the inductor current will have settled to the Norton current for the circuitry connected prior to the switch movement. Because the inductor current cannot change instantaneously, it will still have this value immediately after the switch movement:

$$i(0^+) = i(0^-) = I_{N0}. \qquad (6.116)$$

If we consider KVL for the right-hand mesh after the switch has moved, then

$$-L\frac{di}{dt} - iR_{N1} = 0. \qquad (6.117)$$

This is a first-order linear homogeneous differential equation. Rearranging,

$$\frac{di}{i} = -\frac{dt}{L/R_{N1}}. \qquad (6.118)$$

We can solve by integrating, with x as a variable of integration for current and y as a variable of integration for time,

$$\int_{I_{N0}}^{i(t)} \frac{dx}{x} = -\frac{1}{L/R_{N1}} \int_0^t dy. \tag{6.119}$$

Integrating,

$$[\ln(x)]\Big|_{I_{N0}}^{i(t)} = -\frac{1}{L/R_{N1}} [y]\Big|_0^t. \tag{6.120}$$

Applying the limits,

$$\ln[i(t)] - \ln[I_{N0}] = -\frac{t}{L/R_{N1}}. \tag{6.121}$$

The difference of two logarithms is the logarithm of the quotient:

$$\ln\left(\frac{i(t)}{I_{N0}}\right) = -\frac{t}{L/R_{N1}}. \tag{6.122}$$

Taking the antilogarithm of each side,

$$\frac{i(t)}{I_{N0}} = \exp\left(\frac{-t}{L/R_{N1}}\right). \tag{6.123}$$

Therefore, the inductor current is described by the exponential function of time:

$$i(t) = I_{N0} \exp\left(\frac{-t}{L/R_{N1}}\right); \quad t \geq 0^+. \tag{6.124}$$

The exponential function may also be expressed as

$$i(t) = I_0 \exp(-t/\tau); \quad t \geq 0^+, \tag{6.125}$$

where $I_0 = I_{N0}$ is the initial current in the inductor and τ is the "time constant" for the circuit, given by $\tau = L/R_{N1}$. **As in the case of the RC circuit, all voltages and currents are exponential functions of time, exhibit the same time constant, and settle after about five time constants.** For example, the inductor voltage may be found by using the inductor voltage relationship:

$$v(t) = L\frac{di}{dt} = L\frac{d}{dt}\left[I_{N0} \exp\left(\frac{-t}{L/R_{N1}}\right)\right]$$

$$= I_{N0}R_{N1} \exp\left(\frac{-t}{L/R_{N1}}\right); \quad t \geq 0^+. \tag{6.126}$$

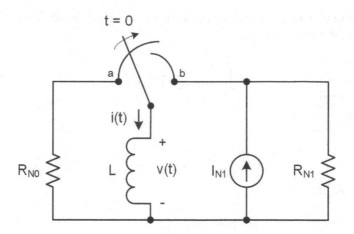

FIGURE 6.23 A switched RL circuit for the consideration of the step response.

6.12   STEP RESPONSE OF AN RL CIRCUIT

Now consider the situation depicted in Figure 6.23, in which an inductor has been connected to a resistor for a long time with the make-before-break switch in position a, and then the switch moves to position b at $t = 0$, causing the inductor current to increase asymptotically to I_{N1} with a parallel resistance R_{N1}.

When we say that the switch has been in position a for a "long time," we mean that the circuit has settled, and therefore all time derivatives have vanished to zero. Therefore, the inductor voltage $L\,di/dt$ will have settled to zero. If we apply KCL with respect to node a at $t = 0^-$, the point in time right before the switch moves,

$$i\left(0^-\right)R_{N0} + L\,di/dt = 0. \tag{6.127}$$

Because the inductor voltage (and therefore the right-hand term) will have settled to zero,

$$i\left(0^-\right) = 0 \tag{6.128}$$

Because the inductor current cannot change instantaneously, it will still have this value immediately after the switch movement:

$$i\left(0^+\right) = i\left(0^-\right) = 0. \tag{6.129}$$

If we consider KCL for node b after the switch has moved, then

$$i - I_{N1} + \frac{L\,di/dt}{R_{N1}} = 0. \tag{6.130}$$

This is a first-order linear differential equation. Rearranging,

$$-\frac{di}{I_{N1} - i} = -\frac{dt}{L/R_{N1}}. \tag{6.131}$$

We can solve by integrating both sides. Using x as a variable of integration for current and y as a variable of integration for time,

$$\int_0^{i(t)} \frac{dx}{I_{N1} - i} = \frac{1}{L / R_{N1}} \int_0^t dy. \tag{6.132}$$

Integrating,

$$-[\ln(I_{N1} - x)]\big|_0^{i(t)} = \frac{1}{L / R_{N1}} [y]\big|_0^t. \tag{6.133}$$

Applying the limits,

$$\ln[I_{N1} - i(t)] + \ln[I_{N1}] = -\frac{t}{L / R_{N1}}. \tag{6.134}$$

The difference of two logarithms is the logarithm of the quotient:

$$\ln\left(\frac{I_{N1} - i(t)}{I_{N1}}\right) = -\frac{t}{L / R_{N1}}. \tag{6.135}$$

Taking the antilogarithm of each side,

$$\frac{I_{N1} - i(t)}{I_{N1}} = \exp\left(\frac{-t}{L / R_{N1}}\right). \tag{6.136}$$

Therefore, the voltage is given by the exponential function of time:

$$i(t) = I_{N1}\left[1 - \exp\left(\frac{-t}{L / R_{N1}}\right)\right]; \quad t \geq 0^+. \tag{6.137}$$

The exponential function may also be expressed as

$$i(t) = I_F[1 - \exp(-t / \tau)]; \quad t \geq 0^+. \tag{6.138}$$

where I_F is the final current in the inductor and τ is the "time constant" for the circuit, given by $\tau = L / R_{N1}$.

6.13   GENERAL CASE OF THE NATURAL AND STEP RESPONSE IN AN RL CIRCUIT

In the general case of a switched RL circuit, the initial and final values for the inductor current may both be non-zero. Then, the total response is the sum of the natural response (associated with the initial current) and the step response (associated with the final value of current). This corresponds to a circuit of the type shown in Figure 6.24. Here, the inductor has been connected for a long time to a two-terminal network with

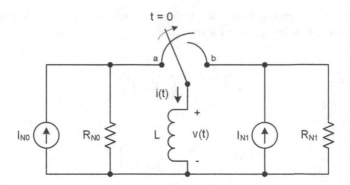

FIGURE 6.24 A switched RL circuit for the consideration of the general case of the natural and step response.

Norton current I_{N0} and Norton resistance R_{N0}, and after $t = 0$ is connected to a two-terminal network with Norton current I_{N1} and Norton resistance R_{N1}. Keep in mind that the overall circuit configuration may not look exactly like the one shown in Figure 6.24, because other switch configurations may be used and it is possible that some circuit elements may be common to the two Norton circuits.

Considering the configuration in Figure 6.24, if we apply KCL to node a at $t = 0^-$, the point in time right before the switch moves,

$$-I_{N0} + \frac{L\, di\, /\, dt}{R_{N0}} + i\left(0^-\right) = 0. \tag{6.139}$$

Because the inductor voltage (and therefore the middle term) will have settled to zero,

$$i\left(0^-\right) = I_{N0}. \tag{6.140}$$

Because the inductor current cannot change instantaneously, it will still have this value immediately after the switch movement:

$$i\left(0^+\right) = i\left(0^-\right) = I_{N0}. \tag{6.141}$$

If we consider KCL for node b after the switch has moved, then

$$i - I_{N1} + \frac{L\, di\, /\, dt}{R_{N1}} = 0. \tag{6.142}$$

Rearranging,

$$\frac{di}{i - I_{N1}} = -\frac{dt}{L\, /\, R_{N1}}. \tag{6.143}$$

We can solve by integrating both sides. Using x as a variable of integration for current and y as a variable of integration for time,

$$\int_{I_{N0}}^{i(t)} \frac{dx}{i-I_{N1}} = -\frac{1}{L/R_{N1}} \int_0^t dy. \tag{6.144}$$

Integrating,

$$[\ln(i-I_{N1})]_{I_{N0}}^{i(t)} = -\frac{1}{L/R_{N1}}[y]_0^t. \tag{6.145}$$

Applying the limits,

$$\ln[i(t)-I_{N1})] - \ln[I_{N0}-I_{N1}] = -\frac{t}{L/R_{N1}}. \tag{6.146}$$

The difference of two logarithms is the logarithm of the quotient:

$$\ln\left(\frac{i(t)-I_{N1}}{I_{N0}-I_{N1}}\right) = -\frac{t}{L/R_{N1}}. \tag{6.147}$$

Taking the antilogarithm of each side,

$$\frac{i(t)-I_{N1}}{I_{N0}-I_{N1}} = \exp\left(\frac{-t}{L/R_{N1}}\right). \tag{6.148}$$

Therefore, the current is given by the exponential function of time:

$$i(t) = I_{N1} + (I_{N0}-I_{N1})\exp\left(\frac{-t}{L/R_{N1}}\right); \quad t \geq 0^+. \tag{6.149}$$

This function may also be expressed as

$$i(t) = I_F + (I_0 - I_F)\exp(-t/\tau); \quad t \geq 0^+. \tag{6.150}$$

where I_0 is the initial current $(I_0 = I_{N0})$, I_F is the final current $(I_F = I_{N1})$, and τ is the "time constant" given by $\tau = L/R_{N1}$. The same form is applicable to any voltage or current in an RL circuit. This is extremely useful because it allows us to immediately write the solution for any such voltage or current as long as the initial value, final value, and time constant are known. For example, the voltage across the inductor may be written as

$$v(t) = V_F + (V_0 - V_F)\exp(-t/\tau); \quad t \geq 0^+, \tag{6.151}$$

where V_0 is the initial voltage, V_F is the final voltage, and τ is the "time constant." It should be noted that the final value of the inductor voltage will always be zero (all time derivatives will settle to zero). Also, the time constant takes on the same value for any voltage or current in an RL circuit.

As in the RC circuit, the overall response can be considered to be the sum of the natural and step response:

$$i(t) = \underbrace{I_0 \exp(-t/\tau)}_{\text{natural response}} + \underbrace{I_F\left[1 - \exp(-t/\tau)\right]}_{\text{step response}}; \quad t \geq 0^+. \tag{6.152}$$

Here, the natural response is associated with the initial value term while the step response is associated with the final value terms.

As an example, consider the RL circuit in Figure 6.25. The switch has been in its starting position for a "long time" before moving at $t = 0$, and we want to determine $v(t)$ and $i(t)$ for $t > 0^+$.

We can use a short-circuit analysis for $t < 0$ (switch open) to find I_{SC0} as shown in Figure 6.26.

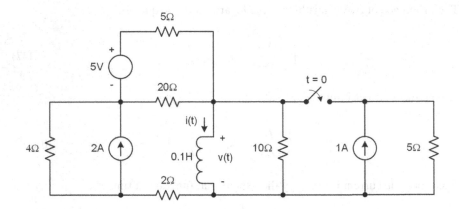

FIGURE 6.25   RL circuit example for consideration of the natural and step response.

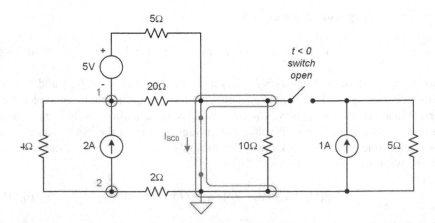

FIGURE 6.26   Short-circuit analysis of the previous circuit for $t < 0$ with respect to the terminals where the inductor had been connected.

Using units of V, A, and Ω,

$$N1 \quad \frac{V_1 - V_2}{4} - 2 + \frac{V_1}{20} + \frac{V_1 + 5}{5} = 0 \tag{6.153}$$

and

$$N2 \quad \frac{V_2 - V_1}{4} + 2 + \frac{V_2}{2} = 0. \tag{6.154}$$

Multiplying by the least common denominators,

$$N1 \quad 5V_1 - 5V_2 - 40 + V_1 + 4V_1 + 20 = 0 \tag{6.155}$$

and

$$V_2 - V_1 + 8 + 2V_2 = 0. \tag{6.156}$$

Collecting like terms,

$$N1 \quad 10V_1 - 5V_2 = 20 \tag{6.157}$$

and

$$N2 \quad -V_1 + 3V_2 = -8. \tag{6.158}$$

In matrix form,

$$\begin{bmatrix} 10 & -5 \\ -1 & 3 \end{bmatrix} \begin{bmatrix} V_1 \\ V_2 \end{bmatrix} = \begin{bmatrix} 20 \\ -8 \end{bmatrix}. \tag{6.159}$$

Solving,

$$V_1 = \frac{\begin{vmatrix} 20 & -5 \\ -8 & 3 \end{vmatrix}}{\begin{vmatrix} 10 & -5 \\ -1 & 3 \end{vmatrix}} = 0.800\,\text{V} \tag{6.160}$$

and

$$V_2 = \frac{\begin{vmatrix} 10 & 20 \\ -1 & -8 \end{vmatrix}}{\begin{vmatrix} 10 & -5 \\ -1 & 3 \end{vmatrix}} = -2.40\,\text{V}. \tag{6.161}$$

Therefore,

$$I_{N0} = I_{SC0} = \frac{V_1}{20\,\Omega} + \frac{V_1 + 5\,\text{V}}{5\,\Omega} = 1.200\,\text{A}. \tag{6.162}$$

Next, we will perform the short-circuit analysis for $t > 0$ with respect to the terminals where the inductor is connected, as shown in Figure 6.27.

Using units of V, A, and Ω,

$$N1 \quad \frac{V_1 - V_2}{4} - 2 + \frac{V_1}{20} + \frac{V_1 + 5}{5} = 0 \tag{6.163}$$

and

$$N2 \quad \frac{V_2 - V_1}{4} + 2 + \frac{V_2}{2} = 0. \tag{6.164}$$

Multiplying by the least common denominators,

$$N1 \quad 5V_1 - 5V_2 - 40 + V_1 + 4V_1 + 20 = 0 \tag{6.165}$$

and

$$N2 \quad V_2 - V_1 + 8 + 2V_2 = 0. \tag{6.166}$$

Collecting like terms,

$$N1 \quad 10V_1 - 5V_2 = 20 \tag{6.167}$$

and

$$N2 \quad -V_1 + 3V_2 = -8. \tag{6.168}$$

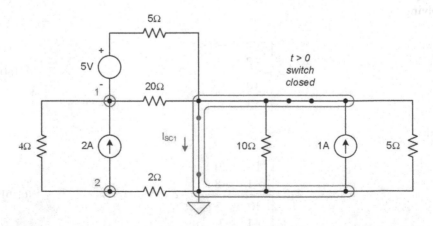

FIGURE 6.27 Short-circuit analysis of the previous circuit for $t > 0$ with respect to the terminals where the inductor is connected.

In matrix form,

$$\begin{bmatrix} 10 & -5 \\ -1 & 3 \end{bmatrix} \begin{bmatrix} V_1 \\ V_2 \end{bmatrix} = \begin{bmatrix} 20 \\ -8 \end{bmatrix}. \tag{6.169}$$

Solving,

$$V_1 = \frac{\begin{vmatrix} 20 & -5 \\ -8 & 3 \end{vmatrix}}{\begin{vmatrix} 10 & -5 \\ -1 & 3 \end{vmatrix}} = 0.800\,\text{V} \tag{6.170}$$

and

$$V_2 = \frac{\begin{vmatrix} 10 & 20 \\ -1 & -8 \end{vmatrix}}{\begin{vmatrix} 10 & -5 \\ -1 & 3 \end{vmatrix}} = -2.40\,\text{V}. \tag{6.171}$$

Therefore,

$$I_{N1} = I_{SC1} = \frac{V_1}{20\,\Omega} + \frac{V_1 + 5\,\text{V}}{5\,\Omega} + 1.00\,\text{A} = 2.200\,\text{A}. \tag{6.172}$$

Finally, we will perform the open-circuit analysis for $t > 0$ with respect to the terminals where the inductor is connected, as shown in Figure 6.28. This will enable the determination of R_{N1} and the time constant.

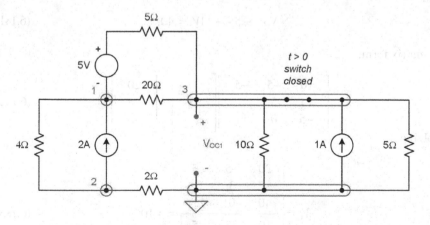

FIGURE 6.28 Open-circuit analysis of the previous circuit for $t > 0$ with respect to the terminals where the inductor is connected.

Using units of V, A, and Ω,

$$N1 \quad \frac{V_1 - V_2}{4} - 2 + \frac{V_1 - V_3}{20} + \frac{V_1 + 5 - V_3}{5} = 0, \tag{6.173}$$

$$N2 \quad \frac{V_2 - V_1}{4} + 2 + \frac{V_2}{2} = 0, \tag{6.174}$$

and

$$N3 \quad \frac{V_3 - V_1}{20} + \frac{V_3 - 5 - V_1}{5} + \frac{V_3}{10} - 1 + \frac{V_3}{5} = 0. \tag{6.175}$$

Multiplying by the least common denominators,

$$N1 \quad 5V_1 - 5V_2 - 40 + V_1 - V_3 + 4V_1 + 20 - 4V_3 = 0, \tag{6.176}$$

$$N2 \quad V_2 - V_1 + 8 + 2V_2 = 0, \tag{6.177}$$

and

$$N3 \quad V_3 - V_1 + 4V_3 - 20 - 4V_1 + 2V_3 - 20 + 4V_3 = 0. \tag{6.178}$$

Collecting like terms,

$$N1 \quad 10V_1 - 5V_2 - 5V_3 = 20, \tag{6.179}$$

$$N2 \quad -V_1 + 3V_2 = -8, \tag{6.180}$$

and

$$N3 \quad -5V_1 + 11V_3 = 40. \tag{6.181}$$

In matrix form,

$$\begin{bmatrix} 10 & -5 & -5 \\ -1 & 3 & 0 \\ -5 & 0 & 11 \end{bmatrix} \begin{bmatrix} V_1 \\ V_2 \\ V_3 \end{bmatrix} = \begin{bmatrix} 20 \\ -8 \\ 40 \end{bmatrix}. \tag{6.182}$$

Solving,

$$V_1 = \frac{\begin{vmatrix} 20 & -5 & -5 \\ -8 & 3 & 0 \\ 40 & 0 & 11 \end{vmatrix}}{\begin{vmatrix} 10 & -5 & -5 \\ -1 & 3 & 0 \\ -5 & 0 & 11 \end{vmatrix}} = 4.10\,\text{V}, \tag{6.183}$$

$$V_2 = \frac{\begin{vmatrix} 10 & 20 & -5 \\ -1 & -8 & 0 \\ -5 & 40 & 11 \end{vmatrix}}{\begin{vmatrix} 10 & -5 & -5 \\ -1 & 3 & 0 \\ -5 & 0 & 11 \end{vmatrix}} = -1.300\,\text{V}, \tag{6.184}$$

and

$$V_3 = \frac{\begin{vmatrix} 10 & -5 & 20 \\ -1 & 3 & -8 \\ -5 & 0 & 40 \end{vmatrix}}{\begin{vmatrix} 10 & -5 & -5 \\ -1 & 3 & 0 \\ -5 & 0 & 11 \end{vmatrix}} = 5.50\,\text{V}. \tag{6.185}$$

Therefore, after movement of the switch,

$$V_{OC1} = V_3 = 5.50\,\text{V} \tag{6.186}$$

and

$$R_{N1} = \frac{V_{OC1}}{I_{SC1}} = \frac{5.50\,\text{V}}{2.20\,\text{A}} = 2.50\,\Omega. \tag{6.187}$$

The initial current is $I_0 = I_{N0} = 1.200\,\text{A}$, the final current is $I_F = I_{N1} = 2.20\,\text{A}$, and the time constant is $\tau = L / R_{N1} = 0.1\,\text{H} / 2.50\,\Omega = 40\,\text{ms}$ so the inductor current is therefore

$$i(t) = I_F + (I_0 - I_F)\exp(-t/\tau)$$
$$= 2.20\,\text{A} - 1.00\,\text{A}\exp(-t/40\,\text{ms}); \quad t \geq 0^+. \tag{6.188}$$

The inductor voltage is

$$v(t) = L\frac{di}{dt} = (0.1\,\text{H})\frac{d}{dt}\{2.20\,\text{A} - 1.00\,\text{A}\exp(-t/40\,\text{ms})\}$$
$$= \frac{(0.1\,\text{H})(-1.00\,\text{A})}{-40\,\text{ms}}\exp\left(\frac{-t}{40\,\text{ms}}\right) = 2.50\,\text{V}\exp\left(\frac{-t}{40\,\text{ms}}\right). \quad t \geq 0^+. \tag{6.189}$$

As expected for the inductor, the final value of the voltage is zero.

To review, see Presentation 6.5 in ebook+.

To test your knowledge, try Quiz 6.5 in ebook+.

6.14   SEQUENTIAL SWITCHING IN FIRST-ORDER CIRCUITS

The general principles presented in the previous sections may be extended to first-order circuits in which there are two or more switching events occurring at different times. In such a case, the initial condition for the second or later switching event is determined by considering the solution for the previous switching event. To illustrate this, we will consider two sequential switching examples: one involving an RL circuit and one involving an RC circuit.

Consider the RL circuit in Figure 6.29. Here, there are two switches: switch one closes at $t = 0$ and then switch two closes at $t = 10\,\text{ms}$.

For $t < 0$, with both switches open as shown in Figure 6.30, there are no sources connected to the inductor, and

$$I_{SC0} = 0. \tag{6.190}$$

Next, we will consider $0^+ \leq t \leq 10\,\text{ms}^-$, for which switch one is closed but switch two is still open as shown in Figure 6.31.

The short-circuit current with respect to the terminals where the inductor is connected is

$$I_{SC1} = \frac{8\,\text{V}}{40\,\Omega} = 0.2\,\text{A}. \tag{6.191}$$

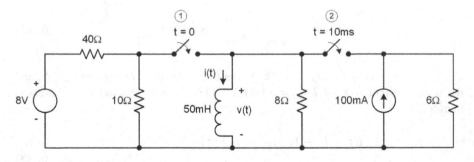

FIGURE 6.29   An RL circuit with two sequential switching events.

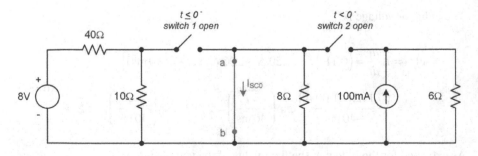

FIGURE 6.30   The previous RL circuit for $t < 0$ (both switches open).

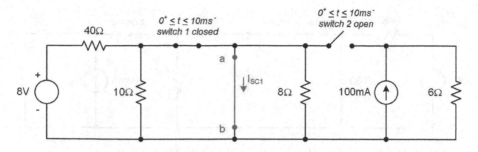

FIGURE 6.31 The previous RL circuit for $0^+ \leq t \leq 10\,\text{ms}^-$ (switch one closed, switch two open).

By the resistance shortcut analysis,

$$R_{N1} = 40\,\Omega \,\|\, 10\,\Omega \,\|\, 8\,\Omega = 4\,\Omega. \tag{6.192}$$

For this time interval (which we will refer to as time interval one), the initial current is $I_{01} = I_{N0} = 0$, the final current is $I_{F1} = I_{N1} = 0.2\,\text{A}$, and the time constant is $\tau_1 = L\,/\,R_{N1} = 50\,\text{mH}\,/\,4\,\Omega = 12.5\,\text{ms}$. The solution for the inductor current during this time interval is therefore

$$i(t) = I_{F1} + (I_{01} - I_{F1})\exp(-t\,/\,\tau_1)$$
$$= \left[0.2 - 0.2\exp(-t\,/\,12.5\,\text{ms})\right]\text{A}; \quad 0^+ \leq t \leq 10\,\text{ms}^- \tag{6.193}$$

We can also solve for the inductor voltage by differentiating

$$v(t) = L\frac{di}{dt} = (50\,\text{mH})\frac{d}{dt}\left[0.2 - 0.2\exp(-t\,/\,12.5\,\text{ms})\right]\text{A}$$

$$= \frac{(50\,\text{mH})(-0.2\,\text{A})}{-12.5\,\text{ms}}\exp\left(\frac{-t}{12.5\,\text{ms}}\right) = 0.8\,\text{V}\exp\left(\frac{-t}{40\,\text{ms}}\right); \quad 0^+ \leq t \leq 10\,\text{ms}^-. \tag{6.194}$$

Before we move on to the second switching event, it is worth noting that the circuit will never settle to the "final values" for time interval one. This is because the second switch will move at $t = 10\,\text{ms}$ whereas it would take about five time constants, or 62.5 ms, for the circuit to settle.

For consideration of the second time interval after the second switching event, $t \geq 10\,\text{ms}^+$, we will use the solution for the first switching event, evaluated at $t = 10\,\text{ms}$ to find the initial condition:

$$I_{02} = i(10\,\text{ms}^+) = i(10\,\text{ms}^-)$$
$$= \left[0.2 - 0.2\exp(-10\,\text{ms}\,/\,12.5\,\text{ms})\right]\text{A} = 0.110\,\text{A}. \tag{6.195}$$

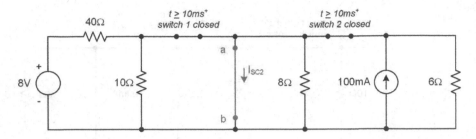

FIGURE 6.32   The previous RL circuit for $t \geq 10\,\mathrm{ms}^{+}$ (both switches closed).

To find the final value of the current for time interval two, we need to find the short-circuit current with respect to the terminals where the inductor is connected, for the situation in which both switches have moved to their final positions as shown in Figure 6.32.

The short-circuit current with respect to the terminals where the inductor is connected is

$$I_{SC2} = \frac{8\,\mathrm{V}}{40\,\Omega} + 100\,\mathrm{mA} = 0.3\,\mathrm{A}. \qquad (6.196)$$

By the resistance shortcut analysis,

$$R_{N2} = 40\,\Omega \,||\, 10\,\Omega \,||\, 8\,\Omega \,||\, 6\,\Omega = 2.4\,\Omega. \qquad (6.197)$$

For the second time interval ($t \geq 10\,\mathrm{ms}^{+}$), the initial current is $I_{01} = 0.110\,\mathrm{A}$ as found above, the final current is $I_{F2} = I_{N2} = 0.3\,\mathrm{A}$, and the time constant is $\tau_2 = L / R_{N2} = 50\,\mathrm{mH} / 2.4\,\Omega = 20.83\,\mathrm{ms}$. The solution for the inductor current during this time interval is therefore

$$i(t) = I_{F2} + (I_{02} - I_{F2})\exp(-(t - t_1)/\tau_2)$$
$$= \left[0.3 - 0.190\exp(-(t - 10\,\mathrm{ms})/20.83\,\mathrm{ms})\right]\mathrm{A}; \quad t \geq 10\,\mathrm{ms}^{+}, \qquad (6.198)$$

where t_1 is the time of the second switching event, which corresponds to the initial value calculated for time interval two.

We can also solve for the inductor voltage by differentiating

$$v(t) = L\frac{di}{dt} = (50\,\mathrm{mH})\frac{d}{dt}\left[0.3 - 0.190\exp(-(t - 10\,\mathrm{ms})/20.83\,\mathrm{ms})\right]\mathrm{A}$$

$$= \frac{(50\,\mathrm{mH})(-0.190\,\mathrm{A})}{-20.83\,\mathrm{ms}}\exp\left(\frac{-(t - 10\,\mathrm{ms})}{20.83\,\mathrm{ms}}\right) = 0.456\,\mathrm{V}\exp\left(\frac{-(t - 10\,\mathrm{ms})}{20.83\,\mathrm{ms}}\right);$$

$$t \geq 10\,\mathrm{ms}^{+}. \qquad (6.199)$$

The overall solution for the current may be expressed in a piecewise fashion:

$$i(t) = \begin{cases} \left[0.2 - 0.2\exp(-t/12.5\,\text{ms})\right]\text{A}; & 0^+ \le t \le 10\,\text{ms}^-; \\ \left[0.3 - 0.190\exp(-(t-10\,\text{ms})/20.83\,\text{ms})\right]\text{A}; & t \ge 10\,\text{ms}^+. \end{cases}$$

$$(6.200)$$

Similarly, the overall voltage solution is

$$v(t) = \begin{cases} \left[0.8\exp(-t/12.5\,\text{ms})\right]\text{V}; & 0^+ \le t \le 10\,\text{ms}^-; \\ \left[0.456\exp(-(t-10\,\text{ms})/20.83\,\text{ms})\right]\text{V}; & t \ge 10\,\text{ms}^+. \end{cases}$$

$$(6.201)$$

A similar approach may be applied to solve an RC circuit with sequential switching. Consider the circuit in Figure 6.33, which involves three switching events: switch one moves at $t = 0$, switch two moves at $t = 20\,\text{ms}$, and switch three moves at $t = 40\,\text{ms}$. Notice that switch two opens whereas the other two switches close; in general, we could have any mix of openings and closures so this must be observed carefully.

For $t \le 0^-$, all switches are in their starting positions as shown in Figure 6.34 (switch one is open, switch two is closed, and switch three is open). We can perform the open-circuit analysis to find the initial capacitor voltage.

Performing the open-circuit analysis with units of V, mA, and $k\Omega$, there are two essential nodes and a single node equation:

$$N1 \quad \frac{V_1}{2} - 10 = 0.$$

$$(6.202)$$

Solving,

$$V_1 = 20\,\text{V}.$$

$$(6.203)$$

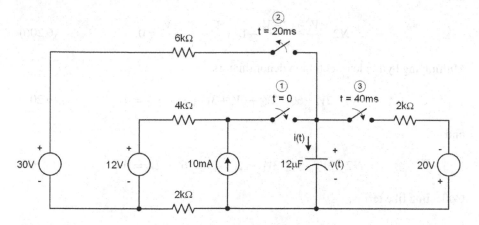

FIGURE 6.33   An RC circuit with three sequential switching events.

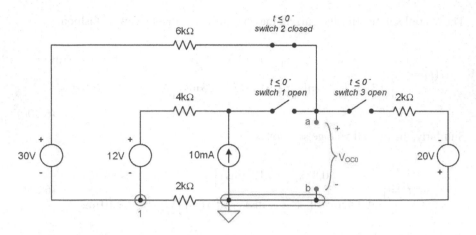

FIGURE 6.34 Open-circuit analysis of the previous RC circuit for $t \leq 0^-$ (switch one open, switch two closed, switch three open) with respect to the terminals where the capacitor had been connected.

By KVL,

$$V_{Th0} = V_{OC0} = V_1 + 30\,\text{V} = 50\,\text{V}. \tag{6.204}$$

Next, we will consider the time interval $0^+ \leq t \leq 20\,\text{ms}^-$, which we will refer to as "time interval one." For this time interval, switches one and two are closed but switch three is open as shown in Figure 6.35.

Using units of V, mA, and kΩ,

$$N1 \quad \frac{V_1 + 30 - V_2}{6} + \frac{V_1}{2} + \frac{V_1 + 12 - V_2}{4} = 0 \tag{6.205}$$

and

$$N2 \quad \frac{V_2 - 12 - V_1}{4} - 10 + \frac{V_2 - 30 - V_1}{6} = 0. \tag{6.206}$$

Multiplying by the least common denominators,

$$N1 \quad 2V_1 + 60 - 2V_2 + 6V_1 + 3V_1 + 36 - 3V_2 = 0 \tag{6.207}$$

and

$$N2 \quad 3V_2 - 36 - 3V_1 - 120 + 2V_2 - 60 - 2V_1 = 0. \tag{6.208}$$

Collecting like terms,

$$N1 \quad 11V_1 - 5V_2 = -96 \tag{6.209}$$

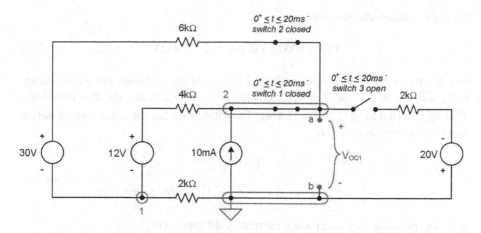

FIGURE 6.35 Open-circuit analysis of the previous RC circuit for $0^+ \le t \le 20\,\mathrm{ms}^-$ (switch one closed, switch two closed, switch three open) with respect to the terminals where the capacitor had been connected.

and

$$N2 \quad -5V_1 + 5V_2 = 216. \tag{6.210}$$

In matrix form,

$$\begin{bmatrix} 11 & -5 \\ -5 & 5 \end{bmatrix} \begin{bmatrix} V_1 \\ V_2 \end{bmatrix} = \begin{bmatrix} -96 \\ 216 \end{bmatrix}. \tag{6.211}$$

Solving,

$$V_1 = \frac{\begin{vmatrix} -96 & -5 \\ 216 & 5 \end{vmatrix}}{\begin{vmatrix} 11 & -5 \\ -5 & 5 \end{vmatrix}} = 20.0\,\mathrm{V}, \tag{6.212}$$

$$V_2 = \frac{\begin{vmatrix} 11 & -96 \\ -5 & 216 \end{vmatrix}}{\begin{vmatrix} 11 & -5 \\ -5 & 5 \end{vmatrix}} = 63.2\,\mathrm{V}. \tag{6.213}$$

Therefore,

$$V_{\mathrm{Th1}} = V_{\mathrm{OC1}} = V_2 = 63.2\,\mathrm{V}. \tag{6.214}$$

By the resistance shortcut analysis,

$$R_{Th1} = 6\,k\Omega \parallel 4\,k\Omega + 2\,k\Omega = 4.4\,k\Omega. \tag{6.215}$$

For this time interval (which we will refer to as time interval one), the initial voltage is $V_{01} = V_{Th0} = 50.0\,V$, the final voltage is $V_{F1} = V_{Th1} = 63.2\,V$, and the time constant is $\tau_1 = R_{Th1}C = (4.4\,k\Omega)(12\,\mu F) = 52.8\,ms$. The solution for the capacitor voltage during this time interval is therefore

$$v(t) = V_{F1} + (V_{01} - V_{F1})\exp(-t/\tau_1)$$
$$= \left[63.2 - 13.2\exp(-t/52.8\,ms)\right]V; \quad 0^+ \leq t \leq 20\,ms^-. \tag{6.216}$$

We can also solve for the capacitor current by differentiating

$$i(t) = C\frac{dv}{dt} = (12\,\mu F)\frac{d}{dt}\left[63.2 - 13.2\exp(-t/52.8\,ms)\right]V$$

$$= \frac{(12\,\mu F)(-13.2\,V)}{-52.8\,ms}\exp(-t/52.8\,ms) \tag{6.217}$$

$$= 3.00\exp(-t/52.8\,ms)\,mA; \quad 0^+ \leq t \leq 20\,ms^-.$$

For consideration of the second time interval, after the second switching event but before the third switching event, $20\,ms^+ \leq t \leq 40\,ms^-$, we will use the solution for the first switching event, evaluated at $t = 20\,ms$ to find the initial condition:

$$v(t) = \left[63.2 - 13.2\exp(-20\,ms/52.8\,ms)\right]V = 54.2\,V. \tag{6.218}$$

To find the final value of the voltage for time interval two, we need to find the open-circuit voltage with respect to the terminals where the capacitor had been connected as shown in Figure 6.36.

Performing the open-circuit analysis with units of V, mA, and kΩ, there are two essential nodes and a single node equation, because the opening of switch two eliminates one essential node:

$$N1 \quad \frac{V_1}{2} - 10 = 0. \tag{6.219}$$

Solving,

$$V_1 = 20\,V, \tag{6.220}$$

and by KVL,

$$V_{Th2} = V_{OC2} = V_1 + 24\,V + (10\,mA)(4\,k\Omega) = 84\,V. \tag{6.221}$$

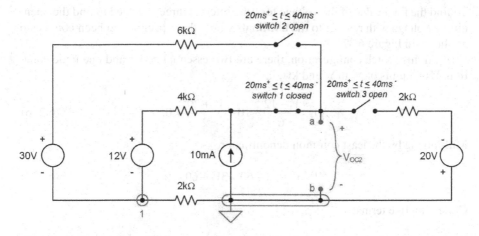

FIGURE 6.36 Open-circuit analysis of the previous RC circuit for $20\,\text{ms}^+ \le t \le 40\,\text{ms}^-$ (switch one closed, switch two open, switch three open) with respect to the terminals where the capacitor had been connected.

By the resistance shortcut analysis,

$$R_{\text{Th}2} = 4\,\text{k}\Omega + 2\,\text{k}\Omega = 6\,\text{k}\Omega. \tag{6.222}$$

For time interval two, the initial voltage is $V_{02} = 54.2\,\text{V}$ as found above, the final voltage is $V_{F2} = V_{\text{Th}2} = 84\,\text{V}$, and the time constant is $\tau_2 = R_{\text{Th}2}C = (6\,\text{k}\Omega)(12\,\mu\text{F}) = 72\,\text{ms}$. The solution for the capacitor voltage during this time interval is therefore

$$
\begin{aligned}
v(t) &= V_{F1} + (V_{01} - V_{F1})\exp(-(t - t_1)/\tau_1) \\
&= \left[84 - 29.8\exp(-(t - 20\,\text{ms})/72\,\text{ms})\right]\text{V}; \quad 20\,\text{ms}^+ \le t \le 40\,\text{ms}^-.
\end{aligned}
\tag{6.223}
$$

We can also solve for the capacitor current by differentiating

$$
\begin{aligned}
i(t) &= C\frac{dv}{dt} = (12\,\mu\text{F})\frac{d}{dt}\left[63.2 - 29.8\exp(-(t - 20\,\text{ms})/72\,\text{ms})\right]\text{V} \\
&= \frac{(12\,\mu\text{F})(-29.8\,\text{V})}{-72\,\text{ms}}\exp(-(t - 20\,\text{ms})/72\,\text{ms}) \\
&= 5.00\exp(-(t - 20\,\text{ms})/72\,\text{ms})\,\text{mA}; \quad 20\,\text{ms}^+ \le t \le 40\,\text{ms}^-.
\end{aligned}
\tag{6.224}
$$

For consideration of the third time interval, after the third switching event, $t \ge 40\,\text{ms}^+$, we will use the solution for the second switching event, evaluated at $t = 40\,\text{ms}$ to find the initial condition:

$$
\begin{aligned}
V_{03} &= v(40\,\text{ms}^+) = v(40\,\text{ms}^-) \\
&= \left[84 - 29.8\exp(-(40\,\text{ms} - 20\,\text{ms})/72\,\text{ms})\right]\text{V} = 61.4\,\text{V}.
\end{aligned}
\tag{6.225}
$$

To find the final value of the voltage for time interval three, we need to find the open-circuit voltage with respect to the terminals where the capacitor had been connected as shown in Figure 6.37.

With this switch configuration, there are two essential nodes and one node equation. Using units of V, mA, and kΩ,

$$N1 \quad \frac{V_1 - 12}{6} - 10 + \frac{V_1 + 20}{2} = 0. \tag{6.226}$$

Multiplying by the least common denominator,

$$N1 \quad V_1 - 12 - 60 + 3V_1 + 60 = 0. \tag{6.227}$$

Collecting like terms,

$$N1 \quad 4V_1 = 12. \tag{6.228}$$

Solving,

$$V_1 = 3.0\,\text{V}. \tag{6.229}$$

Therefore,

$$V_{\text{Th3}} = V_{\text{OC3}} = V_1 = 3.0\,\text{V}. \tag{6.230}$$

By the resistance shortcut analysis,

$$R_{\text{Th3}} = \left(4\,\text{k}\Omega + 2\,\text{k}\Omega\right)\|\,2\,\text{k}\Omega = 1.5\,\text{k}\Omega. \tag{6.231}$$

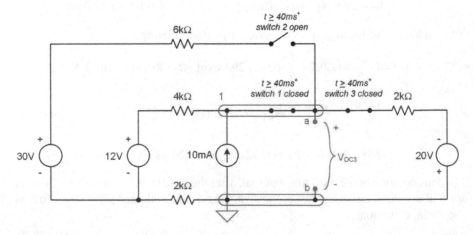

FIGURE 6.37 Open-circuit analysis of the previous RC circuit for $t \geq 40\,\text{ms}^-$ (switch one closed, switch two open, switch three closed) with respect to the terminals where the capacitor had been connected.

For time interval three, the initial voltage is $V_{03} = 61.4\,\text{V}$ as found above, the final voltage is $V_{F3} = V_{\text{Th3}} = 3.0\,\text{V}$, and the time constant is $\tau_3 = R_{\text{Th3}}C = (1.5\,\text{k}\Omega)(12\,\mu\text{F}) = 18.0\,\text{ms}$. The solution for the capacitor voltage during this time interval is therefore

$$v(t) = V_{F1} + (V_{01} - V_{F1})\exp(-(t-t_3)/\tau_1)$$

$$= \left[3.0 + 58.4\exp(-(t-40\,\text{ms})/18.0\,\text{ms})\right]\text{V}; \quad t \geq 40\,\text{ms}^+. \tag{6.232}$$

We can also solve for the capacitor current by differentiating

$$i(t) = C\frac{dv}{dt} = (12\,\mu\text{F})\frac{d}{dt}\left[3.0 + 58.4\exp(-(t-40\,\text{ms})/18.0\,\text{ms})\right]\text{V}$$

$$= \frac{(12\,\mu\text{F})(58.4\,\text{V})}{-18.0\,\text{ms}}\exp(-(t-40\,\text{ms})/18.0\,\text{ms}) \tag{6.233}$$

$$= 38.9\exp(-(t-40\,\text{ms})/18.0\,\text{ms})\,\text{mA}; \quad t \geq 40\,\text{ms}^+.$$

The overall solution for the current may be expressed in a piecewise fashion:

$$v(t) = \begin{cases} \left[63.2 - 13.2\exp(-t/52.8\,\text{ms})\right]\text{V}; & 0^+ \leq t \leq 20\,\text{ms}^-; \\ \left[84 - 29.8\exp(-(t-20\,\text{ms})/72\,\text{ms})\right]\text{V}; & 20\,\text{ms}^+ \leq t \leq 40\,\text{ms}^-; \\ \left[3.0 + 58.4\exp(-(t-40\,\text{ms})/18.0\,\text{ms})\right]\text{V}; & t \geq 40\,\text{ms}^+. \end{cases}$$

$$\tag{6.234}$$

Similarly, the overall current solution is

$$i(t) = \begin{cases} 3.00\exp(-t/52.8\,\text{ms})\,\text{mA}; & 0^+ \leq t \leq 20\,\text{ms}^-; \\ 5.00\exp(-(t-20\,\text{ms})/72\,\text{ms})\,\text{mA}; & 20\,\text{ms}^+ \leq t \leq 40\,\text{ms}^-; \\ 41.5\exp(-(t-40\,\text{ms})/18.0\,\text{ms})\,\text{mA}; & t \geq 40\,\text{ms}^+. \end{cases}$$

$$\tag{6.235}$$

To review, see Presentation 6.6 in ebook+

To test your knowledge, try Quiz 6.6 in ebook+.

6.15 SUMMARY

The **capacitor** is a two-terminal device comprising two metal plates separated by an insulating material. As a consequence of Gauss' law, an external current, called the **displacement current**, flows to charge and discharge the plates and is given by $i = C\,dv/dt$, where i is the capacitor current, v is the capacitor voltage, and C is the capacitance in **Farads**. Hence, the capacitor voltage may not change instantaneously.

The capacitor voltage may be found by integrating the current: $v = V_0 + \dfrac{1}{C}\displaystyle\int_0^t i\,dy$, where V_0 is the initial voltage across the capacitor. The energy w stored in a capacitor is proportional to the square of the voltage: $w = \dfrac{1}{2}Cv^2$. When two or more capacitors are connected in parallel, the equivalent capacitance is the sum of the individual capacitances. When two or more capacitors are connected in series, the equivalent capacitance is the reciprocal of the sum of the reciprocals of the individual capacitances.

The **inductor** is a two-terminal device comprising a coil of wire wrapped around a magnetic medium (which may be air). According to the Faraday law, the voltage across an inductor is given by $v = L\,di\,/\,dt$, where v is the voltage, i is the current, and L is the inductance in **Henries**. Hence, the inductor current may not change instantaneously. The inductor current may be found by integrating the voltage: $i = I_0 + \dfrac{1}{L}\displaystyle\int_0^t v\,dy$, where I_0 is the initial voltage across the capacitor. The energy w stored in an inductor is proportional to the square of the current: $w = \dfrac{1}{2}Li^2$. When inductors are connected in series, the equivalent inductance is the sum of the individual inductances. When inductors are connected in parallel, the equivalent inductance is the reciprocal of the sum of the reciprocals of the individual inductances.

An **RC circuit** is a first-order circuit containing a capacitor, a switch, and sources and resistances. Determination of the capacitor voltage or current as a function of time after movement of the switch requires solution of a first-order differential equation. The solution for the capacitor voltage is of the form $v(t) = V_F + (V_0 - V_F)\exp(-t\,/\,\tau)$. The initial voltage V_0 is determined by the Thevenin voltage for the circuitry connected prior to the movement of the switch: $V_0 = V_{Th0}$. The final voltage V_F is determined by the Thevenin voltage for the circuitry connected after the movement of the switch: $V_F = V_{Th1}$. The time constant is given by $\tau = R_{Th1}C$, where R_{Th1} is the Thevenin resistance of the circuitry connected after movement of the switch and C is the capacitance. Once the capacitor voltage is known, the current can be readily found by differentiation: $i(t) = C\,dv(t)\,/\,dt$.

An **RL circuit** is a first-order circuit containing an inductor, a switch, sources, and resistances. Determination of the inductor current or voltage as a function of time after movement of the switch requires solution of a first-order differential equation. The solution for the inductor current is of the form $i(t) = I_F + (I_0 - I_F)\exp(-t\,/\,\tau)$. The initial current I_0 is determined by the Norton current for the circuitry connected prior to the movement of the switch: $I_0 = I_{N0}$. The final current I_F is determined by the Norton current for the circuitry connected after the movement of the switch: $I_F = I_{N1}$. The time constant is given by $\tau = L\,/\,R_{N1}$, where R_{N1} is the Norton resistance of the circuitry connected after movement of the switch and L is the inductance. Once the inductor current is known, the voltage can be readily found by differentiation: $v(t) = L\,di(t)\,/\,dt$.

The concepts developed above may be applied to **sequential switching problems**, in which an RC or RL circuit contains two or more switches which move at different times. The solution for the first switching event may be found in the same way as described above. For any subsequent switching event, the solution is of the general form $v(t) = V_F + (V_0 - V_F)\exp(-(t - t_n)\,/\,\tau)$ or $i(t) = I_F + (I_0 - I_F)\exp(-(t - t_n)\,/\,\tau)$, in which the initial condition is found by evaluating the solution for the previous switching event at the time of the present switching event ($t = t_n$), the final condition is evaluated by determination of the Thevenin voltage or the Norton current which exists after the present switching event, and the time constant is found as $\tau = R_{\text{Thn}}C$ or $\tau = L / R_{Nn}$, using the Thevenin or Norton resistance which exists after the current switching event.

PROBLEMS

Problem 6.1. For the circuit shown in Figure P6.1, find $v_C(0^-)$, $v_C(0^+)$, $v_C(\infty)$, $i_C(0^-)$, $i_C(0^+)$, $i_C(\infty)$, $v_C(t)$ for $t \geq 0^+$ and $i_C(t)$ for $t \geq 0^+$. Determine the initial and final energy stored in the capacitor. The switch has been in its starting position for a long time before moving at the time indicated.

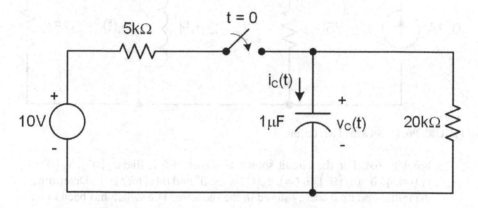

FIGURE P6.1   Switched RC circuit.

Problem 6.2. For the circuit shown in Figure P6.2, find $v_C(0^-)$, $v_C(0^+)$, $v_C(\infty)$, $i_C(0^-)$, $i_C(0^+)$, $i_C(\infty)$, $v_C(t)$ for $t \geq 0^+$ and $i_C(t)$ for $t \geq 0^+$. Determine the initial and final energy stored in the capacitor. The switch has been in its starting position for a long time before moving at the time indicated.

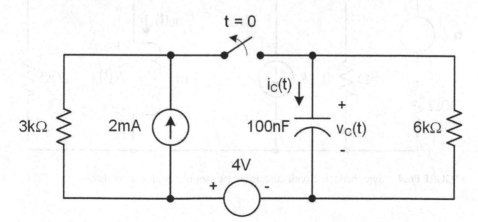

FIGURE P6.2   Switched RC circuit containing two sources and two resistors.

Problem 6.3. For the circuit shown in Figure P6.3, find $v_L(0^-)$, $v_L(0^+)$, $v_L(\infty)$, $i_L(0^-)$, $i_L(0^+)$, $i_L(\infty)$, $v_L(t)$ for $t \geq 0^+$ and $i_L(t)$ for $t \geq 0^+$. Determine the initial and final energy stored in the inductor. The switch has been in its starting position for a long time before moving at the time indicated.

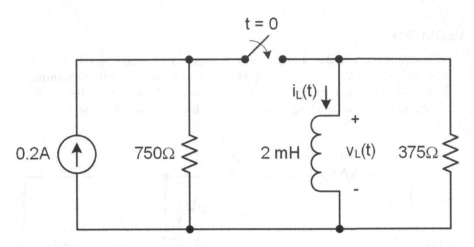

FIGURE P6.3   Switched RL circuit.

Problem 6.4. For the circuit shown in Figure P6.4, find $v_L(0^-)$, $v_L(0^+)$, $v_L(\infty)$, $i_L(0^-)$, $i_L(0^+)$, $i_L(\infty)$, $v_L(t)$ for $t \geq 0^+$ and $i_L(t)$ for $t \geq 0^+$. Determine the initial and final energy stored in the inductor. The switch has been in its starting position for a long time before moving at the time indicated.

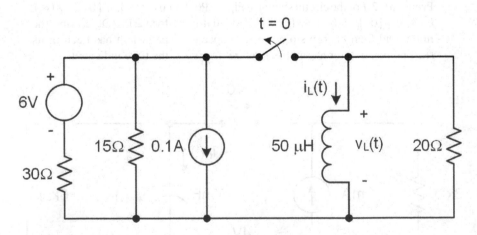

FIGURE P6.4   Switched RL circuit containing two sources and three resistors.

Problem 6.5. For the circuit shown in Figure P6.5, find $v_C(t)$, $i_C(t)$, $i_x(t)$, and $v_y(t)$ for $t \geq 0^+$. The switch has been in its starting position for a long time before moving at the time indicated.

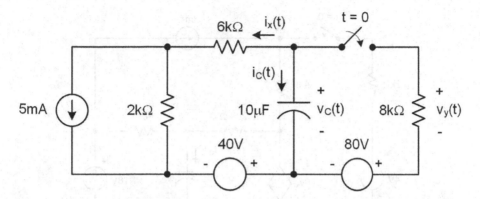

FIGURE P6.5   Switched RC circuit containing three sources and three resistors.

Problem 6.6. For the circuit shown in Figure P6.6, find $v_C(t)$, $i_C(t)$, and $i_x(t)$ for $t \geq 0^+$. The switch has been in its starting position for a long time before moving at the time indicated.

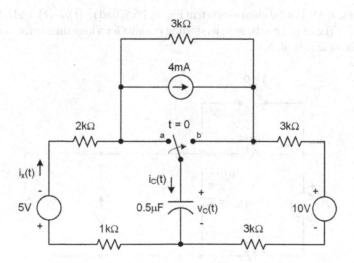

FIGURE P6.6   Switched RC circuit containing three sources, five resistors, and a single-pole, double-throw switch.

Problem 6.7. For the circuit shown in Figure P6.7, determine $v_C(t)$ and $i_C(t)$ for $t \geq 0^+$. The switch has been in its starting position for a long time before moving at the time indicated.

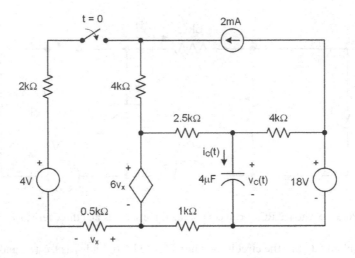

FIGURE P6.7   Switched RC circuit containing mixed sources and six resistors.

Problem 6.8. For the circuit shown in Figure P6.8, find $v_C(t)$, $i_C(t)$, and $i_x(t)$ for $t \geq 0^+$. The switch has been in its starting position for a long time before moving at the time indicated.

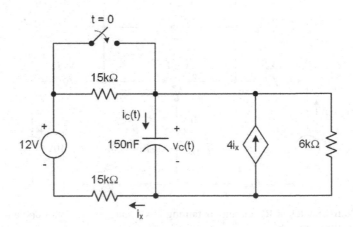

FIGURE P6.8   Switched RC circuit containing an independent source, a CCCS, and three resistors.

Problem 6.9. For the circuit shown in Figure P6.9, find $v_L(t)$, $i_L(t)$, and $i_R(t)$ for $t \geq 0^+$. The switch has been in its starting position for a long time before moving at the time indicated.

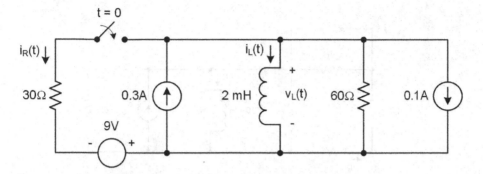

FIGURE P6.9 Switched RL circuit containing three sources and two resistors.

Problem 6.10. For the circuit shown in Figure P6.10, find $v_L(t)$, $i_L(t)$, and $i_x(t)$ for $t \geq 0^+$. The switch has been in its starting position for a long time before moving at the time indicated.

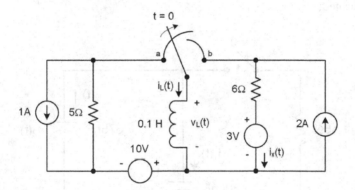

FIGURE P6.10 Switched RL circuit containing four sources, two resistors, and a make-before-break switch.

Problem 6.11. For the circuit of Figure P6.11, determine $v_L(t)$, $i_L(t)$, and $i_x(t)$ for $t \geq 0^+$. The switch has been in its starting position for a long time before moving at the time indicated.

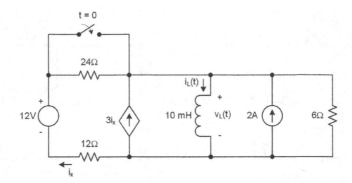

FIGURE P6.11   Switched RL circuit containing mixed sources and three resistors.

Problem 6.12. The circuit shown in Figure P6.12 has zero initial stored energy. Determine $v_1(t)$, $v_2(t)$, $i_1(t)$, $i_2(t)$, and $i_3(t)$ for $t \geq 0^+$. The switch has been in its starting position for a long time before moving at the time indicated.

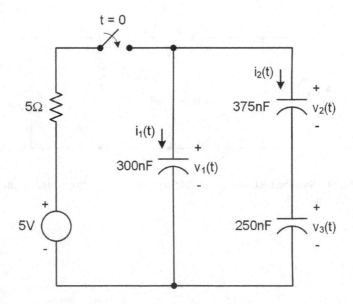

FIGURE P6.12   Switched RC circuit involving three capacitors.

Problem 6.13. For the circuit of Figure P6.13, determine $v_1(t)$, $v_2(t)$, and $i_R(t)$ for $t \geq 0^+$. The switches have been in their starting positions for a long time before moving at the time indicated.

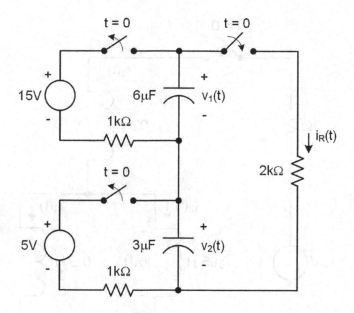

FIGURE P6.13 Switched RC circuit involving three switches which move simultaneously.

Problem 6.14. The circuit shown in Figure P6.14 has zero initial stored energy. Determine $v_1(t)$, $v_2(t)$, $i_1(t)$, $i_2(t)$, and $i_3(t)$ for $t \geq 0^+$. The switch has been in its starting position for a long time before moving at the time indicated.

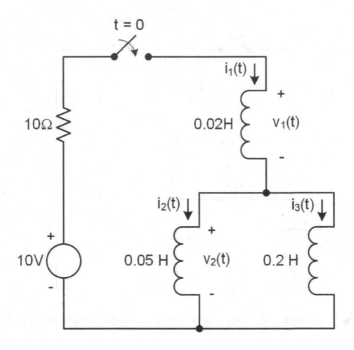

FIGURE P6.14   Switched RL circuit involving three inductors.

Problem 6.15. Find $v(t)$, $i_1(t)$, and $i_2(t)$ for $t \geq 0^+$ in the circuit of Figure P6.15. Is there energy trapped in the circuit after it settles? If so, how much? The switch has been in its starting position for a long time before moving at the time indicated.

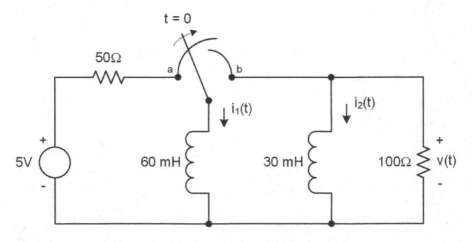

FIGURE P6.15   Switched RL circuit involving two inductors.

Problem 6.16. For the circuit shown in Figure P6.16, find $v_C(t)$ and $i_C(t)$ for $t \geq 0^+$. The switches have been in their starting positions for a long time before moving at the times indicated.

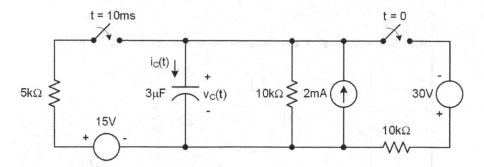

FIGURE P6.16   Sequentially-switched RC circuit.

Problem 6.17. For the circuit shown in Figure P6.17, find $v_C(t)$ and $i_C(t)$ for $t \geq 0^+$. The switches have been in their starting positions for a long time before moving at the times indicated.

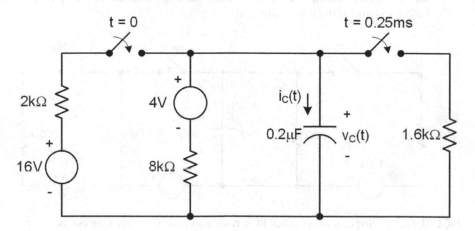

FIGURE P6.17   Sequentially-switched RC circuit involving two sources, three resistors, and two switches.

Problem 6.18. For the circuit shown in Figure P6.18, find $v_L(t)$ and $i_L(t)$ for $t \geq 0^+$. The switches have been in their starting positions for a long time before moving at the times indicated.

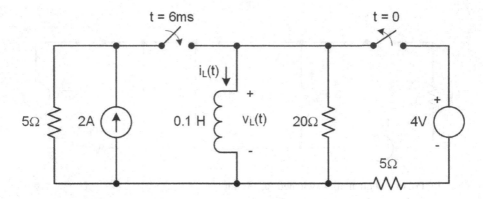

FIGURE P6.18 Sequentially-switched RL circuit.

Problem 6.19. For the circuit shown in Figure P6.19, find $v_C(t)$ and $i_C(t)$ for $t \geq 0^+$. The switches have been in their starting positions for a long time before moving at the times indicated.

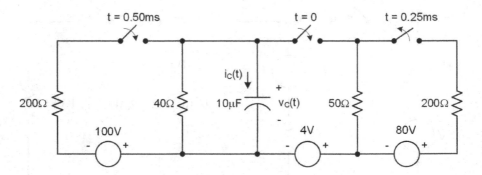

FIGURE P6.19 Sequentially-switched RC circuit involving three switching events.

Problem 6.20. For the circuit shown in Figure P6.20, find $v_L(t)$ and $i_L(t)$ for $t \geq 0^+$. The switches have been in their starting position for a long time before moving at the times indicated.

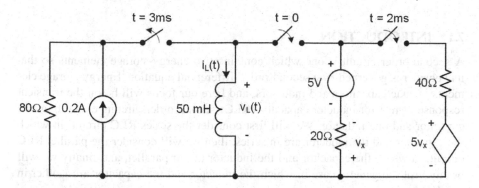

FIGURE P6.20 Sequentially-switched RL circuit involving mixed sources, three resistors, and three switches.

7 Second-Order Circuits

7.1 INTRODUCTION

A second-order circuit is one which contains two energy-storage elements so that its behavior is governed by a second-order differential equation. Energy-storage elements include capacitors and inductors, and here our focus will be on the transient response of resistor-inductor-capacitor (RLC) second-order circuits containing one capacitor and one inductor. We will first consider the series RLC circuit, in which the capacitor and the inductor are in series; then we will consider the parallel RLC circuit, in which the capacitor and the inductor are in parallel, and finally we will briefly visit the general case in which the inductor and the capacitor are neither in parallel nor in series.

7.2 NATURAL RESPONSE OF A SERIES RLC CIRCUIT

Consider the series RLC circuit in Figure 7.1. We will assume that the initial current in the inductor is I_0 and the initial voltage across the capacitor is V_0. (There are numerous switching configurations capable of establishing these initial conditions, and we will consider some of these, but we omit them here for simplicity.) We want to determine $i(t)$ and $v(t)$ for $t \geq 0^+$. If we allow the circuit to settle, both the inductor current and the capacitor voltage will settle to zero: $i(\infty) = 0$ and $v(\infty) = 0$.

To determine the transient response of this circuit while it is settling, we need to set up and solve a second-order homogeneous linear differential equation. To do this, we start by using Kirchhoff's voltage law (KVL):

$$Ri + L\frac{di}{dt} + V_0 + \frac{1}{C}\int_0^t i\,dy = 0. \tag{7.1}$$

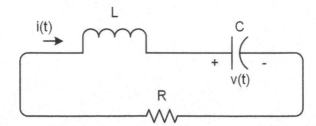

FIGURE 7.1 A series RLC circuit for consideration of the natural response.

  DOI: 10.1201/9781003408529-7

Differentiating with respect to time,

$$R\frac{di}{dt} + L\frac{d^2i}{dt^2} + \frac{i}{C} = 0. \tag{7.2}$$

Rearranging and normalizing with respect to the second-order term,

$$\frac{d^2i}{dt^2} + \frac{R}{L}\frac{di}{dt} + \frac{1}{LC}i = 0. \tag{7.3}$$

We will assume solutions of the form $i(t) = Ae^{st}$. (This should seem plausible, given that a linear combination of the function and its first two derivatives is equal to zero, but we defer to a course on differential equations for a more rigorous justification.) Using our assumed form for the solution,

$$s^2Ae^{st} + \frac{R}{L}sAe^{st} + \frac{1}{LC}Ae^{st} = 0. \tag{7.4}$$

Dividing by Ae^{st} we obtain a quadratic in s, which we call the **characteristic equation** for the series RLC circuit:

$$s^2 + \frac{R}{L}s + \frac{1}{LC} = 0. \tag{7.5}$$

Applying the quadratic formula, we can find the two roots of the characteristic equation:

$$s_1, s_2 = -\frac{R}{2L} \pm \sqrt{\left(\frac{R}{2L}\right)^2 - \frac{1}{LC}}. \tag{7.6}$$

This may be rewritten as

$$s_1, s_2 = -\alpha \pm \sqrt{\alpha^2 - \omega_0^2}, \tag{7.7}$$

where $\alpha = R/(2L)$ is the **neper frequency** and $\omega_0 = 1/\sqrt{LC}$ is the **resonant frequency** for the series RLC circuit.

There are three important cases with regard to the nature of the roots and the response of the circuit. If $\alpha > \omega_0$, the roots are real and distinct, and the solution is given by a linear combination of two exponentials. This is the **overdamped case**:

$$i(t) = A_1e^{s_1t} + A_2e^{s_2t}, \quad t \geq 0^+, \tag{7.8}$$

where A_1 and A_2 are coefficients determined by applying the initial conditions. If $\alpha = \omega_0$, the roots are real and equal, $s_1 = s_2 = \alpha$, and one of the exponentials must be multiplied by t to render two independent solutions. This is the **critically-damped case**:

$$i(t) = D_1te^{-\alpha t} + D_2e^{-\alpha t}, \quad t \geq 0^+. \tag{7.9}$$

where D_1 and D_2 are coefficients determined by applying initial conditions.

Finally, if $\omega_0 > \alpha$ then the roots are complex conjugates: $s_1, s_2 = -\alpha \pm j\sqrt{\omega_0^2 - \alpha^2} = -\alpha \pm j\omega_d$. This is the **underdamped** case. We could treat the solution as a linear combination of two exponentials having complex exponents and complex coefficients, but it will be more useful to rewrite the form of the solution with real exponents and real coefficients. We start by rewriting the expression for $i(t)$ by using the properties of exponents.

$$i(t) = A_1 e^{s_1 t} + A_2 e^{s_2 t}$$

$$= A_1 e^{(-\alpha + j\omega_d)t} + A_2 e^{(-\alpha - j\omega_d)t}$$

$$= A_1 e^{-\alpha t} e^{j\omega_d t} + A_2 e^{-\alpha t} e^{-j\omega_d t}, \quad t \geq 0^+. \tag{7.10}$$

Now, we can make use of Euler's identity: $e^{j\theta} = \cos\theta + j\sin\theta$, yielding

$$i(t) = A_1 e^{-\alpha t}\left[\cos(\omega_d t) + j\sin(\omega_d t)\right] + A_2 e^{-\alpha t}\left[\cos(\omega_d t) + j\sin(\omega_d t)\right]$$

$$= (A_1 + A_2)e^{-\alpha t}\cos(\omega_d t) + j(A_1 - A_2)e^{-\alpha t}\sin(\omega_d t)$$

$$= B_1 e^{-\alpha t}\cos(\omega_d t) + B_2 e^{-\alpha t}\sin(\omega_d t), \quad t \geq 0^+. \tag{7.11}$$

In the final step, we have assumed that A_1 and A_2 are complex conjugates so that $(A_1 + A_2)$ and $j(A_1 - A_2)$ are both real quantities; this is necessarily true because the currents and voltages in real circuits are real, not complex. Therefore we have simplified the final expression by using the real coefficients $B_1 = (A_1 + A_2)$ and $B_2 = j(A_1 - A_2)$. Thus, the **underdamped** solution is a linear combination of two damped sinusoids:

$$i(t) = B_1 e^{-\alpha t}\cos(\omega_d t) + B_2 e^{-\alpha t}\sin(\omega_d t), \quad t \geq 0^+, \tag{7.12}$$

where $\omega_d = \sqrt{\omega_0^2 - \alpha^2}$ is the damped radian frequency.

The three cases of the solution for $i(t)$ are summarized in Table 7.1; the solutions for $v_L(t)$ and $v_C(t)$ have the same general forms but with different coefficients. These general results will be further explained with examples in the following sections.

TABLE 7.1

Three Cases of Natural Response for a Series RLC Circuit

Series RLC Circuit: $\alpha = R/(2L)$, $\omega_0 = 1/\sqrt{LC}$.

$\alpha > \omega_0$ Overdamped	$i(t) = A_1 e^{s_1 t} + A_2 e^{s_2 t}, \quad t \geq 0^+$	$s_1, s_2 = -\alpha \pm \sqrt{\alpha^2 - \omega_0^2}$
$\alpha = \omega_0$ Critically-damped	$i(t) = D_1 t e^{-\alpha t} + D_2 e^{-\alpha t}, \quad t \geq 0^+$	$s_1, s_2 = -\alpha$
$\alpha < \omega_0$ Underdamped	$i(t) = B_1 e^{-\alpha t}\cos(\omega_d t) + B_2 e^{-\alpha t}\sin(\omega_d t), \quad t \geq 0^+$	$s_1, s_2 = -\alpha \pm j\omega_d$ $\omega_d = \sqrt{\omega_0^2 - \alpha^2}$

7.3   OVERDAMPED NATURAL RESPONSE OF A SERIES RLC CIRCUIT

Consider the natural response of the series RLC circuit in Figure 7.2. Suppose that the initial conditions given in the figure have been established by switched circuitry which is omitted from the diagram for simplicity. Whereas the inductor current cannot change instantaneously, $i(0^+) = i(0^-) = i(0)$. Similarly, the capacitor voltage may not change instantaneously so $v_C(0^+) = v_C(0^-) = v_C(0)$.

For this series circuit, the neper frequency is

$$\alpha = \frac{R}{2L} = \frac{500\,\Omega}{2(50\,\text{mH})} = 5,000\,\text{rad/s}, \tag{7.13}$$

and the resonant frequency is

$$\omega_0 = \frac{1}{\sqrt{(1.25\,\mu\text{F})(50\,\text{mH})}} = 4,000\,\text{rad/s}. \tag{7.14}$$

Because $\alpha > \omega_0$, the roots of the characteristic equation are real and distinct:

$$\begin{aligned} s_1, s_2 &= -\alpha \pm \sqrt{\alpha^2 - \omega_0^2} \\ &= \left(-5,000 \pm \sqrt{5,000^2 - 4,000^2}\right)\text{rad/s} = -2,000, -8,000\,\text{rad/s}. \end{aligned} \tag{7.15}$$

This is an **overdamped** circuit and the solution is given by a linear combination of two exponentials:

$$i(t) = A_1 e^{s_1 t} + A_2 e^{s_2 t} = A_1 e^{-2,000t} + A_2 e^{-8,000t}; \quad t \geq 0^+. \tag{7.16}$$

The units of s_1 and s_2 have not been shown explicitly, but it should be remembered that each has units of rad/s and therefore the argument of the exponent is unitless if time is expressed in s.

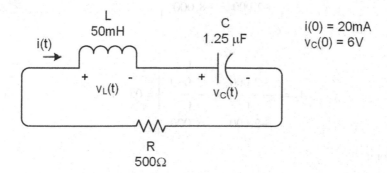

FIGURE 7.2   An overdamped series RLC circuit for consideration of the natural response.

To find the coefficients, we need to apply the initial conditions. The numerical value of the initial inductor current is $i(0^+) = I_0 = 20\,\text{mA}$. From the general form of the solution, evaluated at $t = 0^+$, $i(0^+) = A_1 + A_2$. Equating these two,

$$A_1 + A_2 = 20\,\text{mA}. \tag{7.17}$$

The initial value of the derivative of the current may be found by consideration of the *inductor voltage*, and this can be found by using KVL and the knowledge of the initial capacitor voltage:

$$\left. \frac{di}{dt} \right|_{t=0^+} = \frac{v_L(0^+)}{L} = \frac{-v_C(0^+) - Ri(0^+)}{L} = \frac{-6\,\text{V} - (500\,\Omega)(20\,\text{mA})}{50\,\text{mH}} = -320\,\text{A/s}.$$

$$\tag{7.18}$$

From the general form of the solution,

$$\left. \frac{di}{dt} \right|_{t=0^+} = s_1 A_1 + s_2 A_1 = -2,000 A_1 - 8,000 A_2. \tag{7.19}$$

Equating these two,

$$-2,000 A_1 - 8,000 A_2 = -320\,\text{A/s}. \tag{7.20}$$

Now, we can solve for the coefficients using equations (7.17) and (7.20):

$$\begin{bmatrix} 1 & 1 \\ -2,000\,\text{rad/s} & -8,000\,\text{rad/s} \end{bmatrix} \begin{bmatrix} A_1 \\ A_2 \end{bmatrix} = \begin{bmatrix} 0.02\,\text{A} \\ -320\,\text{A/s} \end{bmatrix}. \tag{7.21}$$

Solving,

$$A_1 = \frac{\begin{vmatrix} 1 & 1 \\ -2,000 & -8,000 \end{vmatrix}}{\begin{vmatrix} 1 & 1 \\ -2,000 & -8,000 \end{vmatrix}} = -0.0267\,\text{A} \tag{7.22}$$

and

$$A_2 = \frac{\begin{vmatrix} 1 & 1 \\ -2,000 & -8,000 \end{vmatrix}}{\begin{vmatrix} 1 & 1 \\ -2,000 & -8,000 \end{vmatrix}} = 0.0467\,\text{A}. \tag{7.23}$$

Therefore,

$$i(t) = \left[-0.0267 e^{-2,000 t} + 0.0467 e^{-8,000 t} \right] \text{A}; \quad t \ge 0^+. \tag{7.24}$$

Once we have solved for the current, which is common to the resistor, inductor, and capacitor, we may find the voltage for each element. For example, the inductor voltage may be found by

$$v_L(t) = L\frac{di}{dt} = (50\,\text{mH})\frac{d}{dt}\left[-0.0267e^{-2,000t} + 0.0467e^{-8,000t}\right]\text{A}$$

$$= \left[2.67e^{-2,000t} - 18.68e^{-8,000t}\right]\text{V}; \quad t \geq 0^+. \tag{7.25}$$

The capacitor voltage could be found by integrating the current but it can also be found using KVL, which may be more convenient:

$$v_C(t) = -iR - L\frac{di}{dt}$$

$$= -(500\,\Omega)\left[-0.0267e^{-2,000t} + 0.0467e^{-8,000t}\right]\text{A}$$

$$-(50\,\text{mH})\frac{d}{dt}\left[-0.0267e^{-2,000t} + 0.0467e^{-8,000t}\right]\text{A} \tag{7.26}$$

$$= \left[10.68e^{-2,000t} - 4.67e^{-8,000t}\right]\text{V}; \quad t \geq 0^+.$$

The transient response (current and capacitor voltage) is plotted in Figure 7.3.

As another example, consider the natural response of the series RLC circuit in Figure 7.4. The make-before-break switch has been in position a for a long time and moves to position b at $t = 0$. Prior to the movement of the switch, all time derivatives would have settled to zero, and therefore the capacitor current would have settled to zero. Because the inductor is in series with the capacitor, its current will also have settled to zero. This means that in a circuit with this configuration, the

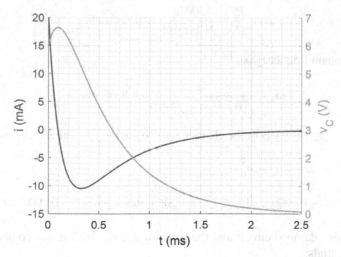

FIGURE 7.3 Transient response (current and capacitor voltage) in the overdamped series RLC circuit of Figure 7.2.

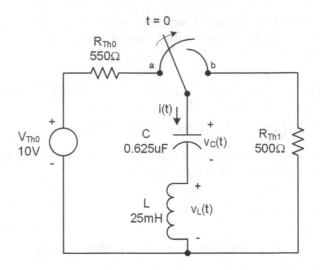

FIGURE 7.4   A switched series RLC circuit for consideration of the natural response.

initial condition for the inductor will always be zero. On the other hand, the voltage across the capacitor will have settled to the value of the Thevenin voltage for the circuitry connected prior to the switch movement, because with the current settled to zero there will be zero voltage across the resistor and zero voltage across the inductor. This same capacitor voltage will exist immediately after the movement of the switch because the capacitor voltage may not change instantaneously. Thus, $v_C(0^+) = v_C(0^-) = V_{Th0} = 10\,\mathrm{V}$.

This is a series RLC circuit because the inductor and the capacitor are in series. The neper frequency is

$$\alpha = \frac{R}{2L} = \frac{500\,\Omega}{2(25\,\mathrm{mH})} = 10,000\,\mathrm{rad/s}, \tag{7.27}$$

and the resonant frequency is

$$\omega_0 = \frac{1}{\sqrt{(0.625\,\mu\mathrm{F})(25\,\mathrm{mH})}} = 8,000\,\mathrm{rad/s}. \tag{7.28}$$

Because $\alpha > \omega_0$, the roots of the characteristic equation are real and distinct:

$$s_1, s_2 = -\alpha \pm \sqrt{\alpha^2 - \omega_0^2}$$

$$= \left(-10,000 \pm \sqrt{10,000^2 - 8,000^2}\right)\mathrm{rad/s} = -4,000, -16,000\,\mathrm{rad/s}. \tag{7.29}$$

This is an overdamped circuit and the solution is given by a linear combination of two exponentials:

$$i(t) = A_1 e^{s_1 t} + A_2 e^{s_2 t} = A_1 e^{-4,000t} + A_2 e^{-16,000t}; \quad t \geq 0^+. \tag{7.30}$$

To find the coefficients, we need to apply the initial conditions. The numerical value of the initial inductor current is zero: $i(0^+) = 0$. From the general form of the solution, evaluated at $t = 0^+$, $i(0^+) = A_1 + A_2$. Equating these two,

$$A_1 + A_2 = 0. \tag{7.31}$$

The initial value of the derivative of the current may be found by consideration of the *inductor voltage*, and this can be found by using KVL and the knowledge of the initial capacitor voltage:

$$\left. \frac{di}{dt} \right|_{t=0^+} = \frac{v_L(0^+)}{L} = \frac{-v_C(0^+) - Ri(0^+)}{L} = \frac{-10\,V - (500\,\Omega)(0)}{25\,mH} = -400\,A/s. \tag{7.32}$$

From the general form of the solution,

$$\left. \frac{di}{dt} \right|_{t=0^+} = s_1 A_1 + s_2 A_1 = -4,000 A_1 - 16,000 A_2. \tag{7.33}$$

Equating these two,

$$-4,000 A_1 - 16,000 A_2 = -400\,A/s. \tag{7.34}$$

Now, we can solve for the coefficients using the two equations involving A_1 and A_2:

$$\begin{bmatrix} 1 & 1 \\ -4,000\,rad/s & -16,000\,rad/s \end{bmatrix} \begin{bmatrix} A_1 \\ A_2 \end{bmatrix} = \begin{bmatrix} 0 \\ -400\,A/s \end{bmatrix}. \tag{7.35}$$

Solving,

$$A_1 = \frac{\begin{vmatrix} 0 & 1 \\ -400\,A/s & -16,000\,rad/s \end{vmatrix}}{\begin{vmatrix} 1 & 1 \\ -4,000\,rad/s & -16,000\,rad/s \end{vmatrix}} = -0.0333\,A \tag{7.36}$$

and

$$A_2 = \frac{\begin{vmatrix} 1 & 0 \\ -4,000\,rad/s & -400\,A/s \end{vmatrix}}{\begin{vmatrix} 1 & 1 \\ -4,000\,rad/s & -16,000\,rad/s \end{vmatrix}} = 0.0333\,A. \tag{7.37}$$

Therefore,

$$i(t) = \left[-0.0333 e^{-4,000t} + 0.0333 e^{-16,000t} \right] A; \quad t \geq 0^+. \tag{7.38}$$

The inductor voltage also exhibits an overdamped form:

$$v_L(t) = L\frac{di}{dt} = (25\,\text{mH})\frac{d}{dt}\left[-0.0333e^{-4,000t} + 0.0333e^{-16,000t}\right]\text{A}$$

$$= \left[3.33e^{-4,000t} - 13.33e^{-16,000t}\right]\text{V}; \quad t \geq 0^+. \tag{7.39}$$

By KVL, the capacitor voltage is

$$v_C(t) = -iR - L\frac{di}{dt}$$

$$= -(500\,\Omega)\left[-0.0333e^{-4,000t} + 0.0333e^{-16,000t}\right]\text{A}$$

$$-(25\,\text{mH})\frac{d}{dt}\left[-0.0333e^{-4,000t} + 0.0333e^{-16,000t}\right]\text{A}$$

$$= \left[13.33e^{-4,000t} - 3.33e^{-16,000t}\right]\text{V}; \quad t \geq 0^+. \tag{7.40}$$

The current and capacitor voltage are plotted in Figure 7.5.

7.4 CRITICALLY-DAMPED NATURAL RESPONSE OF A SERIES RLC CIRCUIT

Now, consider the circuit of Figure 7.6, it is similar to the previous example except that the resistor value R_{Th1} has been changed. Once again $i(0^+) = i(0^-) = 0$ but $v_C(0^+) = v_C(0^-) = V_{\text{Th0}} = 10\,\text{V}$.

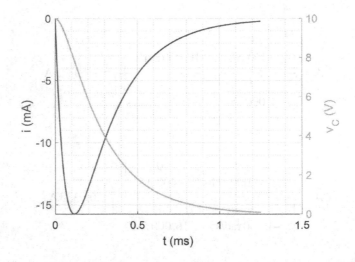

FIGURE 7.5 Transient response (current and capacitor voltage) in the overdamped series RLC circuit of Figure 7.4.

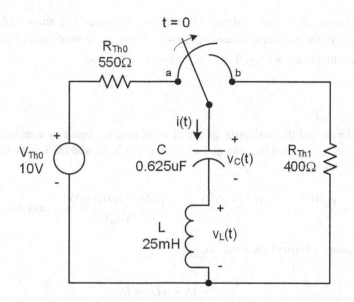

FIGURE 7.6 A critically-damped series RLC circuit for consideration of the natural response.

This is a series RLC circuit; the neper frequency is

$$\alpha = \frac{R}{2L} = \frac{400\,\Omega}{2(25\,\text{mH})} = 8,000\,\text{rad/s}, \tag{7.41}$$

and the resonant frequency is

$$\omega_0 = \frac{1}{\sqrt{(0.625\,\mu\text{F})(25\,\text{mH})}} = 8,000\,\text{rad/s}. \tag{7.42}$$

Because $\alpha = \omega_0$, the roots of the characteristic equation are real and equal:

$$s_1, s_2 = -\alpha \pm \sqrt{\alpha^2 - \omega_0^2}$$

$$= \left(-8,000 \pm \sqrt{8,000^2 - 8,000^2}\right)\text{rad/s}$$

$$= -8,000, -8,000\,\text{rad/s}. \tag{7.43}$$

This is therefore a **critically-damped** circuit and the solution is given by

$$i(t) = D_1 t e^{-\alpha t} + D_2 e^{-\alpha t} = D_1 t e^{-8,000t} + D_2 e^{-8,000t}; \quad t \geq 0^+. \tag{7.44}$$

To find the coefficients, we must apply the initial conditions. The numerical value of the initial inductor current is zero: $i(0^+) = 0$. From the general form of the solution, evaluated at $t = 0^+$, $i(0^+) = D_2$. Equating these two,

$$D_2 = 0. \tag{7.45}$$

The initial value of the derivative of the current may be found by consideration of the *inductor voltage*, and this can be found by using KVL and the knowledge of the initial capacitor voltage:

$$\left.\frac{di}{dt}\right|_{t=0^+} = \frac{v_L(0^+)}{L} = \frac{-v_C(0^+)-Ri(0^+)}{L} = \frac{-10\,\text{V}-(400\,\Omega)(0)}{25\,\text{mH}} = -400\,\text{A/s}. \tag{7.46}$$

From the general form of the solution,

$$\left.\frac{di}{dt}\right|_{t=0^+} = D_1 - \alpha D_2 = D_1. \tag{7.47}$$

Equating these two,

$$D_1 = -400\,\text{A/s}. \tag{7.48}$$

Therefore,

$$i(t) = (-400\,\text{A/s})te^{-8,000t}; \quad t \geq 0^+. \tag{7.49}$$

The inductor voltage also exhibits a critically-damped form:

$$v_L(t) = L\frac{di}{dt} = (25\,\text{mH})\frac{d}{dt}\left[(-400\,\text{A/s})te^{-8,000t}\right]$$

$$= \left[(80,000\,\text{V/s})te^{-8,000t} - 10\,\text{V}e^{-8,000t}\right]; \quad t \geq 0^+. \tag{7.50}$$

By KVL, the capacitor voltage is

$$v_C(t) = -iR - L\frac{di}{dt}$$

$$= -(400\,\Omega)(-400\,\text{A/s})te^{-8,000t}$$

$$-(25\,\text{mH})\frac{d}{dt}\left[(-400\,\text{A/s})te^{-8,000t}\right]$$

$$= \left[(80,000\,\text{V/s})te^{-8,000t} + 10\,\text{V}e^{-8,000t}\right]; \quad t \geq 0^+. \tag{7.51}$$

It is interesting to note that, although the second coefficient is zero in the current expression, the inductor voltage and capacitor voltage expressions both have two non-zero coefficients.

The transient response (current and capacitor voltage) is plotted in Figure 7.7. Visually, it is difficult to distinguish the critically-damped and overdamped responses.

7.5 UNDERDAMPED NATURAL RESPONSE OF A SERIES RLC CIRCUIT

Now consider the circuit in Figure 7.8; it is similar to the circuit of the previous example except that the resistor value R_{Th1} has been changed. Once again $i(0^+) = i(0^-) = 0$ but $v_C(0^+) = v_C(0^-) = V_{Th0} = 10\,\text{V}$.

For this series RLC circuit, the neper frequency is

$$\alpha = \frac{R}{2L} = \frac{50\,\Omega}{2(25\,\text{mH})} = 1,000\,\text{rad/s} \tag{7.52}$$

and the resonant frequency is

$$\omega_0 = \frac{1}{\sqrt{(0.625\,\mu\text{F})(25\,\text{mH})}} = 8,000\,\text{rad/s}. \tag{7.53}$$

Because $\alpha < \omega_0$, the roots of the characteristic equation are complex conjugates and the solution has an **underdamped** form with

$$\omega_d = \sqrt{8,000^2 - 1,000^2}\;\text{rad/s} = 7,937\,\text{rad/s}. \tag{7.54}$$

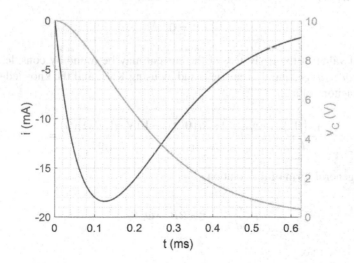

FIGURE 7.7 Transient response (current and capacitor voltage) in the critically-damped series RLC circuit of Figure 7.6.

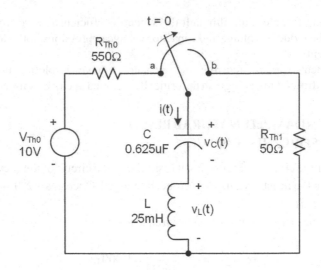

FIGURE 7.8   An underdamped series RLC circuit for consideration of the natural response.

The solution is of the form:

$$i(t) = B_1 e^{-\alpha t} \cos(\omega_d t) + B_2 e^{-\alpha t} \sin(\omega_d t)$$
$$= B_1 e^{-1,000t} \cos(7,937t) + B_2 e^{-1,000t} \sin(7,937t); \quad t \geq 0^+. \tag{7.55}$$

To find the coefficients, we will apply the initial conditions. The numerical value of the initial inductor current is zero: $i(0^+) = 0$. From the general form of the solution, evaluated at $t = 0^+$, $i(0^+) = B_1$. Equating these two,

$$B_1 = 0. \tag{7.56}$$

The initial value of the derivative of the current may be found by consideration of the *inductor voltage*, and this can be found by using KVL and the knowledge of the initial capacitor voltage:

$$\left. \frac{di}{dt} \right|_{t=0^+} = \frac{v_L(0^+)}{L} = \frac{-v_C(0^+) - Ri(0^+)}{L} = \frac{-10\,\text{V} - (50\,\Omega)(0)}{25\,\text{mH}} = -400\,\text{A/s}. \tag{7.57}$$

From the general form of the solution,

$$\left. \frac{di}{dt} \right|_{t=0^+} = -\alpha B_1 + \omega_d B_2. \tag{7.58}$$

Equating these two,

$$-\alpha B_1 + \omega_d B_2 = -400\,\text{A/s}, \tag{7.59}$$

so

$$B_2 = \frac{-400 \text{ A/s} + \alpha B_1}{\omega_d} = \frac{-400 \text{ A/s} + (1,000 \text{ rad/s})(0)}{7,937 \text{ rad/s}} = -0.0504 \text{ A.} \qquad (7.60)$$

Therefore,

$$i(t) = -0.0504 \text{ A} \, e^{-1,000t} \sin(7,937t); \quad t \geq 0^+. \qquad (7.61)$$

The inductor voltage also exhibits an underdamped form:

$$v_L(t) = L\frac{di}{dt} = (25 \text{ mH})\frac{d}{dt}\left[-0.0504 \text{ A} \, e^{-1,000t} \sin(7,937t)\right]$$

$$= \left[-10.0e^{-1,000t} \cos(7,937t) + 1.26e^{-1,000t} \sin(7,937t)\right]\text{V}; \quad t \geq 0^+. \qquad (7.62)$$

By KVL, the capacitor voltage is

$$v_C(t) = -iR - L\frac{di}{dt}$$

$$= -(50 \, \Omega)(-0.0504 \text{ A})e^{-1,000t} \sin(7,937t)$$

$$-(25 \text{ mH})\frac{d}{dt}\left[(-0.0504) \text{ A} \, e^{-1,000t} \sin(7,937t)\right] \qquad (7.63)$$

$$= \left[-10.0e^{-1,000t} \cos(7,937t) + 1.26e^{-1,000t} \sin(7,937t)\right]\text{V}; \quad t \geq 0^+.$$

It is interesting to note that, although the first coefficient is zero in the current expression, the inductor voltage and capacitor voltage expressions both have two non-zero coefficients.

The current and capacitor voltage are plotted in Figure 7.9, and it can be seen that each exhibits the form of a decaying sinusoid. This is further illustrated in Figure 7.10, which shows the current transient along with its envelope $\pm 0.0504 \text{ A} \exp(-\alpha t)$.

7.6 STEP AND NATURAL RESPONSE OF A SERIES RLC CIRCUIT

Consider the series RLC circuit in Figure 7.11. We will assume that the switch has been in position a for a "long time" before moving to position b. If the circuit has been allowed to settle before the movement of the switch, all time derivatives will have settled to zero. Therefore, the current will have settled to zero, and because of the inductor this current may not change instantaneously when the switch is moved. Hence, $i(0^+) = i(0^-) = 0$. It follows that the capacitor voltage will have settled to the Thevenin voltage for the circuit connected prior to the movement of the switch, and the capacitor voltage may not change instantaneously. Therefore, $v_C(0^+) = v_C(0^-) = V_{Th0}$.

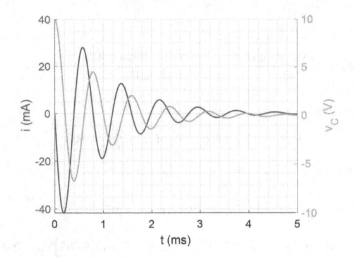

FIGURE 7.9 Transient response (current and capacitor voltage) in the underdamped series RLC circuit of Figure 7.8.

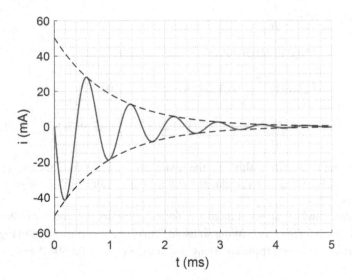

FIGURE 7.10 Current in the underdamped series RLC circuit of Figure 7.8 along with its exponential envelope $\pm 0.0504\,\text{A}\exp(-\alpha t)$.

To determine the transient response of this circuit after the movement of the switch, we need to set up and solve a second-order homogeneous linear differential equation. To do this, we start by using KVL:

$$Ri + L\frac{di}{dt} + V_0 + \frac{1}{C}\int_0^t i\,dy - V_{\text{Th1}} = 0. \qquad (7.64)$$

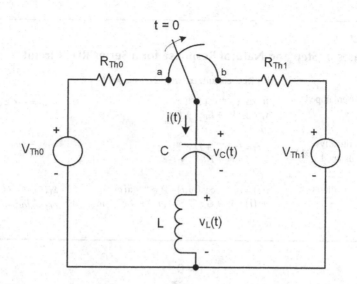

FIGURE 7.11 A series RLC circuit for consideration of the step and natural response.

Differentiating with respect to time,

$$R\frac{di}{dt} + L\frac{d^2i}{dt^2} + \frac{i}{C} = 0. \tag{7.65}$$

Rearranging and normalizing with respect to the second-order term,

$$\frac{d^2i}{dt^2} + \frac{R}{L}\frac{di}{dt} + \frac{1}{LC}i = 0. \tag{7.66}$$

This is identical to the differential equation we obtained when considering the natural response. Hence, the solution for $i(t)$ will have one of the same forms (overdamped, critically-damped, or underdamped) as found when considering the natural response. On the other hand, the solution for the capacitor voltage will have one of the same forms plus a constant to account for a non-zero final value. These forms of the solution are summarized in Table 7.2.

As an example, consider the circuit in Figure 7.12.

For this series RLC circuit, the neper frequency is

$$\alpha = \frac{R}{2L} = \frac{12\,\Omega}{2(10\,\text{mH})} = 600\,\text{rad/s} \tag{7.67}$$

and the resonant frequency is

$$\omega_0 = \frac{1}{\sqrt{(25\,\mu\text{F})(10\,\text{mH})}} = 2,000\,\text{rad/s}. \tag{7.68}$$

TABLE 7.2

Three Cases of Step and Natural Response for a Series RLC Circuit

Series RLC circuit: $\alpha = R/(2L)$, $\omega_0 = 1/\sqrt{LC}$.

$\alpha > \omega_0$	Overdamped	$i(t) = A_1 e^{s_1 t} + A_2 e^{s_2 t}$; $v_C(t) = V_F + E_1 e^{s_1 t} + E_2 e^{s_2 t}$; $t \geq 0^+$	$s_1, s_2 = -\alpha \pm \sqrt{\alpha^2 - \omega_0^2}$
$\alpha = \omega_0$	Critically-damped	$i(t) = D_1 t e^{-\alpha t} + D_2 e^{-\alpha t}$; $v_C(t) = V_F + F_1 t e^{-\alpha t} + F_2 e^{-\alpha t}$ $t \geq 0^+$	$s_1, s_2 = -\alpha$
$\alpha < \omega_0$	Underdamped	$i(t) = B_1 e^{-\alpha t} \cos(\omega_d t) + B_2 e^{-\alpha t} \sin(\omega_d t)$; $v_C(t) = V_F + G_1 e^{-\alpha t} \cos(\omega_d t) + G_2 e^{-\alpha t} \sin(\omega_d t)$ $t \geq 0^+$	$s_1, s_2 = -\alpha \pm j\omega_d$ $\omega_d = \sqrt{\omega_0^2 - \alpha^2}$

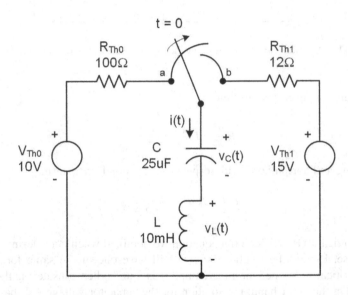

FIGURE 7.12   An underdamped series RLC circuit for consideration of the step and natural response.

Because $\alpha < \omega_0$, the roots of the characteristic equation are complex conjugates and the solution has an **underdamped** form with

$$\omega_d = \sqrt{2,000^2 - 600^2} \text{ rad/s} = 1,908 \text{ rad/s}. \tag{7.69}$$

The solution for the capacitor voltage is of the form:

$$v_C(t) = V_F + E_1 e^{-\alpha t} \cos(\omega_d t) + E_2 e^{-\alpha t} \sin(\omega_d t)$$

$$= V_F + E_1 e^{-600t} \cos(1,908t) + E_2 e^{-600t} \sin(1,908t); \quad t \geq 0^+, \tag{7.70}$$

where $V_F = V_{Th1} = 15\,\text{V}$. To find the coefficients E_1 and E_2, we will apply the initial conditions. We have established that the initial capacitor voltage is $v_C(0^+) = v_C(0^-) = V_{Th0} = 10\,\text{V}$. From the general form of the solution, evaluated at $t = 0^+$, $v_C(0^+) = V_F + E_1$. Equating these two and solving,

$$E_1 = V_{Th0} - V_F = V_{Th0} - V_{Th1} = 10\,\text{V} - 15\,\text{V} = -5\,\text{V}. \tag{7.71}$$

The initial value of the derivative of the capacitor voltage may be found by consideration of the initial current:

$$\left.\frac{dv_C}{dt}\right|_{t=0^+} = \frac{i(0^+)}{C} = \frac{0}{25\,\mu\text{F}} = 0. \tag{7.72}$$

From the general form of the solution,

$$\left.\frac{dv_C}{dt}\right|_{t=0^+} = -\alpha E_1 + \omega_d E_2. \tag{7.73}$$

Equating these two,

$$-\alpha E_1 + \omega_d E_2 = 0, \tag{7.74}$$

so

$$E_2 = \frac{0 + \alpha E_1}{\omega_d} = \frac{0 + (600\,\text{rad/s})(-5\,\text{V})}{1,908\,\text{rad/s}} = -1.572\,\text{V}. \tag{7.75}$$

Therefore,

$$v_C(t) = V_F + E_1 e^{-600t}\cos(1,908t) + E_2 e^{-600t}\sin(1,908t)$$

$$\left[15 - 5e^{-600t}\cos(1,908t) + 1.572e^{-600t}\sin(1,908t)\right]\text{V}; \quad t \geq 0^+. \tag{7.76}$$

The current may be found by differentiation:

$$i(t) = C\frac{dv_C}{dt} = (25\,\mu\text{F})\frac{d}{dt}\left[15 - 5e^{-600t}\cos(1,908t) + 1.572e^{-600t}\sin(1,908t)\right]\text{V}$$

$$= \left[+0.262e^{-600t}\sin(1,908t)\right]\text{A}; \quad t \geq 0^+. \tag{7.77}$$

So whereas the capacitor voltage exhibits a non-zero final value, the current does not. The current and capacitor voltage are plotted in Figure 7.13.

To review, see Presentation 7.1 in ebook+.

To test your knowledge, try Quiz 7.1 in ebook+.

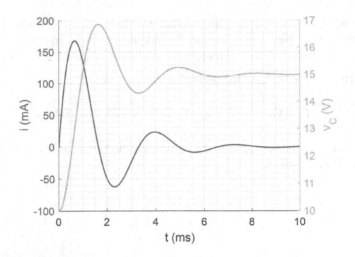

FIGURE 7.13 Step and natural response (current and capacitor voltage) for the under-damped series RLC circuit in Figure 7.12.

7.7 NATURAL RESPONSE OF A PARALLEL RLC CIRCUIT

Now, we will shift our focus to parallel RLC circuits, in which the inductor and the capacitor are connected in parallel. We will find that many of the considerations are similar to those for series circuits, but with key differences in the determination of α, initial conditions, and final conditions. Suppose that the parallel RLC circuit in Figure 7.14 exhibits an initial inductor current of I_0 and an initial capacitor voltage of V_0. (There are switching configurations capable of establishing these initial conditions, but the switching circuitry is omitted here for simplicity.) We want to determine $i_L(t)$ and $v(t)$ for $t \geq 0^+$. If we allow the circuit to settle, both the inductor current and the capacitor voltage will settle to zero: $i_L(\infty) = 0$ and $v(\infty) = 0$.

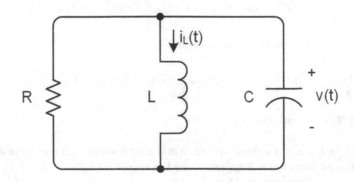

FIGURE 7.14 A parallel RLC circuit for consideration of the natural response.

To determine the transient response of this circuit while it is settling, we need to set up and solve a second-order homogeneous linear differential equation. To do this, we start by using Kirchhoff's current law (KCL):

$$\frac{v}{R} + I_0 + \frac{1}{L}\int_0^t v\,dy + C\frac{dv}{dt} = 0. \tag{7.78}$$

Differentiating with respect to time,

$$\frac{1}{R}\frac{dv}{dt} + \frac{v}{L} + C\frac{d^2v}{dt^2} = 0. \tag{7.79}$$

Rearranging and normalizing with respect to the second-order term,

$$\frac{d^2v}{dt^2} + \frac{1}{RC}\frac{dv}{dt} + \frac{1}{LC}v = 0. \tag{7.80}$$

As before, we will assume solutions of the form $v(t) = Ae^{st}$. Using this assumed form,

$$s^2 Ae^{st} + \frac{1}{RC}sAe^{st} + \frac{1}{LC}Ae^{st} = 0. \tag{7.81}$$

Dividing by Ae^{st}, we obtain the **characteristic equation** for the parallel RLC circuit:

$$s^2 + \frac{1}{RC}s + \frac{1}{LC} = 0. \tag{7.82}$$

Applying the quadratic formula, we can find the two roots of the characteristic equation:

$$s_1, s_2 = -\frac{1}{2RC} \pm \sqrt{\left(\frac{1}{2RC}\right)^2 - \frac{1}{LC}}. \tag{7.83}$$

This may be rewritten as

$$s_1, s_2 = -\alpha \pm \sqrt{\alpha^2 - \omega_0^2}, \tag{7.84}$$

where $\alpha = 1/(2RC)$ is the **neper frequency** and $\omega_0 = 1/\sqrt{LC}$ is the **resonant frequency** for the parallel RLC circuit. Whereas the resonant frequency equation is given by the same expression as the series case, the neper frequency is not.

Once again there are three cases with regard to the nature of the roots and the response of the circuit. These are the **overdamped**, **critically-damped**, and **underdamped** cases summarized in Table 7.3. Although the table gives the forms of the solution for the voltage, which is common to all three elements and therefore a fundamental quantity, it should be recognized that the individual currents in the circuit will exhibit the same general forms.

TABLE 7.3

Three Cases of Natural Response for a Parallel RLC Circuit

Parallel RLC circuit: $\alpha = 1/(2RC)$, $\omega_0 = 1/\sqrt{LC}$.

$\alpha > \omega_0$	Overdamped	$v(t) = A_1 e^{s_1 t} + A_2 e^{s_2 t}$, $\quad t \geq 0^+$	$s_1, s_2 = -\alpha \pm \sqrt{\alpha^2 - \omega_0^2}$
$\alpha = \omega_0$	Critically-damped	$v(t) = D_1 t e^{-\alpha t} + D_2 e^{-\alpha t}$, $\quad t \geq 0^+$	$s_1, s_2 = -\alpha$
$\alpha < \omega_0$	Underdamped	$v(t) = B_1 e^{-\alpha t} \cos(\omega_d t) + B_2 e^{-\alpha t} \sin(\omega_d t)$, $\quad t \geq 0^+$	$s_1, s_2 = -\alpha \pm j\omega_d$
			$\omega_d = \sqrt{\omega_0^2 - \alpha^2}$

7.8   STEP AND NATURAL RESPONSE OF A PARALLEL RLC CIRCUIT

Now consider the parallel RLC circuit in Figure 7.15, and suppose the switch has been in position a for a long time before moving to position b. If the circuit settled with the switch in position a, all time derivatives will have settled to zero. This implies that $v = L di_L / dt$ will have settled to zero before the movement of the switch, and the voltage may not change instantaneously because of the capacitor, so $v(0^+) = v(0^-) = 0$. Also, $i_C = C dv / dt$ will have settled to zero before the movement of the switch so $i_C(0^-) = 0$. However, the capacitor current may change instantaneously so the value of $i_C(0^+)$ may not be zero and must be determined by using KCL. Given these considerations, the inductor current will have settled to the value of the Norton current source connected prior to the switch movement, and the inductor current may not change instantaneously so the same value of current will exist right after the switch moves: $i_L(0^+) = i_L(0^-) = I_{N0}$. If we allow the circuit to resettle with the switch in position b, similar considerations apply and $i_L(\infty) = I_{N1}$.

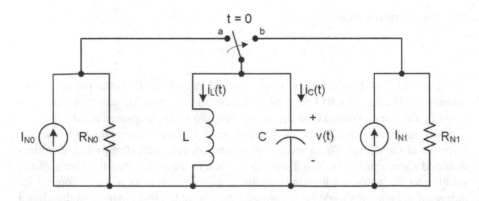

FIGURE 7.15   A parallel RLC circuit for consideration of the step and natural response.

To determine the transient response of this circuit after the switch has moved to position b, we need to set up and solve a second-order homogeneous linear differential equation. To do this, we start by using KCL:

$$I_{N0} + \frac{1}{L}\int_0^t v\,dy + C\frac{dv}{dt} - I_{N1} + \frac{v}{R_{N1}} = 0. \tag{7.85}$$

Differentiating with respect to time,

$$\frac{1}{R}\frac{dv}{dt} + \frac{v}{L} + C\frac{d^2v}{dt^2} = 0. \tag{7.86}$$

Rearranging and normalizing with respect to the second-order term,

$$\frac{d^2v}{dt^2} + \frac{1}{RC}\frac{dv}{dt} + \frac{1}{LC}v = 0. \tag{7.87}$$

As before, we will assume solutions of the form $v(t) = Ae^{st}$. Using this assumed form,

$$s^2Ae^{st} + \frac{1}{RC}sAe^{st} + \frac{1}{LC}Ae^{st} = 0. \tag{7.88}$$

Dividing by Ae^{st}, we obtain the **characteristic equation**:

$$s^2 + \frac{1}{RC}s + \frac{1}{LC} = 0. \tag{7.89}$$

This is the same as the characteristic equation we obtained when solving for the natural response. Therefore, the solutions will have the same forms, apart from a constant to account for the final value of the inductor current. The three forms of the solution are summarized in Table 7.4.

TABLE 7.4

Three Cases of Step and Natural Response for a Parallel RLC Circuit

Series RLC circuit: $\alpha = 1/(2RC$, $\omega_0 = 1/\sqrt{LC}$.

$\alpha > \omega_0$	Overdamped	$v(t) = A_1e^{s_1t} + A_2e^{s_2t}$; $i_L(t) = I_F + E_1e^{s_1t} + E_2e^{s_2t}$; $t \geq 0^+$	$s_1, s_2 = -\alpha \pm \sqrt{\alpha^2 - \omega_0^2}$
$\alpha = \omega_0$	Critically-damped	$v(t) = D_1te^{-\alpha t} + D_2e^{-\alpha t}$; $i_L(t) = I_F + F_1te^{-\alpha t} + F_2e^{-\alpha t}$; $t \geq 0^+$	$s_1, s_2 = -\alpha$
$\alpha < \omega_0$	Underdamped	$v(t) = B_1e^{-\alpha t}\cos(\omega_d t) + B_2e^{-\alpha t}\sin(\omega_d t)$; $i_L(t) = I_F + G_1e^{-\alpha t}\cos(\omega_d t) + G_2e^{-\alpha t}\sin(\omega_d t)$; $t \geq 0^+$	$s_1, s_2 = -\alpha \pm j\omega_d$ $\omega_d = \sqrt{\omega_0^2 - \alpha^2}$

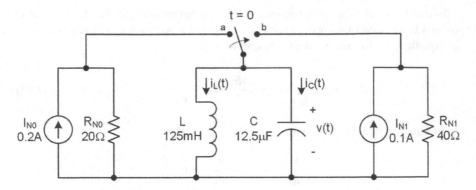

FIGURE 7.16 An overdamped parallel RLC circuit for consideration of the step and natural response.

As an example, consider the circuit in Figure 7.16. The initial value of inductor current is $i_L(0^+) = i_L(0^-) = I_{N0} = 0.2\,\text{A}$, and the final value of inductor current will be $i_L(\infty) = I_{N1} = 0.1\,\text{A}$. The initial voltage will be zero, because the inductor will settle to zero voltage prior to the switch movement and the parallel capacitor prevents the voltage from changing instantaneously. The final voltage will also be zero because the inductor will resettle to zero volts after the switch has been moved.

The neper frequency is

$$\alpha = \frac{1}{2R_{N1}C} = \frac{1}{2(40\,\Omega)(12.5\,\mu\text{F})} = 1{,}000\,\text{rad/s} \tag{7.90}$$

and the resonant frequency is

$$\omega_0 = \frac{1}{\sqrt{(125\,\text{mH})(12.5\,\mu\text{F})}} = 800\,\text{rad/s}. \tag{7.91}$$

Because $\alpha > \omega_0$, the roots of the characteristic equation are real and distinct:

$$s_1, s_2 = -\alpha \pm \sqrt{\alpha^2 - \omega_0^2}$$

$$= \left(-1{,}000 \pm \sqrt{1{,}000^2 - 800^2}\right)\text{rad/s} = -400, -1{,}600\,\text{rad/s}. \tag{7.92}$$

This is an **overdamped** circuit and the solution is of the form:

$$i_L(t) = I_F + E_1 e^{s_2 t} + E_2 e^{s_2 t} = 0.1\,\text{A} + E_1 e^{-400t} + E_2 e^{-1{,}600t};\; t > 0^+. \tag{7.93}$$

To find the coefficients, we need to apply the initial conditions. The numerical value of the initial inductor current is $i_L(0^+) = 0.2\,\text{A}$. From the general form of the solution, evaluated at $t = 0^+$, $i(0^+) = I_F + E_1 + E_2$. Equating these two,

$$E_1 + E_2 = i_L(0^+) - I_F = 0.2\,\text{A} - 0.1\,\text{A} = 0.1\,\text{A}. \tag{7.94}$$

The initial value of the derivative of the current may be found by consideration of the *inductor voltage*, and this can be found by using KVL and the knowledge of the initial capacitor voltage:

$$\frac{di_L}{dt}\bigg|_{t=0^+} = \frac{v_L(0^+)}{L} = \frac{0}{L} = 0. \tag{7.95}$$

From the general form of the solution,

$$\frac{di_L}{dt}\bigg|_{t=0^+} = s_1E_1 + s_2E_2 = -400E_1 - 1,600E_2. \tag{7.96}$$

Equating these two,

$$-400E_1 - 1,600E_2 = 0. \tag{7.97}$$

Now, we can solve for the coefficients using the two equations involving A_1 and A_2:

$$\begin{bmatrix} 1 & 1 \\ -400\,\text{rad/s} & -1,600\,\text{rad/s} \end{bmatrix} \begin{bmatrix} E_1 \\ E_2 \end{bmatrix} = \begin{bmatrix} 0.1\,\text{A} \\ 0 \end{bmatrix}. \tag{7.98}$$

Solving,

$$E_1 = \frac{\begin{vmatrix} 0.1\,\text{A} & 1 \\ 0 & -1,600\,\text{rad/s} \end{vmatrix}}{\begin{vmatrix} 1 & 1 \\ -400\,\text{rad/s} & -1,600\,\text{rad/s} \end{vmatrix}} = 0.1333\,\text{A} \tag{7.99}$$

and

$$E_2 = \frac{\begin{vmatrix} 1 & 0.1\,\text{A} \\ -400\,\text{rad/s} & 0 \end{vmatrix}}{\begin{vmatrix} 1 & 1 \\ -400\,\text{rad/s} & -1,600\,\text{rad/s} \end{vmatrix}} = -0.0333\,\text{A}. \tag{7.100}$$

Therefore,

$$i_L(t) = \left[0.1 + 0.1333e^{-400t} - 0.0333e^{-1,600t}\right]\text{A}; \quad t \geq 0^+. \tag{7.101}$$

The inductor voltage also exhibits an overdamped form:

$$v(t) = L\frac{di_L}{dt} = (125\,\text{mH})\frac{d}{dt}\left[0.1 + 0.1333e^{-400t} - 0.0333e^{-1,600t}\right]\text{A}$$

$$= \left[-6.665e^{-400t} + 6.665e^{-1,600t}\right]\text{V}; \quad t \geq 0^+. \tag{7.102}$$

The inductor current and voltage are plotted in Figure 7.17.

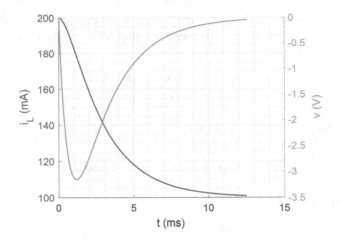

FIGURE 7.17 Inductor current and voltage for the overdamped parallel RLC circuit in Figure 7.16.

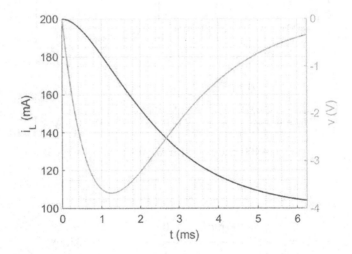

FIGURE 7.18 A critically-damped parallel RLC circuit for consideration of the step and natural response.

As a second example, consider the circuit in Figure 7.18. This is similar to the previous example except that the value of R_{N1} has been modified.

The neper frequency is

$$\alpha = \frac{1}{2R_{N1}C} = \frac{1}{2(50\,\Omega)(12.5\,\mu F)} = 800\,\text{rad/s} \qquad (7.103)$$

and the resonant frequency is

$$\omega_0 = \frac{1}{\sqrt{(125\,\text{mH})(12.5\,\mu F)}} = 800\,\text{rad/s}. \qquad (7.104)$$

Because $\alpha = \omega_0$, the roots of the characteristic equation are real and equal:

$$s_1, s_2 = -\alpha \pm \sqrt{\alpha^2 - \omega_0^2}$$

$$= \left(-800 \pm \sqrt{800^2 - 800^2}\right) \text{rad/s} = -800 - 800 \text{ rad/s.}$$

(7.105)

This is therefore a **critically-damped** circuit and the solution is of the form:

$$i_L(t) = I_F + F_1 t e^{-\alpha t} + F_2 e^{-\alpha t}$$

$$= 0.1 \text{A} + F_1 t e^{-800t} + F_2 e^{-800t}; \quad t \geq 0^+.$$

(7.106)

To find the coefficients, we need to apply the initial conditions. The numerical value of the initial inductor current is $i_L(0^+) = 0.2 \text{ A}$. From the general form of the solution, evaluated at $t = 0^+$, $i(0^+) = I_F + F_2$. Equating these two and solving,

$$F_2 = i_L(0^+) - I_F = 0.2 \text{A} - 0.1 \text{A} = 0.1 \text{A.}$$

(7.107)

The initial value of the derivative of the inductor current may be found by consideration of the *inductor voltage*:

$$\left.\frac{di_L}{dt}\right|_{t=0^+} = \frac{v_L(0^+)}{L} = \frac{0}{L} = 0.$$

(7.108)

From the general form of the solution,

$$\left.\frac{di_L}{dt}\right|_{t=0^+} = F_1 - \alpha F_2 = F_1 - (800 \text{ rad/s})(0.1 \text{A}).$$

(7.109)

Solving,

$$F_1 = 80 \text{ A/s.}$$

(7.110)

Therefore,

$$i_L(t) = \left[0.1 \text{A} + (80 \text{ A/s}) t e^{-800t} + (0.1 \text{A}) e^{-800t}\right]; \quad t \geq 0^+.$$

(7.111)

The inductor voltage also exhibits a critically-damped form:

$$v(t) = L\frac{di}{dt} = (125 \text{ mH})\frac{d}{dt}\left[0.1 \text{A} + (80 \text{ A/s}) t e^{-800t} + (0.1 \text{A}) e^{-800t}\right]$$

$$= \left[(-8,000 \text{ V/s}) t e^{-800t}\right]; \quad t \geq 0^+.$$

(7.112)

The inductor current and voltage are shown in Figure 7.19.

As a third example, consider the circuit in Figure 7.20. Here, the value of R_{N1} has been further modified to render an underdamped circuit. (Notice that in the

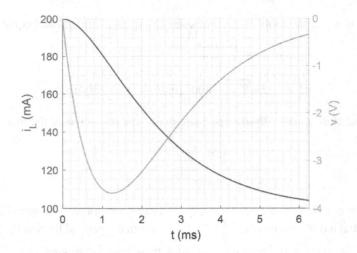

FIGURE 7.19 Inductor current and voltage transients for the critically-damped parallel RLC circuit in Figure 7.18.

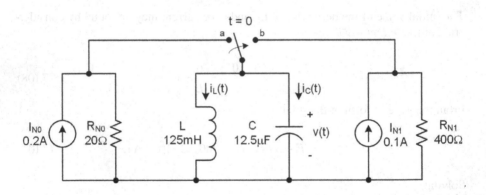

FIGURE 7.20 An underdamped parallel RLC circuit for consideration of the step and natural response.

parallel RLC circuit, an underdamped response corresponds to larger values of R_{N1}. This contrasts with the case of the series RLC circuit.) The initial inductor current will be $I_0 = i_L(0^+) = i_L(0^-) = I_{N0} = 0.2\,\text{A}$, and the final inductor current will be $I_F = i_L(\infty) = I_{N1} = 0.1\,\text{A}$.

The neper frequency is

$$\alpha = \frac{1}{2R_{N1}C} = \frac{1}{2(400\,\Omega)(12.5\,\mu\text{F})} = 100\,\text{rad/s}, \qquad (7.113)$$

and the resonant frequency is

$$\omega_0 = \frac{1}{\sqrt{(125\,\text{mH})(12.5\,\mu\text{F})}} = 800\,\text{rad/s}. \tag{7.114}$$

Because $\alpha < \omega_0$, the roots of the characteristic equation are complex conjugates.

$$s_1, s_2 = -\alpha \pm j\sqrt{\omega_0^2 - \alpha^2} = -\alpha \pm j\omega_d,$$

where

$$\omega_d = \sqrt{800^2 - 100^2}\,\text{rad/s} = 794\,\text{rad/s}. \tag{7.115}$$

This is therefore an **underdamped** circuit and the solution is of the form:

$$i_L(t) = I_F + G_1 e^{-\alpha t}\cos(\omega_d t) + G_2 e^{-\alpha t}\sin(\omega_d t);$$
$$\tag{7.116}$$
$$= 0.1\,\text{A} + G_1 e^{-100t}\cos(794t) + G_2 e^{-100t}\sin(794t); \quad t \geq 0^+.$$

To find the coefficients, we need to apply the initial conditions. The numerical value of the initial inductor current is $i_L(0^+) = 0.2\,\text{A}$. From the general form of the solution, evaluated at $t = 0^+$, $i(0^+) = I_F + G_1$. Equating these two and solving,

$$G_1 = i_L(0^+) - I_F = 0.2\,\text{A} - 0.1\,\text{A} = 0.1\,\text{A}. \tag{7.117}$$

The initial value of the derivative of the inductor current may be found by consideration of the *inductor voltage*:

$$\left.\frac{di_L}{dt}\right|_{t=0^+} = \frac{v_L(0^+)}{L} = \frac{0}{L} = 0. \tag{7.118}$$

From the general form of the solution,

$$\left.\frac{di_L}{dt}\right|_{t=0^+} = -\alpha G_1 + \omega_d G_2 = -(100\,\text{rad/s})(0.1\,\text{A}) + (794\,\text{rad/s})G_2. \tag{7.119}$$

Solving,

$$G_2 = 0.0125\,\text{A}. \tag{7.120}$$

Therefore,

$$i_L(t) = \left[0.1 + 0.1e^{-100t}\cos(794t) + 0.0125e^{-100t}\sin(794t)\right]\text{A}; \quad t \geq 0^+. \tag{7.121}$$

The inductor voltage also exhibits an underdamped form:

$$v(t) = L\frac{di}{dt} = \left(125\,\text{mH}\right)\frac{d}{dt}\left[0.1 + 0.1e^{-100t}\cos\left(794t\right) + 0.0125e^{-100t}\sin\left(794t\right)\right]\text{A}$$

$$= \left[-10.16e^{-100t}\sin\left(794t\right)\right]\text{V}; \quad t \geq 0^{+}.$$

(7.122)

The inductor current and voltage are plotted in Figure 7.21.

To review, see Presentations 7.2, 7.3, and 7.4 in ebook+.

To test your knowledge, try Quizzes 7.2, 7.3, and 7.4 in ebook+. To put your knowledge to practice, try Laboratory Exercise 7.1 in ebook+.

7.9   GENERAL RLC CIRCUIT

Some second-order RLC circuits may be neither parallel nor series in character; that is, the inductor and the capacitor are neither in parallel nor in series. One example is shown in Figure 7.22. Here, the switch has been open for a long time and closes at $t = 0$. In order to solve for i and v for $t \geq 0^{+}$, it is necessary to derive and solve the second-order differential equation.

In order to derive the differential equation, we disable the independent sources and consider the position of the switch for $t \geq 0^{+}$. Here, there is only one independent voltage source to disable, and we disable it by setting it to zero volts or a short. (If there had been an independent current source, we would disable it by setting it to zero amperes or an open circuit.) This results in the circuit of Figure 7.23.

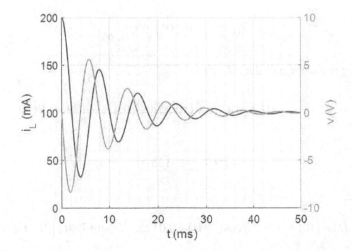

FIGURE 7.21   Inductor current and voltage transients for the underdamped parallel RLC circuit of Figure 7.20.

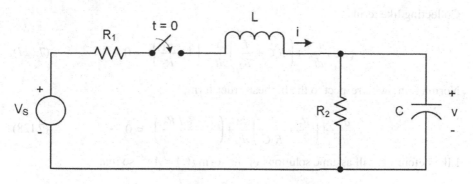

FIGURE 7.22 A general RLC circuit which is neither parallel and series in character.

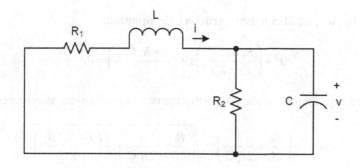

FIGURE 7.23 A general RLC circuit which has been simplified.

By KVL,

$$iR_1 + L\frac{di}{dt} + v = 0. \tag{7.123}$$

By KCL,

$$i = \frac{v}{R_2} + C\frac{dv}{dt}. \tag{7.124}$$

Differentiating the equation for i,

$$\frac{di}{dt} = \frac{1}{R_2}\frac{dv}{dt} + C\frac{d^2v}{dt^2}. \tag{7.125}$$

Substituting equations (7.123) and (7.124) into equation (7.122),

$$R_1\left[\frac{v}{R_2} + C\frac{dv}{dt}\right] + L\left[\frac{1}{R_2}\frac{dv}{dt} + C\frac{d^2v}{dt^2}\right] + v = 0. \tag{7.126}$$

Collecting like terms,

$$(LC)\frac{d^2v}{dt^2} + \left(R_1C + \frac{L}{R_2}\right)\frac{dv}{dt} + \left(1 + \frac{R_1}{R_2}\right)v = 0. \tag{7.127}$$

Normalizing with respect to the highest-order term,

$$\frac{d^2v}{dt^2} + \left(\frac{R_1}{L} + \frac{1}{R_2C}\right)\frac{dv}{dt} + \left(\frac{1 + R_1/R_2}{LC}\right)v = 0. \tag{7.128}$$

Like before we will assume solutions of the form $i(t) = Ae^{st}$, so that

$$s^2Ae^{st} + \left(\frac{R_1}{L} + \frac{1}{R_2C}\right)sAe^{st} + \left(\frac{1 + R_1/R_2}{LC}\right)Ae^{st} = 0. \tag{7.129}$$

Dividing by Ae^{st}, we obtain the **characteristic equation**:

$$s^2 + \left(\frac{R_1}{L} + \frac{1}{R_2C}\right)s + \left(\frac{1 + R_1/R_2}{LC}\right) = 0. \tag{7.130}$$

Applying the quadratic formula, we can find the two roots of the characteristic equation:

$$s_1, s_2 = -\left(\frac{R_1}{2L} + \frac{1}{2R_2C}\right) \pm \sqrt{\left(\frac{R_1}{2L} + \frac{1}{2R_2C}\right)^2 - \left(\frac{1 + R_1/R_2}{LC}\right)}. \tag{7.131}$$

This may be rewritten as

$$s_1, s_2 = -\alpha \pm \sqrt{\alpha^2 - \omega_0^2}, \tag{7.132}$$

where the **neper frequency** is

$$\alpha = \left(\frac{R_1}{2L} + \frac{1}{2R_2C}\right) \tag{7.133}$$

and the resonant frequency is

$$\omega_0 = \sqrt{\left(\frac{1 + R_1/R_2}{LC}\right)}. \tag{7.134}$$

There are three cases of the solution (overdamped, critically-damped, and under-damped) which have similar forms to those found before except that the neper frequency and resonant frequency are calculated differently.

It is interesting to consider two limiting cases of this circuit. First, consider what happens if R_1 approaches zero:

$$\lim_{R_1 \to 0} \alpha = \lim_{R_1 \to 0} \left(\frac{R_1}{2L} + \frac{1}{2R_2C}\right) = \frac{1}{2R_2C} \tag{7.135}$$

and

$$\lim_{R_1 \to 0} \omega_0 = \lim_{R_1 \to 0} \sqrt{\left(\frac{1 + R_1 / R_2}{LC}\right)} = \sqrt{\frac{1}{LC}}. \tag{7.136}$$

This corresponds to a simple parallel RLC circuit as expected for the case in which R_1 has been replaced by a short. On the other hand, if we consider the limiting case in which R_2 approaches infinity,

$$\lim_{R_2 \to \infty} \alpha = \lim_{R_2 \to \infty} \left(\frac{R_1}{2L} + \frac{1}{2R_2 C}\right) = \frac{R_1}{2L} \tag{7.137}$$

and

$$\lim_{R_2 \to \infty} \omega_0 = \lim_{R_2 \to \infty} \sqrt{\left(\frac{1 + R_1 / R_2}{LC}\right)} = \sqrt{\frac{1}{LC}}. \tag{7.138}$$

This corresponds to a simple series RLC circuit as expected for the case in which R_2 has been replaced by an open circuit.

To review, see Presentation 7.5 in ebook+.

To test your knowledge, try Quiz 7.5 in ebook+.

7.10 SUMMARY

In this chapter, we considered the transient response for switched second-order circuits. A second-order circuit contains two energy-storage elements, such as an inductor and a capacitor, two capacitors, or two inductors, and its transient behavior is described by a second-order differential equation. We considered some important cases of second-order circuits here, including the series RLC circuit, the parallel RLC circuit, and the general RLC circuit which may be neither series nor parallel.

First, we analyzed the **natural response of a series RLC circuit** as shown in Figure 7.24.

The initial inductor current is assumed to be I_0 and the initial capacitor voltage is assumed to be V_0. Both final values will be zero. The differential equation describing the behavior for $t \geq 0^+$ is

$$\frac{d^2 i}{dt^2} + \frac{R}{L}\frac{di}{dt} + \frac{1}{LC} i = 0. \tag{7.139}$$

We assume solutions of the form $i(t) = Ae^{st}$, leading to the **characteristic equation**:

$$s^2 + \frac{R}{L} s + \frac{1}{LC} = 0. \tag{7.140}$$

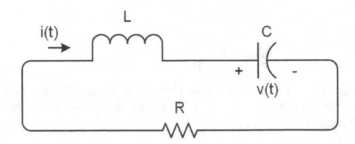

FIGURE 7.24   A simple series RLC circuit without sources.

The roots are

$$s_1, s_2 = -\alpha \pm \sqrt{\alpha^2 - \omega_0^2},\tag{7.141}$$

where $\alpha = R/(2L)$ is the **neper frequency** and $\omega_0 = 1/\sqrt{LC}$ is the **resonant frequency** for the series RLC circuit. There are three cases of the solution for $t \geq 0^+$:

$$i(t) = \begin{cases} A_1 e^{s_1 t} + A_2 e^{s_2 t}, & (\text{overdamped, } \alpha > \omega_0); \\ D_1 t e^{-\alpha t} + D_2 e^{-\alpha t}, & (\text{critically-damped,} \alpha = \omega_0); \\ B_1 e^{-\alpha t} \cos(\omega_d t) + B_2 e^{-\alpha t} \sin(\omega_d t), & (\text{underdamped,} \alpha < \omega_0). \end{cases}$$

$$\tag{7.142}$$

In each case, the two coefficients are found by consideration of the initial conditions, and in the underdamped case the **damped radian frequency** is $\omega_d = \sqrt{\omega_0^2 - \alpha^2}$.

Next, we considered the **step and natural response of the series RLC** circuit using the switched network in Figure 7.25.

In this circuit configuration, the initial capacitor voltage is $V_0 = V_{Th0}$ and the final capacitor voltage is $V_F = V_{Th1}$, but the initial and final inductor current are both zero. When solving for the current, the differential equation and the characteristic equation are the same as in the case of the natural response, so the forms of the solution are the same. The solutions for the capacitor voltage are similar but with an added constant representing the final value. For $t \geq 0^+$:

$$v_C(t) = \begin{cases} V_F + E_1 e^{s_1 t} + E_2 e^{s_2 t}, & (\text{overdamped, } \alpha > \omega_0); \\ V_F + F_1 t e^{-\alpha t} + F_2 e^{-\alpha t} & (\text{critically-damped,} \alpha = \omega_0); \\ V_F + G_1 e^{-\alpha t} \cos(\omega_d t) + G_2 e^{-\alpha t} \sin(\omega_d t), & (\text{underdamped,} \alpha < \omega_0). \end{cases}$$

$$\tag{7.143}$$

In each case, the two coefficients are found by consideration of the initial conditions, and in the underdamped case the **damped radian frequency** is $\omega_d = \sqrt{\omega_0^2 - \alpha^2}$.

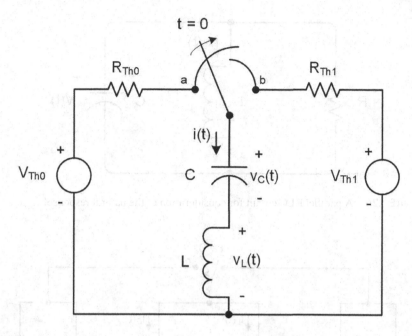

FIGURE 7.25 A switched series RLC circuit for consideration of the step and natural response.

Third, we analyzed the **natural response of a parallel RLC circuit** as shown in Figure 7.26.

The initial inductor current is assumed to be I_0 and the initial capacitor voltage is assumed to be V_0. Both final values will be zero. The differential equation describing the behavior for $t \geq 0^+$ is

$$\frac{d^2v}{dt^2} + \frac{1}{RC}\frac{dv}{dt} + \frac{1}{LC}v = 0. \tag{7.144}$$

We assume solutions of the form $v(t) = Ae^{st}$, leading to the **characteristic equation**:

$$s^2 + \frac{1}{RC}s + \frac{1}{LC} = 0. \tag{7.145}$$

The roots are

$$s_1, s_2 = -\alpha \pm \sqrt{\alpha^2 - \omega_0^2}, \tag{7.146}$$

where $\alpha = 1/(2RC)$ is the **neper frequency** and $\omega_0 = 1/\sqrt{LC}$ is the **resonant frequency** for the parallel RLC circuit. There are three cases of the solution for $t \geq 0^+$:

$$v(t) = \begin{cases} A_1 e^{s_1 t} + A_2 e^{s_2 t}, & \text{(overdamped, } \alpha > \omega_0); \\ D_1 t e^{-\alpha t} + D_2 e^{-\alpha t}, & \text{(critically-damped, } \alpha = \omega_0); \\ B_1 e^{-\alpha t}\cos(\omega_d t) + B_2 e^{-\alpha t}\sin(\omega_d t), & \text{(underdamped, } \alpha < \omega_0). \end{cases} \tag{7.147}$$

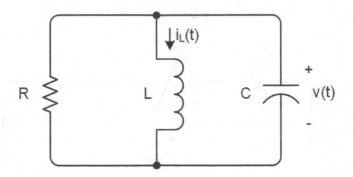

FIGURE 7.26 A parallel RLC circuit for consideration of the natural response.

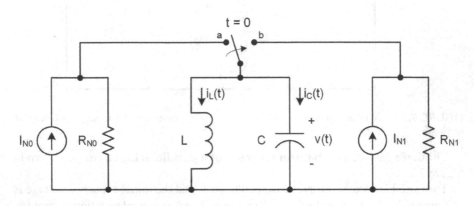

FIGURE 7.27 A parallel RLC circuit for consideration of the step and natural response.

In each case, the two coefficients are found by consideration of the initial conditions, and in the underdamped case the **damped radian frequency** is $\omega_d = \sqrt{\omega_0^2 - \alpha^2}$.

Fourth, we considered the **step and natural response of the parallel RLC** circuit using the switched network in Figure 7.27.

In this circuit configuration, the initial inductor current is $I_0 = I_{N0}$ and the final inductor current is $I_F = I_{N1}$, but the initial and final capacitor voltage are both zero. When solving for the voltage $v(t)$, the differential equation and the characteristic equation are the same as in the case of the natural response, so the forms of the solution are the same. The solutions for the inductor current are similar but with an added constant representing the final value. For $t \geq 0^+$:

$$
i_L(t) = \begin{cases} I_F + E_1 e^{s_1 t} + E_2 e^{s_2 t}, & \text{(overdamped, } \alpha > \omega_0); \\ I_F + F_1 t e^{-\alpha t} + F_2 e^{-\alpha t}, & \text{(critically-damped, } \alpha = \omega_0); \\ I_F + G_1 e^{-\alpha t} \cos(\omega_d t) + G_2 e^{-\alpha t} \sin(\omega_d t), & \text{(underdamped, } \alpha < \omega_0). \end{cases}
$$

$$(7.148)$$

In each case, the two coefficients are found by consideration of the initial conditions, and in the underdamped case the **damped radian frequency** is $\omega_d = \sqrt{\omega_0^2 - \alpha^2}$.

More generally, the initial conditions must be found by consideration of the particular switching configuration, which may be different from those shown in Figures 7.25 and 7.27. Also, some RLC circuits are neither parallel nor series in character, so it becomes necessary to derive and solve the differential equation rather than use one of the predetermined solutions obtained above.

To evaluate your mastery of Chapters 6 and 7, solve Example Exam 7.1 in ebook+ or Example Exam 7.2 in ebook+.

PROBLEMS

Problem 7.1. For the circuit shown in Figure P7.1, find $v_L(0^-)$, $v_L(0^+)$, $v_L(\infty)$, $v_C(0^-)$, $v_C(0^+)$, $v_C(\infty)$, $i(0^-)$, $i(0^+)$, $i(\infty)$, and $di/dt|_{t=0^+}$. Determine $v_L(t)$, $v_C(t)$, and $i(t)$ for $t \geq 0^+$. Calculate the initial and final energy stored in the capacitor. Assume that the switch has been in its starting position for a long time before moving at $t = 0$.

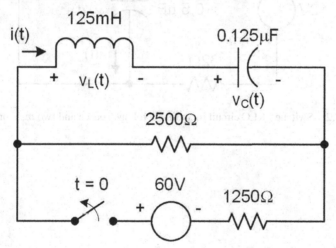

FIGURE P7.1 Switched series RLC circuit.

Problem 7.2. For the circuit shown in Figure P7.2, find $v_L(0^-)$, $v_L(0^+)$, $v_L(\infty)$, $v_C(0^-)$, $v_C(0^+)$, $v_C(\infty)$, $i(0^-)$, $i(0^+)$, $i(\infty)$, and $di/dt|_{t=0^+}$. Determine $v_L(t)$, $v_C(t)$, and $i(t)$ for $t \geq 0^+$. Assume that the switch has been in its starting position for a long time before moving at $t = 0$.

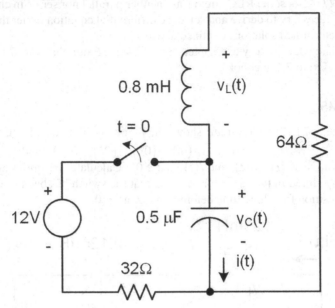

FIGURE P7.2   Switched RLC circuit including a voltage source and two resistors.

Problem 7.3. For the circuit shown in Figure P7.3, determine $v_L(t)$, $v_C(t)$, and $i(t)$ for $t \geq 0^+$. Assume that the switch has been in its starting position for a long time before moving at $t = 0$.

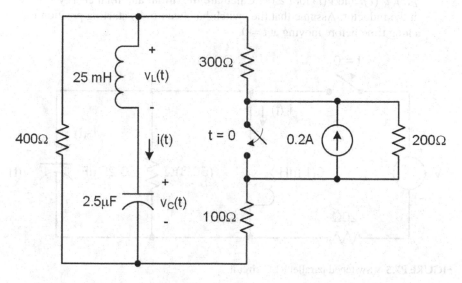

FIGURE P7.3   Switched RLC circuit including a current source and four resistors.

Problem 7.4. For the circuit shown in Figure P7.4, determine $v_L(t)$, $v_C(t)$, and $i(t)$ for $t \geq 0^+$. Assume that the switch has been in its starting position for a long time before moving at $t = 0$.

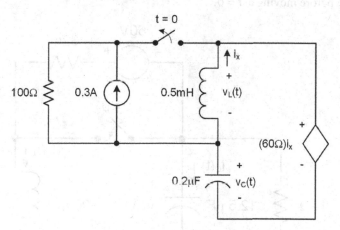

FIGURE P7.4   Switched RLC circuit including mixed sources.

Problem 7.5. For the circuit shown in Figure P7.5, find $i_L(0^-)$, $i_L(0^+)$, $i_L(\infty)$, $i_C(0^-)$, $i_C(0^+)$, $i_C(\infty)$, $v(0^-)$, $v(0^+)$, $v(\infty)$, and $dv/dt|_{t=0^+}$. Determine $i_L(t)$, $i_C(t)$, and $v(t)$ for $t \geq 0^+$. Calculate the initial and final energy stored in the inductor. Assume that the switch has been in its starting position for a long time before moving at $t = 0$.

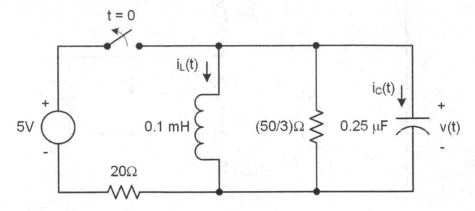

FIGURE P7.5 Switched parallel RLC circuit.

Problem 7.6 For the circuit shown in Figure P7.6, find $i_L(0^-)$, $i_L(0^+)$, $i_L(\infty)$, $i_C(0^-)$, $i_C(0^+)$, $i_C(\infty)$, $v(0^-)$, $v(0^+)$, $v(\infty)$, and $dv/dt|_{t=0^+}$. Determine $i_L(t)$, $i_C(t)$, and $v(t)$ for $t \geq 0^+$. Calculate the initial and final energy stored in the inductor. Assume that the switch has been in its starting position for a long time before moving at $t = 0$.

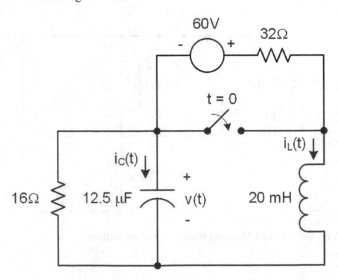

FIGURE P7.6 Switched RLC circuit which includes a voltage source and two resistors.

Problem 7.7. For the circuit shown in Figure P7.7, find $i_L\left(0^-\right)$, $i_L\left(0^+\right)$, $i_L\left(\infty\right)$, $i_C\left(0^-\right)$, $i_C\left(0^+\right)$, $i_C\left(\infty\right)$, $v\left(0^-\right)$, $v\left(0^+\right)$, $v\left(\infty\right)$, and $dv/dt|_{t=0^+}$. Determine $i_L\left(t\right)$, $i_C\left(t\right)$, and $v\left(t\right)$ for $t \geq 0^+$. Calculate the initial and final energy stored in the inductor. Assume that the switch has been in its starting position for a long time before moving at $t = 0$.

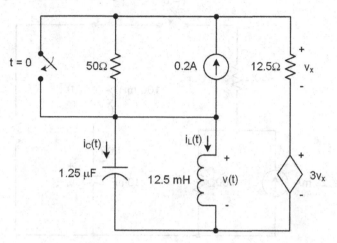

FIGURE P7.7   Switched parallel RLC circuit which includes mixed sources and resistors.

Problem 7.8. For the circuit shown in Figure P7.8, find $i_L\left(t\right)$, $i_C\left(t\right)$, and $v\left(t\right)$ for $t \geq 0^+$. Assume that the switch has been in its starting position for a long time before moving at $t = 0$.

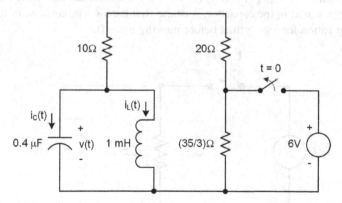

FIGURE P7.8   Switched parallel RLC circuit which includes a voltage source and three resistors.

Problem 7.9. For the circuit shown in Figure P7.9, find $v_L(0^-)$, $v_L(0^+)$, $v_L(\infty)$, $v_C(0^-)$, $v_C(0^+)$, $v_C(\infty)$, $i(0^-)$, $i(0^+)$, $i(\infty)$, and $di/dt|_{t=0^+}$. Determine $v_L(t)$, $v_C(t)$, and $i(t)$ for $t \geq 0^+$. Assume that the switch has been in its starting position for a long time before moving at $t = 0$.

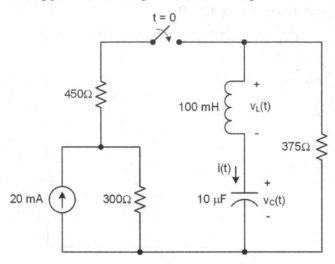

FIGURE P7.9   Switched series RLC circuit which includes a current source and three resistors.

Problem 7.10. For the circuit shown in Figure P7.10, find $v_L(0^-)$, $v_L(0^+)$, $v_L(\infty)$, $v_C(0^-)$, $v_C(0^+)$, $v_C(\infty)$, $i(0^-)$, $i(0^+)$, $i(\infty)$, and $di/dt|_{t=0^+}$. Determine $v_L(t)$, $v_C(t)$, and $i(t)$ for $t \geq 0^+$. Calculate the initial and final energy stored in the capacitor. Assume that the switch has been in its starting position for a long time before moving at $t = 0$.

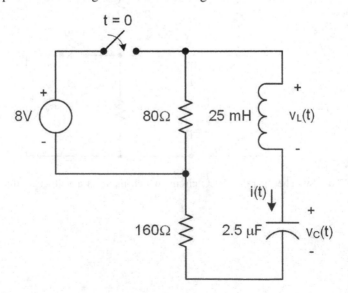

FIGURE P7.10   Switched series RLC circuit which includes a voltage source and a resistor.

Problem 7.11. For the circuit shown in Figure P7.11, determine $i_L(t)$, $i_C(t)$, and $v(t)$ for $t \geq 0^+$. Assume that the switch has been in its starting position for a long time before moving at $t = 0$.

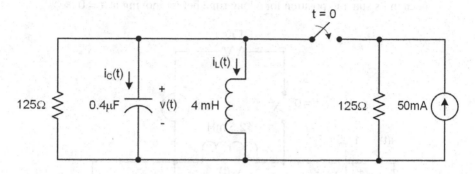

FIGURE P7.11   Switched parallel RLC circuit which includes a current source and two resistors.

Problem 7.12. For the circuit shown in Figure P7.12, find $i_L(t)$, $i_C(t)$, and $v(t)$ for $t \geq 0^+$. Calculate the initial and final energy stored in the inductor and in the capacitor. Assume that the switch has been in its starting position for a long time before moving at $t = 0$.

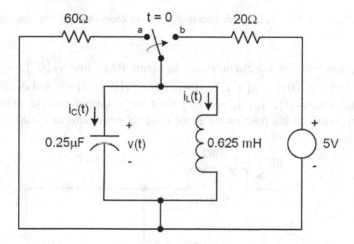

FIGURE P7.12   Parallel RLC circuit involving a single-pole, double-throw switch.

Problem 7.13. For the circuit shown in Figure P7.13, find $v_L(0^-)$, $v_L(0^+)$, $v_L(\infty)$, $v_C(0^-)$, $v_C(0^+)$, $v_C(\infty)$, $i(0^-)$, $i(0^+)$, $i(\infty)$, and $di/dt|_{t=0^+}$. Determine $v_L(t)$, $v_C(t)$, and $i(t)$ for $t \geq 0^+$. Assume that the switch has been in its starting position for a long time before moving at $t = 0$.

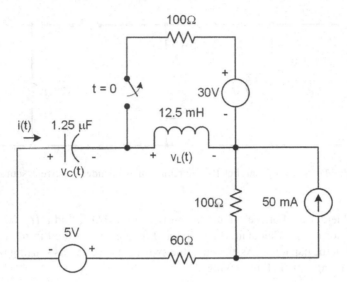

FIGURE P7.13 RLC circuit which becomes series in character after the movement of a switch.

Problem 7.14. For the circuit shown in Figure P7.14, find $v_L(0^-)$, $v_L(0^+)$, $v_L(\infty)$, $v_C(0^-)$, $v_C(0^+)$, $v_C(\infty)$, $i(0^-)$, $i(0^+)$, $i(\infty)$, and $di/dt|_{t=0^+}$. Determine $v_L(t)$, $v_C(t)$, and $i(t)$ for $t \geq 0^+$. Assume that the switch has been in its starting position for a long time before moving at $t = 0$.

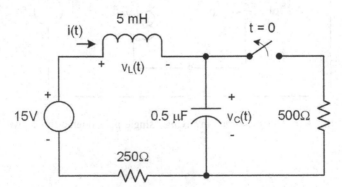

FIGURE P7.14 Switched RLC circuit which changes in character after the movement of the switch.

Problem 7.15. For the circuit shown in Figure P7.15, find $v_L(0^-)$, $v_L(0^+)$, $v_L(\infty)$, $v_C(0^-)$, $v_C(0^+)$, $v_C(\infty)$, $i(0^-)$, $i(0^+)$, $i(\infty)$, and $di/dt|_{t=0^+}$. Determine $v_L(t)$, $v_C(t)$, and $i(t)$ for $t \geq 0^+$. Assume that the switch has been in its starting position for a long time before moving at $t = 0$.

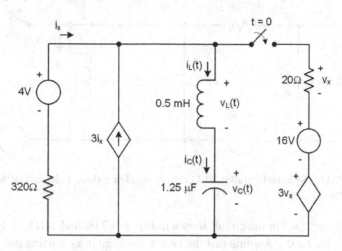

FIGURE P7.15   Switched series RLC circuit involving mixed sources.

Problem 7.16. For the circuit shown in Figure P7.16, find $v_L(t)$, $v_C(t)$, and $i(t)$ for $t \geq 0^+$. Assume that the switch has been in its starting position for a long time before moving at $t = 0$.

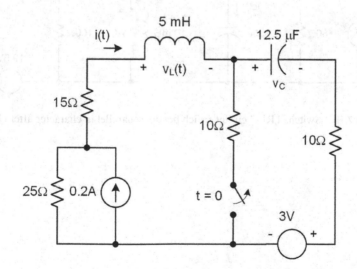

FIGURE P7.16   Switched RLC circuit involving two sources and four resistors, which becomes series in character after the movement of a switch.

Problem 7.17. For the circuit shown in Figure P7.17, find $i_L(t)$, $i_C(t)$, and $v(t)$ for $t \geq 0^+$. Assume that the switch has been in its starting position for a long time before moving at $t = 0$.

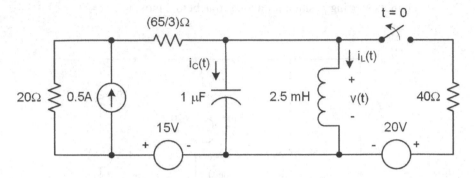

FIGURE P7.17 Switched parallel RLC circuit involving three independent sources and three resistors.

Problem 7.18. For the circuit shown in Figure P7.18, find $i_L(t)$, $i_C(t)$, and $v(t)$ for $t \geq 0^+$. Assume that the switch has been in its starting position for a long time before moving at $t = 0$

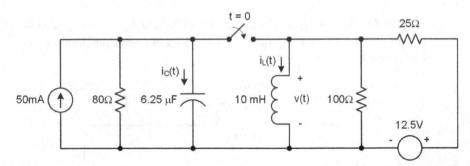

FIGURE P7.18 Switched RLC circuit which becomes parallel in character after closure of a switch.

Problem 7.19. For the circuit shown in Figure P7.19, find $i_L(t)$, $i_C(t)$, and $v(t)$ for $t \geq 0^+$. Assume that the switch has been in its starting position for a long time before moving at $t = 0$.

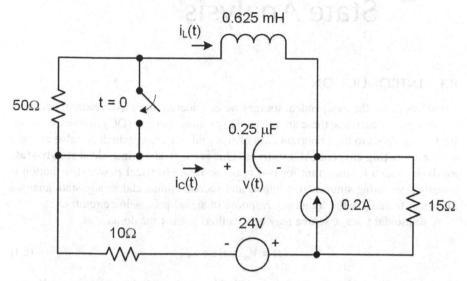

FIGURE P7.19 Switched RLC circuit which changes character after movement of a switch.

Problem 7.20. For the circuit shown in Figure P7.20, find $i_L(t)$, $i_C(t)$, and $v(t)$ for $t \geq 0^+$. Assume that the switch has been in its starting position for a long time before moving at $t = 0$.

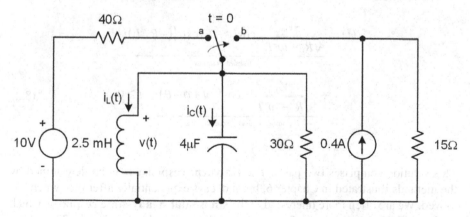

FIGURE P7.20 Switched parallel RLC circuit involving a single-pole, double-throw switch.

8 Sinusoidal Steady-State Analysis

8.1 INTRODUCTION

Up to this point, the independent sources we considered provided constant, unchanging voltages or currents; these are referred to as direct current (DC) sources. Now, we turn our analysis to the important case of sinusoidal sources, which are also referred to as **alternating current (AC) sources**. This is the field of **sinusoidal steady-state analysis**, which is important for two reasons: first, electrical power distribution is usually done using sinusoidal voltages, and second, sinusoidal steady-state analysis allows us to analyze the frequency response of signal-processing circuitry.

A sinusoidal voltage source may be specified in the time domain as

$$v(t) = V_m \cos(\omega t + \phi), \tag{8.1}$$

where V_m is the amplitude of the sinusoid in V, ω is the radial frequency in rad/s, and ϕ is the phase angle in rad. Other quantities of interest are the frequency f in Hz, where $f = \omega / (2\pi)$, and the period T in s, given by $T = 1 / f$.

Consider the switched application of a sinusoidal voltage source to the resistor-inductor (RL) circuit shown in Figure 8.1.

The solution is

$$i(t) = \underbrace{\frac{-V_m}{\sqrt{R^2 + \omega^2 L^2}} \cos(\phi - \theta) \exp\{(-R / L)t\}}_{\text{transient response}}$$

$$+ \underbrace{\frac{V_m}{\sqrt{R^2 + \omega^2 L^2}} \cos(\omega t + \phi - \theta)}_{\text{sinusoidal steady-state response}}; \quad t \ge 0^+. \tag{8.2}$$

This solution comprises two parts. The **transient response** may be determined by the methods illustrated in Chapter 6, but it decays exponentially after the switch has moved. We are often more interested in the **sinusoidal steady-state response**, which is **sinusoidal with the same frequency as the source in the circuit**. This chapter will describe phasor and impedance methods used to determine the sinusoidal steady-state response. We will see that this approach simplifies the solution by allowing us to solve algebraic equations instead of differential equations.

DOI: 10.1201/9781003408529-8

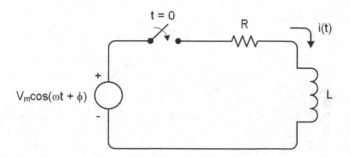

FIGURE 8.1   Switched application of a sinusoidal voltage source to an RL circuit.

8.2   REVIEW OF COMPLEX NUMBERS

In order to solve AC circuits algebraically, without the need for differential equations, we will make extensive use of complex numbers and complex mathematics. It should be understood that this represents a system of shortcuts and that we are using complex quantities to represent real circuit elements, real currents, and real voltages. Nonetheless, working with complex quantities is a critical part of sinusoidal steady-state analysis so we will review the mathematics of complex numbers.

A complex number $A + jB$ involves a real component A and an imaginary component, jB, where j is the imaginary unit,[1] $j = \sqrt{-1}$. This complex number may be shown on the complex plane using Cartesian (rectangular) coordinates or polar coordinates of the form $R\angle\theta$, as shown in Figure 8.2.

Using Euler's relationship,[2] we can convert from rectangular coordinates to polar coordinates[3]:

$$\underbrace{A + jB}_{\text{rectangular coordinates}} \rightarrow \underbrace{\left(A^2 + B^2\right)^{1/2} e^{\tan^{-1}(B/A)} = Re^{j\theta}}_{\text{polar coordinates}}. \tag{8.3}$$

We can also convert from polar coordinates to rectangular coordinates:

$$\underbrace{Re^{j\theta}}_{\text{polar coordinates}} \rightarrow \underbrace{R\cos\theta + jR\sin\theta = A + jB}_{\text{rectangular coordinates}}. \tag{8.4}$$

Often, we will express complex numbers in polar form using the shorthand notation $R\angle\theta$; it should be recognized that this means the same thing as $Re^{j\theta}$.

When adding complex quantities,

$$(A + jB) + (C + jD) = (A + C) + j(B + D) \tag{8.5}$$

and

$$A\angle\alpha + B\angle\beta = (A\cos\alpha + B\cos\beta) + j(A\sin\alpha + B\sin\beta). \tag{8.6}$$

When multiplying complex quantities,

$$(A + jB)(C + jD) = (AC - BD) + j(BC + AD) \tag{8.7}$$

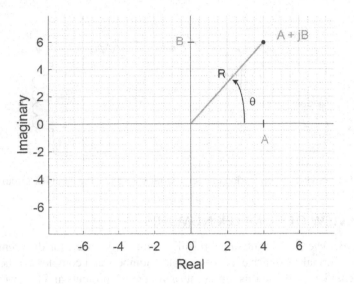

FIGURE 8.2   A complex quantity $A + jB$ shown in the complex plane.

and

$$(A\angle\alpha)(B\angle\beta) = AB\angle(\alpha+\beta). \tag{8.8}$$

When dividing complex quantities,

$$\frac{A+jB}{C+jD} = \frac{(A+jB)}{(C+jD)}\frac{(C-jD)}{(C-jD)} = \frac{(AC+BD)+j(BC-AD)}{C^2+D^2} \tag{8.9}$$

and

$$\frac{A\angle\alpha}{B\angle\beta} = (A/B)\angle(\alpha-\beta). \tag{8.10}$$

Finally, we note that $1/j = -j$, because

$$\frac{1}{j} = \left(\frac{1}{j}\right)\left(\frac{j}{j}\right) = \frac{j}{-1} = -j. \tag{8.11}$$

8.3   PHASORS

Consider a sinusoidal voltage in the time domain given by

$$v(t) = V_m \cos(\omega t + \phi). \tag{8.12}$$

If we imagine a vector of length V_m emanating from the origin of the complex plane[4] with a starting angle of ϕ with respect to the horizontal axis and spinning counterclockwise at an angular velocity ω, then $v(t)$ may be considered to be the projection of this spinning vector on the real axis. If we freeze this spinning vector at $t = 0$, this results in a phasor representation of the sinusoid in the complex plane. For example, consider the following three sinusoidal voltages:

$$v_a(t) = 12\,\text{V}\cos\left[\left(400\,\text{rad/s}\right)t + \pi/6\right],$$

$$v_b(t) = 10\,\text{V}\cos\left[\left(400\,\text{rad/s}\right)t + \pi/4\right], \qquad (8.13)$$

$$v_c(t) = 8\,\text{V}\cos\left[\left(400\,\text{rad/s}\right)t + \pi/3\right].$$

These time-domain voltages are plotted in Figure 8.3.

These same three sinusoids may be represented as complex phasors, which can be shown in the complex plane. The complex phasors, denoted as bold uppercase letters, are

$$V_a = 12\,\text{V}\exp\left(j\pi/6\right) = 12\,\text{V}\left[\cos(\pi/6) + j\sin(\pi/6)\right] = (10.39 + j6.00)\,\text{V},$$

$$V_b = 10\,\text{V}\exp\left(j\pi/4\right) = 10\,\text{V}\left[\cos(\pi/4) + j\sin(\pi/4)\right] = (7.07 + j7.07)\,\text{V}, \quad (8.14)$$

$$V_c = 8\,\text{V}\exp\left(j\pi/3\right) = 8\,\text{V}\left[\cos(\pi/3) + j\sin(\pi/3)\right] = (4.00 + j6.92)\,\text{V}.$$

In shorthand polar notation, we can represent these phasors as

$$V_a = 12\,\text{V}\angle\pi/6,$$

$$V_b = 10\,\text{V}\angle\pi/4, \qquad (8.15)$$

$$V_c = 8\,\text{V}\angle\pi/3.$$

It is important to be familiar with all of these equivalent ways for expressing a phasor, and to be equally comfortable with rectangular forms and polar forms. The three phasors described above may be plotted in the complex plane as shown in Figure 8.4. We can represent sinusoidal currents as phasors in much the same way. However, it must be emphasized that voltages and currents are real, not complex, and the complex phasor is simply a convenient way to represent the amplitude and phase of a real sinusoid!

8.4 IMPEDANCES

When solving for the sinusoidal steady-state response of a circuit, we first transform the circuit from the **time domain** to the **phasor domain** (also referred to as the frequency domain). To do this, we represent sources using complex phasors while passive components (resistors, inductors, and capacitors and their combinations) are represented by complex impedances.

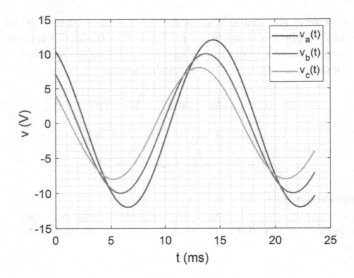

FIGURE 8.3 Three sinusoidal voltages represented in the time domain.

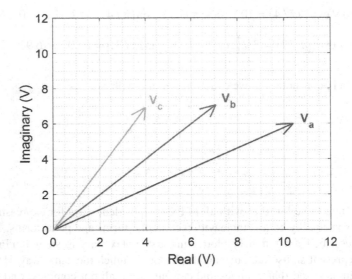

FIGURE 8.4 Three sinusoidal voltages represented as phasors.

8.4.1 IMPEDANCE OF A RESISTOR

First, consider the case of a resistor with an applied sinusoidal voltage as shown in Figure 8.5a.

Suppose the time-domain voltage is given by

$$v(t) = V_m \cos(\omega t + \phi). \tag{8.16}$$

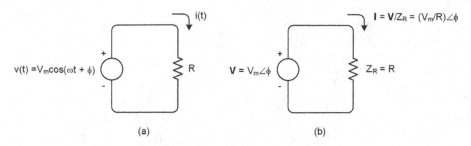

FIGURE 8.5 A resistor with a sinusoidal voltage applied: (a) time domain and (b) phasor domain.

By Ohm's law, the time-domain current is

$$i(t) = v(t) / R = (V_m / R)\cos(\omega t + \phi). \qquad (8.17)$$

As shown in Figure 8.5b, the phasor quantities are

$$V = V_m \angle \phi \qquad (8.18)$$

and

$$I = (V_m / R) \angle \phi. \qquad (8.19)$$

The impedance of the resistor is the ratio of the phasor voltage to the phasor current:

$$Z_R = V / I = [V_m \angle \phi] / [(V_m / R) \angle \phi] = R \angle 0 = R. \qquad (8.20)$$

The impedance of a resistor is equal to its resistance and has zero phase angle; therefore, the current and voltage for a resistor are in phase (at the same phase angle).

Figure 8.6 shows the time-domain voltage and current for a resistor in the case of $v(t) = 10\,\text{V}\cos(400t + \pi / 12)$ and $R = 1.5\,\Omega$. It can be seen that the voltage and current are in phase. Figure 8.7 shows the phasor voltage and current for this case, showing that the two phasors have the same angle or phase.

8.4.2 IMPEDANCE OF AN INDUCTOR

Next, consider an inductor with an applied sinusoidal current as shown in Figure 8.8a.
Suppose the time-domain current is given by

$$i(t) = I_m \cos(\omega t + \phi). \qquad (8.21)$$

The time-domain voltage is

$$v(t) = L\frac{di}{dt} = -\omega L I_m \sin(\omega t + \phi). \qquad (8.22)$$

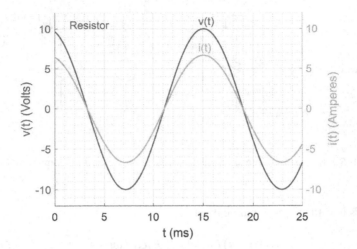

FIGURE 8.6 Example time-domain voltage and current for a resistor. The time-domain voltage is $v(t) = 10V\cos(400t + \pi/12)$ and the resistance is $R = 1.5\Omega$.

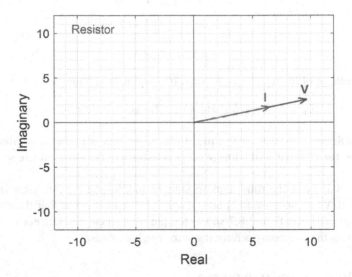

FIGURE 8.7 Example phasor-domain voltage and current for a resistor. The phasor voltage is $V = 10V\angle\pi/12$, and the impedance of the resistor is $Z_R = 1.5\Omega$.

It is customary to describe sinusoidal voltages and currents in terms of cosines, not sines, and the phase angles are always given with respect to the cosine function. In order to express the voltage using a cosine, we use the trigonometric identity $\sin(\alpha) = -\cos(\alpha + \pi/2)$ resulting in

$$v(t) = \omega L I_m \cos(\omega t + \phi + \pi/2). \tag{8.23}$$

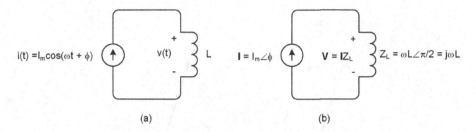

FIGURE 8.8 An inductor with a sinusoidal current applied: (a) time domain and (b) phasor domain.

Thus, the phasor quantities (see Figure 8.8b) are

$$I = I_m \angle \phi \qquad (8.24)$$

and

$$V = \omega L I_m \angle (\phi + \pi / 2). \qquad (8.25)$$

The impedance of the inductor is the ratio of the phasor voltage to the phasor current:

$$Z_L = V / I = [\omega L I_m \angle (\phi + \pi / 2)] / [I_m \angle \phi] = \omega L \angle \pi / 2 = j\omega L. \qquad (8.26)$$

The impedance of an inductor is equal to jωL; it has a phase angle of π/2 so the current lags the voltage by π/2 radians.

Figure 8.9 shows the time-domain voltage and current for an inductor in the case of $v(t) = 10 \text{ V} \cos(400t + \pi / 12)$ and $L = 3.75 \text{ mH}$. It can be seen that the **current lags the voltage by a phase angle of** $\pi / 2$. (In the time domain, the current peaks one-quarter period **after** the voltage.) Figure 8.10 shows the phasor voltage and current for this case. Recognizing that increasing angle represents counterclockwise rotation of a phasor, we see that the current phasor lags the voltage phasor by $\pi / 2$.

8.4.3 IMPEDANCE OF A CAPACITOR

Finally, we can consider a capacitor with an applied sinusoidal voltage as shown in Figure 8.11a.

Suppose the time-domain voltage is given by

$$v(t) = V_m \cos(\omega t + \phi). \qquad (8.27)$$

The time-domain current is

$$i(t) = C \frac{dv}{dt} = -\omega C V_m \sin(\omega t + \phi). \qquad (8.28)$$

In order to express the current using a cosine function, we note that $\sin(\alpha) = -\cos(\alpha + \pi / 2)$ so

$$i(t) = \omega C V_m \cos(\omega t + \phi + \pi / 2). \qquad (8.29)$$

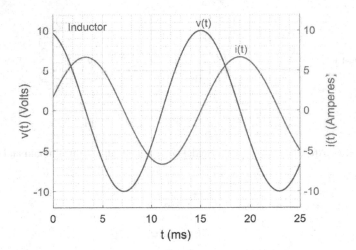

FIGURE 8.9 Example time-domain voltage and current for an inductor. The time-domain voltage is $v(t) = 10Vcos(400t + \pi / 12)$, and the inductance is $L = 3.75\ mH$.

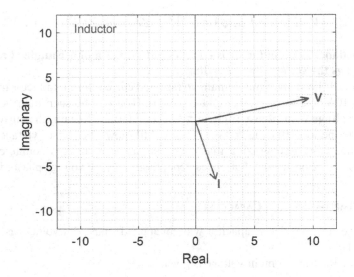

FIGURE 8.10 Example phasor-domain voltage and current for an inductor. The phasor voltage is $V = 10V\angle\pi / 12$, and the impedance of the inductor is $Z_L = j1.5\Omega$.

Thus, the phasor quantities (see Figure 8.11b) are

$$I = \omega C V_m \angle(\phi + \pi / 2) \tag{8.30}$$

and

$$V = V_m \angle\phi. \tag{8.31}$$

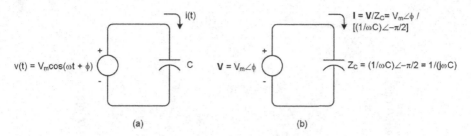

FIGURE 8.11 A capacitor with a sinusoidal voltage applied: (a) time domain and (b) phasor domain.

The impedance of the capacitor is the ratio of the phasor voltage to the phasor current:

$$Z_C = V / I = [V_m \angle \phi] / [\omega C V_m \angle(\phi + \pi / 2)]$$
$$= 1 / (\omega C) \angle - \pi / 2 = 1 / (j\omega C). \tag{8.32}$$

The impedance of a capacitor is equal to 1/(jωC); it has a phase angle of −π/2 so the current leads the voltage by π/2 radians.

Figure 8.12 shows the time-domain voltage and current for a capacitor in the case of $v(t) = 10\,\text{V}\cos(400t + \pi / 12)$ and $C = 5/3\,\text{mF}$. It can be seen that the **current leads the voltage by a phase angle of** $\pi / 2$. (In the time domain, the current peaks one-quarter period **before** the voltage.) Figure 8.13 shows the phasor voltage and current for this case. Recognizing that increasing angle represents counterclockwise rotation of a phasor, we see that the current phasor leads the voltage phasor by $\pi / 2$.

8.4.4 SERIES IMPEDANCES

Kirchhoff's laws and Ohm's law apply in the phasor domain, and we can analyze the case of series impedances in a manner analogous to that used for series resistors. Consider the case of three series impedances shown in Figure 8.14.

By Kirchhoff's voltage law,

$$-V_s + IZ_1 + IZ_2 + IZ_3 = 0. \tag{8.33}$$

Solving for the phasor current,

$$I = \frac{V_s}{Z_1 + Z_2 + Z_3}, \tag{8.34}$$

and the equivalent impedance for the series combination is

$$Z_{eq} = \frac{V_s}{I} = Z_1 + Z_2 + Z_3. \tag{8.35}$$

We can extend this to any number of series impedances, showing that series impedances combine like series resistances, and the equivalent impedance for the series combination is the sum of the individual impedances. Because Kirchhoff's laws and

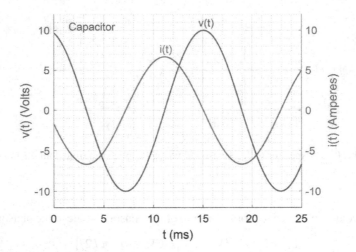

FIGURE 8.12 Example time-domain voltage and current for a capacitor. The time-domain voltage is $v(t) = 10V\cos(400t + \pi/12)$, and the capacitance is $C = 5/3\text{mF}$.

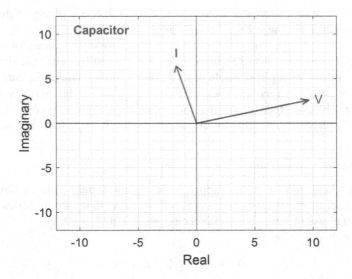

FIGURE 8.13 Example phasor-domain voltage and current for a capacitor. The phasor voltage is $V = 10V\angle\pi/12$, and the impedance of the capacitor is $Z_C = -j1.5\Omega$.

Ohm's law apply in the phasor domain, the same is true of the voltage-divider rule. In the case of three series impedances shown here, the use of the voltage-divider rule yields

$$V_1 = V_s \frac{Z_1}{Z_1 + Z_2 + Z_3}, \tag{8.36}$$

$$V_2 = V_s \frac{Z_2}{Z_1 + Z_2 + Z_3}, \tag{8.37}$$

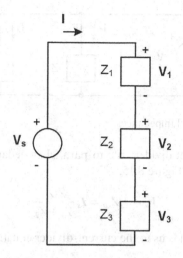

FIGURE 8.14   Three impedances in series.

and

$$V_3 = V_s \frac{Z_3}{Z_1 + Z_2 + Z_3}. \tag{8.38}$$

8.4.5   PARALLEL IMPEDANCES

Another important situation involves the parallel connection of two or more impedances, as shown in Figure 8.15.

Applying Kirchhoff's current law to the top node,

$$-I_s + I_1 + I_2 = 0 \tag{8.39}$$

or

$$I_s = I_1 + I_2. \tag{8.40}$$

By Ohm's law,

$$I_s = \frac{V_s}{Z_1} + \frac{V_s}{Z_2}. \tag{8.41}$$

The equivalent impedance for the parallel combination is

$$Z_{eq} = \frac{V_s}{I_s} = \left(\frac{1}{Z_1} + \frac{1}{Z_2} \right)^{-1}. \tag{8.42}$$

In other words, parallel impedances combine like parallel resistors. **The equivalent
impedance for two parallel impedances is equal to the reciprocal of the sum of
their reciprocals**. In the particular case of two parallel impedances, the equivalent
impedance is equal to the product divided by the sum:

$$Z_{eq} = \frac{Z_1 Z_2}{Z_1 + Z_2}. \tag{8.43}$$

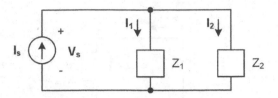

FIGURE 8.15   Two parallel impedances.

We can apply the current-divider rule to parallel impedances in a phasor-domain circuit. For the circuit in Figure 8.15,

$$V_s = I_s Z_{eq} = I_s \frac{Z_1 Z_2}{Z_1 + Z_2}. \tag{8.44}$$

The use of Ohm's law leads us to the current-divider equations:

$$I_1 = I_s \frac{Z_2}{Z_1 + Z_2} \tag{8.45}$$

and

$$I_2 = I_s \frac{Z_1}{Z_1 + Z_2}. \tag{8.46}$$

A statement of the current-divider rule is as follows: when two impedances are connected in parallel, the phasor current in one of the impedances is equal to the total current phasor multiplied by the impedance in the other branch and divided by the sum of the impedances.

We can generalize the current-divider rule to three or more impedances in parallel by combining all of the impedances but one, rendering a system with two parallel impedances.

8.4.6   COMBINATIONS OF SERIES AND PARALLEL IMPEDANCES

We can extend the concepts of the previous two sections to more complex situations involving series and parallel combinations of impedances. Consider the circuit in Figure 8.16, shown both in the time domain (8.16a) and in the phasor domain (8.16b).

Here, C and R_1 are in parallel, R_2 and L are in parallel, and these two combinations are then connected in series. The equivalent impedance is therefore

$$Z_{eq} = \frac{R_1 / j\omega C}{R_1 + 1 / j\omega C} + \frac{j\omega L R_2}{j\omega L + R_2}. \tag{8.47}$$

Both the real and imaginary parts of the equivalent impedance are seen to be strong functions of frequency in Figure 8.17. The real component of the impedance plateaus at $10\,\Omega$ in the middle range of frequencies. Moreover, the imaginary part exhibits a zero crossing at $\sim 50,000\,\text{rad/s}$, so the impedance is real (purely resistive) at this radial frequency.

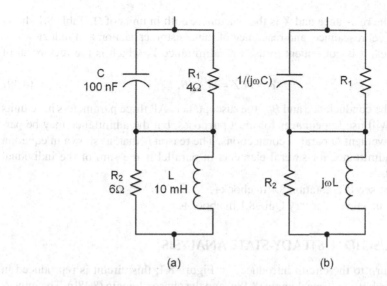

FIGURE 8.16 A series/parallel connection of four impedances: (a) time domain and (b) phasor domain.

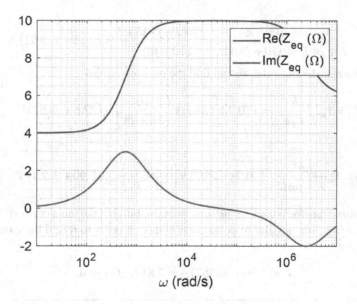

FIGURE 8.17 Real and imaginary parts of the equivalent impedance for the series/parallel connection of Figure 8.16.

8.4.7 IMPEDANCE AND ADMITTANCE

In general, a complex impedance can be expressed as

$$Z = R + jX, \tag{8.48}$$

where R is the resistance and X is the reactance, both in units of Ω. Table 8.1 shows the impedance, resistance, and reactance of the resistor, capacitor, and inductor.

Sometimes, it is convenient to use the admittance Y, which is the reciprocal of impedance:

$$Y = 1/Z = G + jB, \tag{8.49}$$

where G is the conductance and B is the susceptance. All three parameters have units of Ω^{-1}. We will use impedances for most purposes, but the admittance may be particularly convenient in parallel connections. The reason is that, as shown in equation (8.42), the admittance for several elements in parallel is the sum of the individual admittances.

To review, see Presentation 8.1 in ebook+.

To test your knowledge, try Quiz 8.1 in ebook+.

8.5 SINUSOIDAL STEADY-STATE ANALYSIS

We now return to the circuit introduced in Figure 8.1; this circuit is reproduced in Figure 8.18 in both the time domain (8.18a) and the phasor domain (8.18b). The impedance of the inductor in the phasor domain is $Z_L = j\omega L = j(200\,\text{rad/s})(40\,\text{mH}) = j8\,\Omega$.

After conversion to the phasor domain, we can find the phasor current by Ohm's law:

$$I = \frac{V_s}{R + j\omega L} = \frac{100\,\text{V}\angle\pi/4}{(10 + j8)\,\Omega} = \frac{(70.7 + j70.7)\,\text{V}}{(10 + j8)\,\Omega} = (7.76 + j0.862)\,\text{A} \tag{8.50}$$

By the voltage-divider rule, we can find the phasor voltages:

$$V_1 = V_s \frac{R}{R + j\omega L} = (70.7 + j70.7)\,V\left(\frac{10}{10 + j8}\right) = (77.6 + j8.62)\,V \tag{8.51}$$

and

$$V_2 = V_s \frac{j\omega L}{R + j\omega L} = (70.7 + j70.7)\,V\left(\frac{j8}{10 + j8}\right) = (-6.90 + j62.1)\,V. \tag{8.52}$$

We can then transform back to the time domain, but first we must convert the phasor quantities from rectangular to polar coordinates, in order to reveal their magnitudes and phases:

$$I = (7.76 + j0.862)\,A = 7.81\,A\angle 0.11\,\text{rad}, \tag{8.53}$$

TABLE 8.1
Impedance, Resistance, and Reactance for the Passive Components

	Impedance	Resistance	Reactance
Resistor	$R + j0$	R	0
Inductor	$0 + j\omega L$	0	ωL
Capacitor	$0 + 1/(j\omega C)$	0	$-1/(\omega C)$

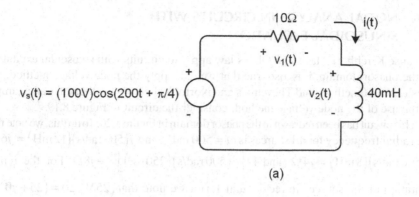

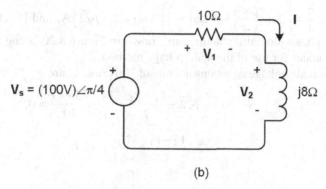

FIGURE 8.18 Example circuit for sinusoidal steady-state analysis: (a) time domain and (b) phasor domain.

$$V_1 = (77.6 + j8.62)\,\text{V} = 78.1\,\text{V}\angle 0.11\,\text{rad}, \tag{8.54}$$

and

$$V_2 = (-6.90 + j62.1)\,\text{V} = 62.5\,\text{V}\angle 1.68\,\text{rad}. \tag{8.55}$$

Now, we can transform back to the time domain:

$$i(t) = 7.81\,\text{A}\cos(200t + 0.11), \tag{8.56}$$

$$v_1(t) = 78.1\,\text{V}\cos(200t + 0.11), \tag{8.57}$$

and

$$v_2(t) = 62.5\,\text{V}\cos(200t + 1.68). \tag{8.58}$$

Thus, by transforming to the phasor domain, we were able to solve algebraically and then transform back to the time domain. If we had solved this circuit directly in the time domain, it would have involved the solution of a differential equation. Whereas we can scale the phasor approach to very complicated circuits by use of the node voltage or mesh current method, a time-domain solution becomes cumbersome for a circuit of even modest complexity.

8.6   NODAL ANALYSIS IN CIRCUITS WITH SINUSOIDAL EXCITATION

Because Kirchhoff's laws and Ohm's law apply to circuits with sinusoidal excitation in the phasor domain, it is also true that we can apply the node voltage method, the mesh current method, and Thevenin's and Norton's theorems to them. As an example of the use of the node voltage method, consider the circuit in Figure 8.19.

This circuit has been redrawn in the phasor domain in Figure 8.20. To do this, we note that the radial frequency for all sources is $\omega = 500\,\text{rad/s}$, and $j(500\,\text{rad/s})(12\,\text{mH}) = j6\,\Omega$, $j(500\,\text{rad/s})(8\,\text{mH}) = j4\,\Omega$, and $1/[j(500\,\text{rad/s})(250\,\mu\text{F})] = -j8\,\Omega$. For the representation of the sources in rectangular form, we note that $(25\,\text{V})\angle 0 = (25 + j0)\,\text{V}$,

$$(2\,\text{A})\angle\frac{\pi}{4} = \left[2\cos\left(\frac{\pi}{4}\right) + j2\sin\left(\frac{\pi}{4}\right)\right]\text{A} = \left(\sqrt{2} + j\sqrt{2}\right)\text{A}, \text{ and } (6\,\text{A})\angle 0 = (6 + j0)\,\text{A}.$$

These phasor-domain quantities are shown in Figure 8.20, along with the labeling of the nodes for use of the node voltage method.

The node voltage equations in units of V, A, and Ω are

$$N1 \quad \sqrt{2} + j\sqrt{2} + \frac{V_1 - V_2}{7} + \frac{V_1 + 25 - V_3}{5 + j10} = 0, \tag{8.59}$$

$$N2 \quad \frac{V_2 - V_1}{7} + \frac{V_2}{-j8} + 6 = 0, \tag{8.60}$$

and

$$N3 \quad \frac{V_3 - 25 - V_1}{5 + j10} - 6 + \frac{V_3}{9} = 0. \tag{8.61}$$

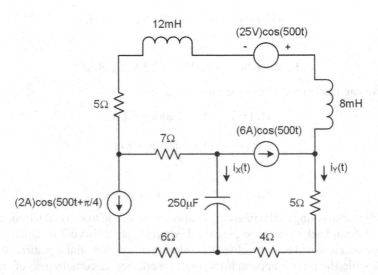

FIGURE 8.19 Example AC circuit for application of the node voltage method, represented in the time domain.

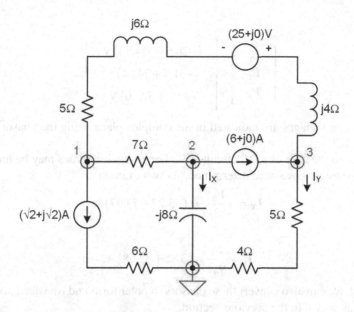

FIGURE 8.20 Example AC circuit for application of the node voltage method, represented in the phasor domain with the nodes labeled.

It will not be helpful to use the least common denominator here; doing so would actually complicate the calculations. Instead, we retain the fractions as they appear when collecting like terms.

$$N1 \quad V_1\left[\frac{1}{7}+\frac{1}{5+j10}\right]+V_2\left[\frac{-1}{7}\right]+V_3\left[\frac{-1}{5+j10}\right]=-\sqrt{2}-j\sqrt{2}-\frac{25}{5+j10}, \quad (8.62)$$

$$N2 \quad V_1\left[\frac{-1}{7}\right]+V_2\left[\frac{1}{7}+\frac{1}{-j8}\right]=-6, \quad (8.63)$$

and

$$N3 \quad V_1\left[\frac{-1}{5+j10}\right]+V_3\left[\frac{1}{5+j10}+\frac{1}{9}\right]=\frac{25}{5+j10}+6. \quad (8.64)$$

In matrix form,

$$\begin{bmatrix} \frac{1}{7}+\frac{1}{5+j10} & \frac{-1}{7} & \frac{-1}{5+j10} \\ \frac{-1}{7} & \frac{1}{7}+\frac{1}{-j8} & 0 \\ \frac{-1}{5+j10} & 0 & \frac{1}{5+j10}+\frac{1}{9} \end{bmatrix} \begin{bmatrix} V_1 \\ V_2 \\ V_3 \end{bmatrix} = \begin{bmatrix} -\sqrt{2}-j\sqrt{2}-\frac{25}{5+j10} \\ -6 \\ \frac{25}{5+j10}+6 \end{bmatrix}.$$

$$(8.65)$$

Solving,

$$\begin{bmatrix} V_1 \\ V_2 \\ V_3 \end{bmatrix} = \begin{bmatrix} (-26.6 + j14.33)\,\text{V} \\ (-31.7 + j42.1)\,\text{V} \\ (34.7 + j23.0)\,\text{V} \end{bmatrix}. \tag{8.66}$$

These phasor voltages are indicated in the complex plane using the phasor diagram in Figure 8.21.

Similar to the case of a DC circuit, all other circuit quantities may be found once the node voltages have been determined. As two examples,

$$I_X = \frac{V_2}{-j8\,\Omega} = (-5.27 - j3.97)\,\text{A} \tag{8.67}$$

and

$$I_Y = \frac{V_3}{9\,\Omega} = (3.85 + j2.58)\,\text{A}. \tag{8.68}$$

If needed, we can also convert these phasors to polar form and transform to the time domain, as shown in the previous section.

We can analyze special cases of the node voltage method in AC circuits in the same manner as in DC circuits. As an example, consider the circuit in Figure 8.22.

This circuit has been redrawn in the phasor domain in Figure 8.23. To do this, we note that the radial frequency of both sources is $\omega = 800\,\text{rad/s}$, and $j(800\,\text{rad/s})(10\,\text{mH}) = j8\,\Omega$, $j(800\,\text{rad/s})(2\,\text{mH}) = j1.6\,\Omega$, and $1/\left[j(800\,\text{rad/s})(100\,\mu\text{F})\right] = -j12.5\,\Omega$. For the representation of the sources in rectangular form, we note that $(20\,\text{V})\angle\pi/2 = (0 + j20)\,\text{V}$ and $(50\,\text{V})\angle 0 = (50 + j0)\,\text{V}$.

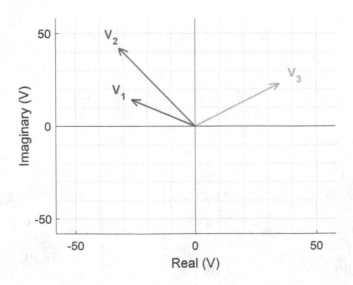

FIGURE 8.21 Phasor diagram showing the node voltages of the circuit in Figure 8.19.

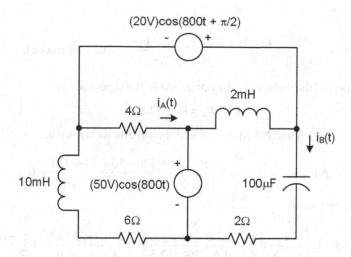

FIGURE 8.22 Example AC circuit for application of the node voltage method, represented in the time domain.

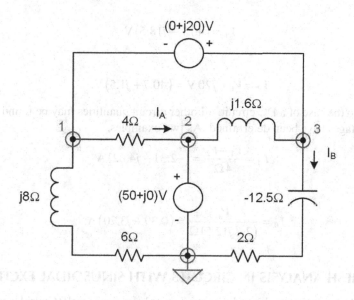

FIGURE 8.23 Example AC circuit for application of the node voltage method, represented in the phasor domain with the nodes labeled.

These phasor-domain quantities are shown in Figure 8.23, along with the labeling of the nodes for use of the node voltage method. It can be seen that there are four essential nodes and two special cases: one known node voltage and one supernode. (Here, this is true for any choice of the reference node!)

The node voltage equations in units of V, A, and Ω are

$$N2 \quad V_2 = 50 + j0; \quad \text{(known node voltage)} \tag{8.69}$$

and

$$N13 \quad \frac{V_1}{6+j8}+\frac{V_1-V_2}{4}+\frac{V_3-V_2}{j1.6}+\frac{V_3}{2-j12.5}=0; \quad \text{(supernode)}. \tag{8.70}$$

The equation of the voltage source contained in the supernode is

$$VS \quad V_3 = V_1 + j20. \tag{8.71}$$

Substituting the VS and N2 equations into the supernode equation,

$$N2N13VS \quad \frac{V_1}{6+j8}+\frac{V_1-50}{4}+\frac{V_1+j20-50}{j1.6}+\frac{V_1+j20}{2-j12.5}=0. \tag{8.72}$$

Collecting like terms,

$$N2N13VS \quad V_1\left[\frac{1}{6+j8}+\frac{1}{4}+\frac{1}{j1.6}+\frac{1}{2-j12.5}\right]=\left[\frac{50}{4}+\frac{-j20+50}{j1.6}+\frac{-j20}{2-j12.5}\right]. \tag{8.73}$$

Solving,

$$V_1 = (40.7 - j18.5)\,\text{V} \tag{8.74}$$

and

$$V_3 = V_1 + j20\,\text{V} = (40.7 + j1.5)\,\text{V}. \tag{8.75}$$

Similar to the case of a DC circuit, all other circuit quantities may be found once the node voltages have been determined. As two examples,

$$I_A = \frac{V_1-V_2}{4\,\Omega} = (-2.31 - j4.62)\,\text{A} \tag{8.76}$$

and

$$I_B = \frac{V_3}{(2-j12.5)\,\Omega} = (0.39 + j3.20)\,\text{A}. \tag{8.77}$$

8.7 MESH ANALYSIS IN CIRCUITS WITH SINUSOIDAL EXCITATION

The mesh current method may be applied to AC circuits as well. Consider the circuit in Figure 8.24, already rendered in the phasor domain, and with the meshes labeled and numbered.

The mesh current equations in units of V, A, and Ω are

$$M13 \quad 5I_1 + j3I_1 - j18I_1 + j4(I_1-I_4)+5(I_3-I_4)$$
$$+4I_3 - j9(I_3-I_2)+7(I_1-I_2)= 0; \quad \text{(supermesh)}; \tag{8.78}$$

$$M2 \quad I_2 = -1 - j2 \quad \text{(known mesh current)}; \tag{8.79}$$

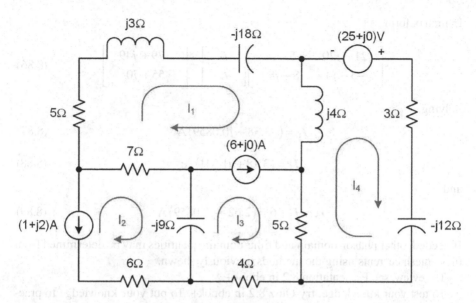

FIGURE 8.24   Example AC circuit for application of the mesh current method, represented in the phasor domain and with the meshes labeled.

and

$$M4 \quad 5(I_4 - I_3) + j4(I_4 - I_1) - 25 + 3I_4 - j12I_4 = 0. \tag{8.80}$$

The equation of the current source contained in the supermesh is

$$CS \quad I_3 - I_1 = 6. \tag{8.81}$$

Substituting the CS and M2 equations into the others,

$$M13M2CS \quad 5I_1 + j3I_1 - j18I_1 + j4(I_1 - I_4) + 5(I_1 + 6 - I_4)$$

$$+ 4(I_1 + 6) - j9\left[(I_1 + 6) - (-1 - j2)\right] + 7\left[I_1 - (-1 - j2)\right] = 0; \tag{8.82}$$

$$M4CS \quad 5\left[I_4 - (I_1 + 6)\right] + j4(I_4 - I_1) - 25 + 3I_4 - j12I_4 = 0. \tag{8.83}$$

Collecting like terms,

$$M13M2CS \quad I_1(21 - j20) + I_4(-5 - j4) = -79 + j49 \tag{8.84}$$

and

$$M4CS \quad I_1(-5 - j4) + I_4(8 - j8) = 55 + j0. \tag{8.85}$$

In matrix form,

$$
\begin{bmatrix} 21-j20 & -5-j4 \\ -5-j4 & 8-j8 \end{bmatrix} \begin{bmatrix} I_1 \\ I_4 \end{bmatrix} = \begin{bmatrix} -79+j49 \\ 55+j0 \end{bmatrix}.
\tag{8.86}
$$

Solving,

$$
I_1 = (-3.38 - j0.0859)\,\text{A};
\tag{8.87}
$$

$$
I_4 = (3.18 + j1.541)\,\text{A};
\tag{8.88}
$$

and

$$
I_3 = I_1 + 6 = (2.62 - j0.0859)\,\text{A}.
\tag{8.89}
$$

If needed, other phasor-domain and time-domain quantities may be determined from these mesh currents using the methods previously shown.

To review, see Presentation 8.2 in ebook+.

To test your knowledge, try Quiz 8.2 in ebook+. To put your knowledge to practice, try Laboratory Exercise 8.1 in ebook+.

8.8   THEVENIN'S AND NORTON'S THEOREMS IN AC CIRCUITS

In this section, we will consider the use of Thevenin's theorem, source transformations, and Norton's theorem in circuits with sinusoidal excitation. It should become clear that many of the considerations are the same as in DC circuits, although it is necessary to use complex quantities; in an AC circuit, the Thevenin voltage is a complex phasor and the Thevenin impedance is generally complex as well.

8.8.1   THEVENIN'S THEOREM

With respect to an AC circuit, Thevenin's theorem states that for any two-terminal network involving impedances and sources, there is a Thevenin equivalent network involving a single voltage source V_{Th} and a single impedance Z_{Th}, as shown in Figure 8.25. Generally, both V_{Th} and Z_{Th} will be complex. The Thevenin circuit is **externally equivalent** to the original circuit, in the sense that if we connect additional circuitry to the two terminals a and b, the voltages and currents in this external circuitry will be the same for the Thevenin equivalent as the original circuit. They are not internally equivalent, however. This means that we cannot use the Thevenin equivalent circuit to directly find voltages and currents inside the original circuit.

If the two-terminal network contains independent sources, we may use the open-circuit and short-circuit conditions to find the Thevenin equivalent circuit. The externally-equivalent networks should exhibit the same open-circuit voltage as shown in Figure 8.26.

Therefore, the Thevenin voltage is equal to the open-circuit voltage for the original circuit.

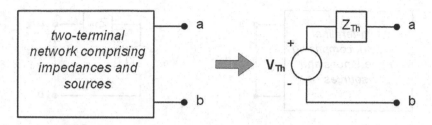

FIGURE 8.25 A network containing impedances and sources along with its Thevenin equivalent.

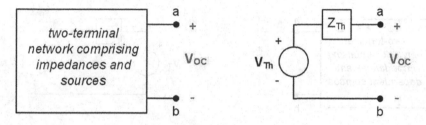

FIGURE 8.26 The open-circuit condition for a network containing impedances and sources along with its Thevenin equivalent.

$$V_{\text{Th}} = V_{\text{OC}}. \tag{8.90}$$

The short-circuit current will also be equal for the two circuits as shown in Figure 8.27, so

$$I_{\text{SC}} = V_{\text{Th}} / Z_{\text{Th}}, \tag{8.91}$$

and the Thevenin impedance may be found from

$$Z_{\text{Th}} = V_{\text{Th}} / I_{\text{SC}} = V_{\text{OC}} / I_{\text{SC}}. \tag{8.92}$$

We will see below that, in a two-terminal network containing only independent sources, we may also find the Thevenin impedance by disabling the sources and determining the equivalent impedance with respect to the terminals a and b. (We will refer to this method as the "impedance shortcut.")

Occasionally, the original circuit may contain no independent sources, either because it contains only impedances or because the sources within it are all dependent sources. If this is the case, both the open-circuit voltage and the short-circuit current will be zero so we may not use their ratio to find the Thevenin resistance. Instead, we can apply a test source as shown in Figure 8.28. We can use either a voltage test source or a current test source, if one is more convenient than the other. In Figure 8.28, we show the application of a voltage test source of value V_{Test}, and the resulting current I_{Test} is to be determined. We could also apply a current test source of value I_{Test}, and then find the resulting voltage V_{Test}. Either way, a simple application of KVL and Ohm's law reveals that

$$Z_{\text{Th}} = (V_{\text{Test}} - V_{\text{Th}}) / I_{\text{Test}}. \tag{8.93}$$

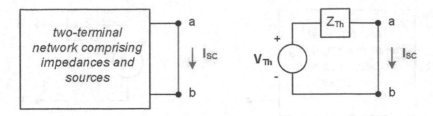

FIGURE 8.27   The short-circuit condition for a network containing impedances and sources along with its Thevenin equivalent.

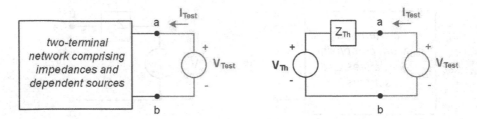

FIGURE 8.28   The application of a test source to a circuit containing resistors and sources as well as its Thevenin equivalent.

In a case for which the Thevenin voltage is zero, this simplifies to

$$Z_{\text{Th}} = V_{\text{Test}} / I_{\text{Test}}. \tag{8.94}$$

As an example, consider the two-terminal circuit in Figure 8.29. It contains only independent sources, and we may use any two of the open-circuit, short-circuit, and impedance shortcut analyses.

For the open-circuit analysis, we will use the labeling of nodes shown in Figure 8.30. Note that although the bottom node is not an essential node for the open-circuit analysis, it will be an essential node for the short-circuit analysis. Hence, we use it as the reference so we can keep the same labeling of nodes for both analyses.

We will start by doing the open-circuit analysis by the node voltage method as shown in Figure 5.6.

In units of V, A, and Ω, the node voltage equations are

$$N1 \quad -4 + \frac{V_1 - V_2}{j2} + \frac{V_1 - V_2}{3 + j4} = 0 \tag{8.95}$$

and

$$N2 \quad \frac{V_2 - V_1}{j2} + \frac{V_2 - V_1}{3 + j4} + \frac{V_2 - \left(20 + j20\sqrt{3}\right)}{3} = 0. \tag{8.96}$$

Collecting like terms,

$$N1 \quad V_1\left[\frac{1}{j2} + \frac{1}{3 + j4}\right] + V_2\left[\frac{-1}{j2} + \frac{-1}{3 + j4}\right] = 4 \tag{8.97}$$

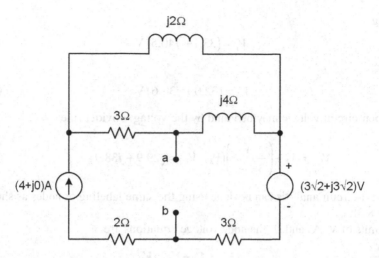

FIGURE 8.29 A two-terminal network containing only independent sources and impedances.

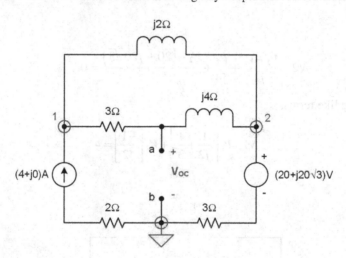

FIGURE 8.30 Open-circuit analysis of a two-terminal network containing only indepen-dent sources and impedances.

and

$$N2 \quad V_1\left[\frac{-1}{j2}+\frac{-1}{3+j4}\right]+V_2\left[\frac{1}{j2}+\frac{1}{3+j4}+\frac{1}{3}\right]=\frac{\left(20+j20\sqrt{3}\right)}{3}. \tag{8.98}$$

In matrix form,

$$\begin{bmatrix} \dfrac{1}{j2}+\dfrac{1}{3+j4} & \dfrac{-1}{j2}+\dfrac{-1}{3+j4} \\[2ex] \dfrac{-1}{j2}+\dfrac{-1}{3+j4} & \dfrac{1}{j2}+\dfrac{1}{3+j4}+\dfrac{1}{3} \end{bmatrix} \begin{bmatrix} V_1 \\[2ex] V_2 \end{bmatrix}=\begin{bmatrix} 4 \\[2ex] \dfrac{\left(20+j20\sqrt{3}\right)}{3} \end{bmatrix}. \tag{8.99}$$

Solving,

$$V_1 = (33.1 + j40.5)\,\text{V} \tag{8.100}$$

and

$$V_2 = (32.0 + j34.6)\,\text{V}. \tag{8.101}$$

The open-circuit voltage may be found by the voltage-divider rule:

$$V_{OC} = V_1 + \left(\frac{3}{3 + j4}\right)(V_2 - V_1) = (29.9 + j38.9)\,\text{V}. \tag{8.102}$$

The short-circuit analysis can be done using the same labeling of nodes as shown in Figure 8.31.

In units of V, A, and Ω, the node voltage equations are

$$N1 \quad -4 + \frac{V_1 - V_2}{j2} + \frac{V_1}{3} = 0 \tag{8.103}$$

and

$$N2 \quad \frac{V_2 - V_1}{j2} + \frac{V_2}{j4} + \frac{V_2 - \left(20 + j20\sqrt{3}\right)}{3} = 0. \tag{8.104}$$

Collecting like terms,

$$N1 \quad V_1\left[\frac{1}{j2} + \frac{1}{3}\right] + V_2\left[\frac{-1}{j2}\right] = 4 \tag{8.105}$$

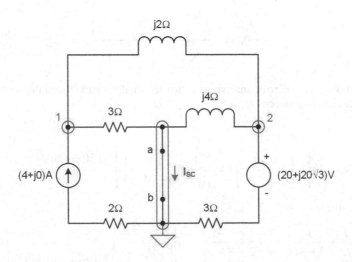

FIGURE 8.31 Short-circuit analysis of a two-terminal network containing only independent sources and impedances.

and

$$N2 \quad V_1\left[\frac{-1}{j2}\right] + V_2\left[\frac{1}{j2} + \frac{1}{j4} + \frac{1}{3}\right] = \frac{\left(20 + j20\sqrt{3}\right)}{3}. \tag{8.106}$$

In matrix form,

$$\begin{bmatrix} \dfrac{1}{j2} + \dfrac{1}{3} & \dfrac{-1}{j2} \\[2mm] \dfrac{-1}{j2} & \dfrac{1}{j2} + \dfrac{1}{j4} + \dfrac{1}{3} \end{bmatrix} \begin{bmatrix} V_1 \\[2mm] V_2 \end{bmatrix} = \begin{bmatrix} 4 \\[2mm] \dfrac{\left(20 + j20\sqrt{3}\right)}{3} \end{bmatrix}. \tag{8.107}$$

Solving,

$$V_1 = \left(14.62 + j17.54\right) \text{V} \tag{8.108}$$

and

$$V_2 = \left(2.92 + j19.29\right) \text{V}. \tag{8.109}$$

The short-circuit current may be found using KCL:

$$I_{\text{SC}} = \frac{V_1}{3\,\Omega} + \frac{V_2}{j4\,\Omega} = \left(9.69 + j5.12\right) \text{A}. \tag{8.110}$$

Now, we can determine the Thevenin equivalent circuit:

$$V_{\text{Th}} = V_{\text{OC}} = \left(29.9 + j38.9\right) \text{V} \tag{8.111}$$

and

$$Z_{\text{Th}} = \frac{V_{\text{OC}}}{I_{\text{SC}}} = \left(4.07 + j1.867\right)\Omega. \tag{8.112}$$

It is important to realize that the impedances for the reactive components ($j2\,\Omega$ and $j4\,\Omega$) have been determined for a particular frequency; therefore, this Thevenin equivalent circuit is only applicable at that frequency. In general, the impedances of inductors and capacitors are frequency dependent so the Thevenin equivalent circuit is frequency dependent.

Because the circuit contains no dependent sources, we may also use the impedance shortcut analysis. To apply this, we disable the independent sources and find the resulting equivalent impedance with respect to the terminals a and b. To disable the sources, we replace the current source with an open and replace the voltage source with a short, as shown in Figure 8.32.

The Thevenin impedance is the equivalent impedance "seen" looking into the terminals a and b; it is

$$Z_{\text{Th}} = 3\,\Omega + \left(3 + j2\right)\Omega \,\|\, j4\,\Omega = \left(4.07 + j1.867\right)\Omega. \tag{8.113}$$

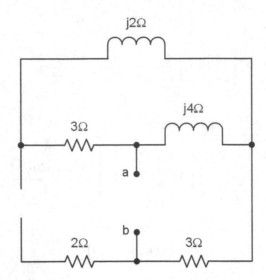

FIGURE 8.32 Impedance shortcut analysis of a two-terminal network containing only independent sources and impedances.

It should be noted that the impedance shortcut may not be used for a circuit containing mixed (dependent and independent) sources; this is because a dependent source contributes to the Thevenin impedance. Fortunately, the open-circuit and short-circuit analyses may be used with mixed sources.

As a second example, consider the circuit in Figure 8.33, which contains a dependent source but no independent sources.

The Thevenin voltage is necessarily zero, but to find the Thevenin impedance we can apply a test source as shown in Figure 8.34. (The same approach is applicable with two or more dependent sources, although the analysis may be more complicated.)

In units of V, A, and Ω, the node voltage equation is

$$N1 \quad \frac{V_1 - V_{\text{Test}}}{3 + j1} + \frac{V_1}{4 - j6} + \frac{V_1 - 3v_x}{3} = 0, \tag{8.114}$$

and the equation of the dependent source is

$$DS \quad v_x = V_1 \left(\frac{4}{4 - j6} \right). \tag{8.115}$$

Combining equations,

$$N1DS \quad \frac{V_1 - V_{\text{Test}}}{3 + j1} + \frac{V_1}{4 - j6} + \frac{V_1}{3} - V_1 \left(\frac{4}{4 - j6} \right) = 0. \tag{8.116}$$

Collecting like terms,

$$N1DS \quad V_1 \left[\frac{1}{3 + j1} + \frac{1}{4 - j6} + \frac{1}{3} - \frac{4}{4 - j6} \right] = \frac{V_{\text{Test}}}{3 + j1}. \tag{8.117}$$

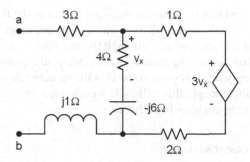

FIGURE 8.33 Two-terminal circuit containing only a dependent source and impedances.

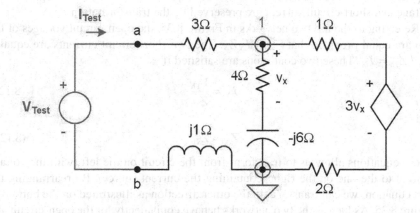

FIGURE 8.34 Test-source analysis of a two-terminal network containing only independent sources and impedances.

Solving,

$$N1DS \quad V_1 = \frac{V_{Test}}{3+j1}\left[\frac{1}{3+j1} - \frac{3}{4-j6} + \frac{1}{3}\right]^{-1}. \tag{8.118}$$

Now, we can find the test current:

$$I_{Test} = \frac{V_{Test} - V_1}{3+j1} = \frac{V_{Test}}{3+j1}\left\{1 - \frac{1}{3+j1}\left[\frac{1}{3+j1} - \frac{3}{4-j6} + \frac{1}{3}\right]^{-1}\right\}. \tag{8.119}$$

The Thevenin impedance is therefore

$$Z_{Th} = \frac{V_{Test}}{I_{Test}} = (3+j1)\left\{1 - \frac{1}{3+j1}\left[\frac{1}{3+j1} - \frac{3}{4-j6} + \frac{1}{3}\right]^{-1}\right\}^{-1} \tag{8.120}$$

$$= (3.79 + j3.66)\,\Omega.$$

It should be recognized that the value of V_{Test} cancels out in the final analysis; therefore, we may leave its value unspecified (as was done here), or we may choose a specific value in volts if it is convenient. Although the test-source analysis is not directly applicable to a circuit containing mixed sources, we could disable the independent sources and then apply a test source to the remaining network to find its Thevenin impedance. Generally, though, this will be less convenient than the use of the open-circuit and short-circuit analyses in such a case.

8.8.2 SOURCE TRANSFORMATIONS

A two-terminal network comprising a voltage source and a series impedance may be transformed to a two-terminal network comprising a current source and a parallel impedance, and these two will be externally equivalent as long as the open-circuit voltage and short-circuit current are preserved by the transformation.

Referring to the top two networks in Figure 8.35, the open-circuit voltages of the two are equal provided that $V_{Th} = I_N Z_N$. The two short-circuit currents are equal if $V_{Th} / Z_{Th} = I_N$. These two conditions are satisfied if

$$I_N = \frac{V_{Th}}{Z_{Th}} \tag{8.121}$$

and

$$Z_N = Z_{Th}. \tag{8.122}$$

These equations allow us to transform from the circuit on the left (with the voltage source) to the one on the right (containing the current source). By rearranging the first equation, we can transform in the other direction as illustrated on the bottom of Figure 8.35. As long as the two networks behave equivalently for the open-circuit and short-circuit conditions, they will behave equivalently when connected to a network of sources and impedances. The two are externally equivalent; that is, all external voltages and currents will be unchanged by transforming from one to the other.

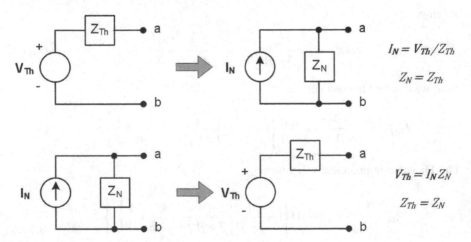

FIGURE 8.35 Source transformations.

8.8.3 Norton's Theorem

With respect to an AC circuit, Norton's theorem states that for any two-terminal network involving impedances and sources, there is a Norton equivalent network involving a single current source I_N and a parallel impedance Z_N, as shown in Figure 8.36. This follows directly from the use of Thevenin's theorem and a source transformation. Therefore, the methods employed to determine the Norton equivalent are the same as those used to find the Thevenin equivalent.

To review, see Presentation 8.3 in ebook+. To test your knowledge, try Quiz 8.3 in ebook+.

8.9 SINUSOIDAL STEADY-STATE POWER

8.9.1 Instantaneous Power, Average Power, and Reactive Power

For an element in a circuit with sinusoidal excitation, such as that shown in Figure 8.37, the instantaneous power is $p(t) = i(t)v(t)$. (The passive sign convention applies the same way as in DC circuits.)

If $v(t) = V_m \cos(\omega t + \phi_v)$ and $i(t) = I_m \cos(\omega t + \phi_i)$, then

$$p(t) = V_m I_m \cos(\omega t + \phi_v)\cos(\omega t + \phi_i). \tag{8.123}$$

If we shift the time reference by $-\phi_i / \omega$, then

$$p(t) = V_m I_m \cos(\omega t + \phi_v - \phi_i)\cos(\omega t). \tag{8.124}$$

We can use the trigonometric identity for the product of two cosines: $\cos\alpha\cos\beta = \dfrac{1}{2}\cos(\alpha - \beta) - \dfrac{1}{2}\cos(\alpha + \beta)$. Let $\alpha = \omega t + \phi_v - \phi_i$ and $\beta = \omega t$, yielding

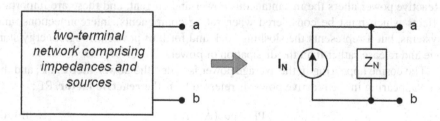

FIGURE 8.36 A network containing impedances and sources along with its Norton equivalent.

FIGURE 8.37 An element in a circuit with sinusoidal excitation.

$$p(t) = \frac{V_m I_m}{2}\cos(\phi_v - \phi_i) + \frac{V_m I_m}{2}\cos(2\omega t + \phi_v - \phi_i). \tag{8.125}$$

Next, we will use the identity $\cos(\alpha + \beta) = \cos\alpha\cos\beta - \sin\alpha\sin\beta$; let $\alpha = \phi_v - \phi_i$ and $\beta = 2\omega t$, yielding

$$p(t) = \frac{V_m I_m}{2}\cos(\phi_v - \phi_i) + \frac{V_m I_m}{2}\cos(\phi_v - \phi_i)\cos(2\omega t)$$
$$- \frac{V_m I_m}{2}\sin(\phi_v - \phi_i)\sin(2\omega t). \tag{8.126}$$

This may be rewritten in the simpler form

$$p(t) = P + P\cos(2\omega t) - Q\sin(2\omega t), \tag{8.127}$$

where

$$P = \frac{V_m I_m}{2}\cos(\phi_v - \phi_i) \tag{8.128}$$

and

$$Q = \frac{V_m I_m}{2}\sin(\phi_v - \phi_i). \tag{8.129}$$

If we average the instantaneous power in (8.127) over any integral number of periods, we obtain P; therefore, P represents the average power for the element under consideration. Because we are using the passive sign convention, a positive value represents power dissipated whereas a negative value shows power developed. The term involving Q averages to zero for any integral number of periods. We therefore refer to Q as reactive power, because it is associated with reactive components of impedance and it results from the component of current which is out of phase with the voltage. Reactive power alters the instantaneous power and current, and these are important effects which must be considered when rating components, interconnections, and systems, but it represents the sloshing back and forth of power through energy storage and release rather than the dissipation of power.

The cosine appearing in the average power is called the **power factor**, PF, and the sine appearing in the reactive power is referred to as the **reactive factor**, RF:

$$\text{PF} = \cos(\phi_v - \phi_i) \tag{8.130}$$

and

$$\text{RF} = \sin(\phi_v - \phi_i). \tag{8.131}$$

Because the cosine is an even function, the PF does not explicitly contain information about the sign of $(\phi_v - \phi_i)$. However, it is customary to specify whether the PF is **leading** or **lagging**. In the case of a lagging PF, the current lags the voltage and $(\phi_v - \phi_i) > 0$. This represents a load with net inductive character. In the case of a leading PF, the current leads the voltage and $(\phi_v - \phi_i) < 0$. This represents the case of a net capacitive load.

For efficient power delivery, it is desirable to maximize the PF (to unity if possible) and minimize the RF (to zero if possible). In a large plant with many machines and motors, the PF will tend to be lagging but may be corrected by banks of capacitors. In a plant with many electronically-controlled furnaces, the PF may be leading and could be corrected by installing an inductor bank.

8.9.2 AVERAGE POWER AND ROOT MEAN SQUARE (rms) VALUES OF VOLTAGE OR CURRENT

If we apply a periodic voltage waveform v to a resistor, the average power P can be found by integrating the instantaneous power $p = v^2 / R$ over a period T of the voltage waveform and dividing by the period:

$$P = \frac{1}{T}\int_0^T p\,dt = \frac{1}{T}\int_0^T \frac{v^2}{R}\,dt = \frac{\frac{1}{T}\int_0^T v^2 dt}{R} = \frac{V_{\text{rms}}^2}{R}, \tag{8.132}$$

where V_{rms} is the root mean square (rms) voltage, also referred to as the effective voltage, and is given by

$$V_{\text{rms}} = \sqrt{\frac{1}{T}\int_0^T v^2 dt}. \tag{8.133}$$

To find the root mean square value, we can remember "rms" and do the operations in reverse order: we **square** the voltage, we take the **mean**, and then we take the **root**.

Now, consider the case of a sinusoidal waveform (which is arguably the most important to us). Suppose the time-domain voltage is given by $v(t) = V_m \cos(\omega t + \phi_v)$. We may integrate over any time interval equal to the period:

$$V_{\text{rms}} = \sqrt{\frac{1}{T}\int_{t_0}^{t_0+T} V_m^2 \cos^2(\omega t + \phi_v)\,dt}. \tag{8.134}$$

If we choose t_0 to offset the phase difference and shift the time reference,

$$V_{\text{rms}} = \sqrt{\frac{1}{T}\int_0^T V_m^2 \cos^2(\omega t)\,dt} = V_m\sqrt{\frac{1}{T}\left[\frac{t}{2} + \frac{1}{4\omega}\sin(2\omega t)\right]_0^T}. \tag{8.135}$$

Because $T = 2\pi / \omega$,

$$V_{\text{rms}} = V_m\sqrt{\frac{1}{T}} = V_m\sqrt{\frac{1}{T}\left[\frac{T}{2} - 0 + \frac{1}{4\omega}\sin\left(\frac{4\pi\omega}{\omega}\right) - \frac{1}{4\omega}\sin(0)\right]} = \frac{V_m}{\sqrt{2}}. \tag{8.136}$$

Therefore, **the root mean square value of a sinusoidal voltage is equal to the amplitude divided by the square root of two**. A similar conclusion holds for a sinusoidal current waveform.

8.9.3 COMPLEX POWER S

The average power in an element with sinusoidal excitation is

$$P = \frac{V_m I_m}{2}\cos(\phi_v - \phi_i) = \frac{V_m}{\sqrt{2}}\frac{I_m}{\sqrt{2}}\cos(\phi_v - \phi_i)$$

$$= V_{\mathrm{rms}} I_{\mathrm{rms}} \cos(\phi_v - \phi_i),$$

(8.137)

and the reactive power is

$$Q = \frac{V_m I_m}{2}\sin(\phi_v - \phi_i) = \frac{V_m}{\sqrt{2}}\frac{I_m}{\sqrt{2}}\sin(\phi_v - \phi_i)$$

$$= V_{\mathrm{rms}} I_{\mathrm{rms}} \sin(\phi_v - \phi_i).$$

(8.138)

We can conveniently express both using a single complex quantity which we refer to as the complex power S:

$$S = P + jQ = V_{\mathrm{rms}} I_{\mathrm{rms}} \cos(\phi_v - \phi_i) + j V_{\mathrm{rms}} I_{\mathrm{rms}} \sin(\phi_v - \phi_i)$$

$$= V_{\mathrm{rms}} I_{\mathrm{rms}} \exp(\phi_v - \phi_i) = V_{\mathrm{rms}} \exp(\phi_v) I_{\mathrm{rms}} \exp(-\phi_i)$$

(8.139)

$$= V_{\mathrm{rms}} I_{\mathrm{rms}}^*.$$

Therefore, **the complex power for a two-terminal element may be determined by multiplying the phasor voltage by the conjugate of the phasor current.** It should be recognized that the passive sign convention applies, so if either reference polarity from Figure 8.37 were reversed it would be necessary to include a minus sign in the complex power equation.

For an impedance, we can utilize Ohm's law to develop another expression for the complex power:

$$S = V_{\mathrm{rms}} I_{\mathrm{rms}}^* = (Z I_{\mathrm{rms}}) I_{\mathrm{rms}}^* = Z|I_{\mathrm{rms}}|^2 = (R + jX)|I_{\mathrm{rms}}|^2.$$

(8.140)

Hence,

$$P = R|I_{\mathrm{rms}}|^2$$

(8.141)

and

$$Q = X|I_{\mathrm{rms}}|^2.$$

(8.142)

If we display the complex power on the complex plane, the resulting right triangle formed by dropping a vertical to the horizontal is referred to as the power triangle. An example is shown in Figure 8.38. The length of the hypotenuse, $|S| = \sqrt{P^2 + Q^2}$,

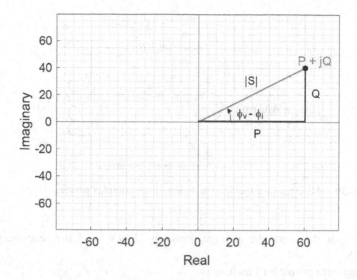

FIGURE 8.38 Complex power and power triangle for a load.

is referred to as the apparent power and is given by the product of the rms voltage amplitude and rms current amplitude: $|S| = V_{rms}I_{rms}$. The other two sides of the triangle have lengths equal to $|P| = V_{rms}I_{rms}\cos(\phi_v - \phi_i)$ and $|Q| = V_{rms}I_{rms}\sin(\phi_v - \phi_i)$. In the example shown, both P and Q are positive, representing a load with a lagging PF, and the power triangle is in the first quadrant. For a load with a leading PF, the power triangle would be in the fourth quadrant. The power triangle for a source developing power would be in the third or fourth quadrant, depending on whether its load was leading or lagging, respectively. The angle $\phi_v - \phi_i$ is referred to as the power angle.

The apparent power, average power, and reactive power all have units of VA. However, it is customary to distinguish between them by using VA for apparent power, W for average power, and VARS (volt-amperes reactive) for the reactive power.

It should be clear now that all complex power calculations involve *rms* values, and we will not always explicitly use an "rms" subscript, although we will use the notations V (rms) and A (rms) for units. Other notations are used as well, so it is important to always be aware of whether *rms* or amplitude values are being used.

As an example of complex power calculation, consider the circuit in Figure 8.39, rendered in the phasor domain.

The phasor current is

$$I = \frac{V}{Z} = \frac{120\angle\pi/4\,\text{V (rms)}}{\left(4\sqrt{3} + j4\right)\Omega} = \frac{120\angle\pi/4\,\text{V (rms)}}{8\angle\pi/6\,\Omega} = 15\angle\pi/12\,\text{A (rms)}. \quad (8.143)$$

The complex power for the $\left(4\sqrt{3} + j4\right)\Omega$ load is

$$S_L = VI^* = \left[120\angle\pi/4\,\text{V (rms)}\right]\left[15\angle-\pi/12\,\text{A (rms)}\right]$$

$$= (1559 + j900)\,\text{VA}. \quad (8.144)$$

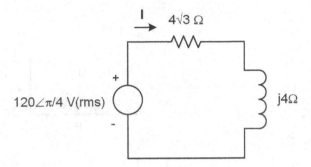

FIGURE 8.39 AC circuit for the determination of the complex power.

Therefore, for the load, the average power is 1559 W and the reactive power is 900 VARS.

The apparent power for this load is

$$|\boldsymbol{S}_L| = VI = \left(120\,\text{V}\,(\text{rms})\right)\left(15\,\text{A}\,(\text{rms})\right) = 1800\,\text{VA}. \tag{8.145}$$

This is also the length of the hypotenuse for the power triangle:

$$|\boldsymbol{S}_L| = \sqrt{P^2 + Q^2} = \sqrt{\left(1559\,\text{VA}\right)^2 + \left(900\,\text{VA}\right)^2} = 1800\,\text{VA}. \tag{8.146}$$

The PF for the load is

$$PF = \cos\left(\phi_v - \phi_i\right) = \cos(\pi/4 - \pi/6) = 0.866 \ \text{(lagging)}; \tag{8.147}$$

and the RF for the load

$$RF = \sin\left(\phi_v - \phi_i\right) = \sin(\pi/4 - \pi/6) = 0.5. \tag{8.148}$$

The apparent power for the source, using the passive sign convention, is

$$\boldsymbol{S}_S = -VI^* = -\left[120\angle\pi/4\,\text{V}\,(\text{rms})\right]\left[15\angle-\pi/12\,\text{A}\,(\text{rms})\right]$$

$$= \left(-1559 - j900\right)\text{VA}. \tag{8.149}$$

Figure 8.40 shows the power triangles for the load and source in this example.

To review, see Presentation 8.4 in ebook+.

To test your knowledge, try Quiz 8.4 in ebook+.

8.10 MAXIMUM POWER TRANSFER IN CIRCUITS WITH SINUSOIDAL EXCITATION

As in the DC case, an important application of Thevenin's theorem (or Norton's theorem) is in solving maximum power transfer problems. Suppose a load impedance Z_L is connected to a two-terminal network involving sources and impedances as shown

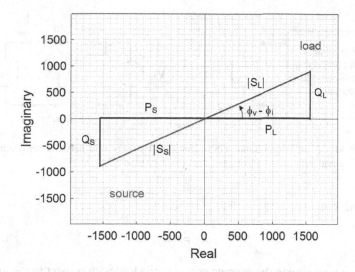

FIGURE 8.40   Power triangles for the load and source of the example circuit

on the left-hand side of Figure 8.41. We can represent the two-terminal network by its Thevenin equivalent as shown on the right of this figure.

We want to know the maximum average power which can be delivered to the load, and what value of load impedance will give rise to this maximum power transfer. This problem can be easily solved using the Thevenin representation for the original circuit.

In order to solve this problem, we recognize that $Z_{Th} = R_{Th} + jX_{Th}$ and $Z_L = R_L + jX_L$. In the following, it is assumed that the Thevenin voltage is an rms value, which will give rise to an rms current. The load current may be found by Ohm's law:

$$I_L = \frac{V_{Th}}{(R_{Th} + R_L) + j(X_{Th} + X_L)}. \tag{8.150}$$

The magnitude of the load current is

$$|I_L| = \frac{|V_{Th}|}{\sqrt{(R_{Th} + R_L)^2 + (X_{Th} + X_L)^2}}, \tag{8.151}$$

and the average power in the load is

$$P_L = |I_L|^2 R_L = \frac{|V_{Th}|^2 R_L}{(R_{Th} + R_L)^2 + (X_{Th} + X_L)^2}. \tag{8.152}$$

In order to maximize the average power in the load with respect to the load reactance, we can take the partial derivative and set it to zero:

$$\partial P_L / \partial X_L = \frac{-2(X_{Th} + X_L)|V_{Th}|^2 R_L}{\left[(R_{Th} + R_L)^2 + (X_{Th} + X_L)^2\right]^2} = 0. \tag{8.153}$$

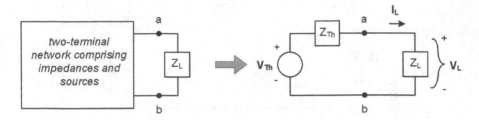

FIGURE 8.41 A load impedance connected to a two-terminal network represented by its Thevenin equivalent.

This is true only if

$$X_L = -X_{Th}.$$ (8.154)

Next, we can maximize the average power in the load with respect to the load resistance, using the same approach:

$$\partial P_L / \partial R_L = \frac{\left[(R_{Th}+R_L)^2+(X_{Th}+X_L)^2\right]|V_{Th}|^2}{\left[(R_{Th}+R_L)^2+(X_{Th}+X_L)^2\right]^2} - \frac{-2(R_{Th}+R_L)|V_{Th}|^2 R_L}{\left[(R_{Th}+R_L)^2+(X_{Th}+X_L)^2\right]^2} = 0.$$

(8.155)

Multiplying through by $\left[(R_{Th}+R_L)^2+(X_{Th}+X_L)^2\right]^2 /|V_{Th}|^2$, we obtain

$$(R_{Th}+R_L)^2+(X_{Th}+X_L)^2-2R_L(R_{Th}+R_L)=0.$$ (8.156)

Expanding,

$$R_{Th}^2+2R_{Th}R_L+R_L^2+(X_{Th}+X_L)^2-2R_LR_{Th}-2R_L^2=0.$$ (8.157)

Simplifying,

$$R_L^2 = R_{Th}^2+(X_{Th}+X_L)^2$$ (8.158)

and

$$R_L = \sqrt{R_{Th}^2+(X_{Th}+X_L)^2}.$$ (8.159)

Therefore, the average power in the load is maximized if $X_L = -X_{Th}$ and $R_L = R_{Th}$, or

$$Z_L = Z_{Th}^*.$$ (8.160)

In this ideal case, the average power in the load will be

$$P_L = \frac{|V_{Th}|^2 R_L}{(R_{Th}+R_L)^2+(X_{Th}+X_L)^2} = \frac{|V_{Th}|^2}{4R_L}.$$ (8.161)

If there are restrictions on the load impedance, then the load reactance should be set as close as possible to $-X_L$, and then the load resistance should be set as close as possible to $\sqrt{R_{Th}^2 + (X_{Th} + X_L)^2}$. For example, for the case in which the load impedance is restricted to be purely resistive (real, not complex), $X_L = 0$ and the load resistance should be chosen to be

$$R_L = \sqrt{R_{Th}^2 + X_{Th}^2}. \tag{8.162}$$

As an example, consider the load connected to the two-terminal network shown in Figure 8.42. We would like to find the value of the complex load impedance, which will maximize the average power delivered to the load, and this value of power. Second, we would like to find the purely-resistive load impedance, which will maximize the average power delivered to the load and the corresponding power.

We start by doing the open-circuit analysis with respect to the terminals where the load impedance is to be connected as shown in Figure 8.43.

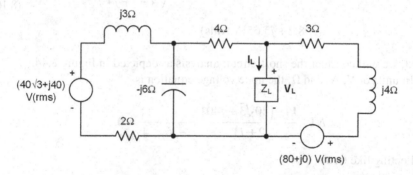

FIGURE 8.42   A load impedance connected to a two-terminal network for the consideration of maximum power transfer.

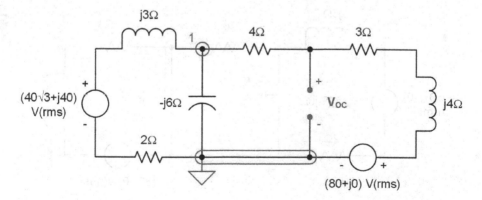

FIGURE 8.43   Open-circuit analysis of a two-terminal network for the consideration of maximum power transfer.

In units of V, A, and Ω, the node voltage equation is

$$N1 \quad \frac{V_1 - \left(40\sqrt{3} + j40\right)}{2 + j3} + \frac{V_1}{-j6} + \frac{V_1 - 80}{7 + j4} = 0. \tag{8.163}$$

Collecting like terms,

$$N1 \quad V_1\left[\frac{1}{2+j3} + \frac{1}{-j6} + \frac{1}{7+j4}\right] = \frac{40\sqrt{3}+j40}{2+j3} + \frac{80}{7+j4}. \tag{8.164}$$

Solving,

$$V_1 = \left(110.6 - j3.30\right)V\,(\text{rms}) \tag{8.165}$$

and by the voltage-divider rule,

$$V_{\text{Th}} = V_{\text{OC}} = \left(80 + j0\right)V\,(\text{rms}) + V_1\left(\frac{3+j4}{4+3+j4}\right) \tag{8.166}$$

$$= \left(98.2 + j5.65\right)V\,(\text{rms}).$$

Next, we will consider the short-circuit analysis as depicted in Figure 8.44.
In units of V, A, and Ω, the node voltage equation is

$$N1 \quad \frac{V_1 - \left(40\sqrt{3} + j40\right)}{2 + j3} + \frac{1}{-j6} + \frac{V_1}{4} = 0. \tag{8.167}$$

Collecting like terms,

$$N1 \quad V_1\left[\frac{1}{2+j3} + \frac{1}{-j6} + \frac{1}{4}\right] = \frac{40\sqrt{3}+j40}{2+j3}. \tag{8.168}$$

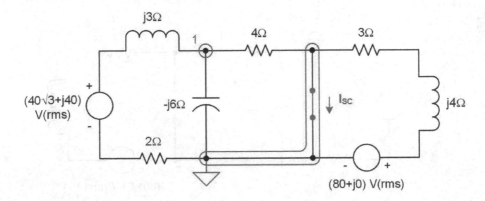

FIGURE 8.44 Short-circuit analysis of a two-terminal network for the consideration of maximum power transfer.

Solving,

$$V_1 = (51.8 - j16.13)\,\text{V (rms)}. \tag{8.169}$$

The short-circuit current is

$$I_{\text{SC}} = \frac{V_1}{4\,\Omega} + \frac{80\,\text{V (rms)}}{(3+j4)\,\Omega} = (22.6 - j16.83)\,\text{A (rms)}. \tag{8.170}$$

The Thevenin impedance is

$$Z_{\text{Th}} = \frac{V_{\text{OC}}}{I_{\text{SC}}} = (2.68 + j2.25)\,\Omega. \tag{8.171}$$

If there are no restrictions on Z_L, then the average power in the load is maximized with $Z_L = Z_{\text{Th}}^* = (2.68 - j2.25)\,\Omega$, and the value of average power in this load impedance is

$$P_L = \frac{|V_{\text{Th}}|^2 R_L}{(R_{\text{Th}} + R_L)^2 + (X_{\text{Th}} + X_L)^2} = \frac{|V_{\text{Th}}|^2}{4R_L} = 904\,\text{W}. \tag{8.172}$$

If on the other hand Z_L is restricted to be real, so that $X_L = 0$, then the average power in the load is maximized with $R_L = \sqrt{R_{\text{Th}}^2 + X_{\text{Th}}^2} = 3.50\,\Omega$, and the average power in the load with this impedance is

$$P_L = \frac{|V_{\text{Th}}|^2 R_L}{(R_{\text{Th}} + R_L)^2 + (X_{\text{Th}})^2} = 784\,\text{W}. \tag{8.173}$$

Figure 8.45 shows both cases; we can see that in the ideal case the average load power peaks at 904 W with $R_L = 2.68\,\Omega$, and in the restricted case the average power peaks at 784 W with $R_L = 3.50\,\Omega$.

To review, see Presentation 8.5 in ebook+.

To test your knowledge, try Quiz 8.5 in ebook+.

8.11 THREE-PHASE CIRCUITS AND SYSTEMS

Power transmission and distribution is usually done using three-phase circuits and systems, and the reason for this is efficiency. To see this, first consider the delivery of AC power to three load impedances using three parallel single-phase systems (all at a common phase angle) as shown in Figure 8.46. Six resistive transmission lines are used, and it is assumed that they all have the same series resistance R_S. Suppose the voltages all have the same magnitude V_M, and the load impedances all are equal: $Z_A = Z_B = Z_C = Z$. Then, the load currents are all equal, $I_A = I_B = I_C = I$, and the total power loss in the series resistances is $P_{\text{loss}} = 6|I|^2 R$.

Now consider the delivery of AC power to the same three load impedances using a three-phase system as shown in Figure 8.47. Only a single return line is needed, so a total of four resistive transmission lines is used, and they are assumed to each have the same series resistance R_S. The phases of the three voltage sources are chosen so that the three return currents will cancel, and the summed current in the single

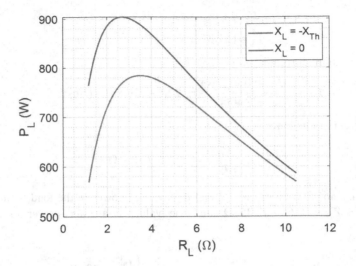

FIGURE 8.45 Average power in the load as a function of load resistance for the example circuit of Figure 8.42, both for the ideal case ($X_L = -X_{Th}$) and for the case in which the load impedance is restricted to be real ($X_L = 0$).

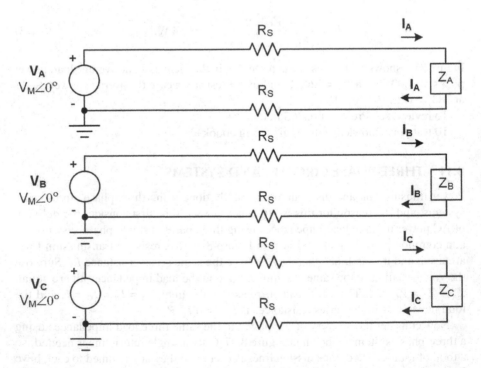

FIGURE 8.46 Delivery of power to three equal load impedances using resistive lines and three single-phase systems.

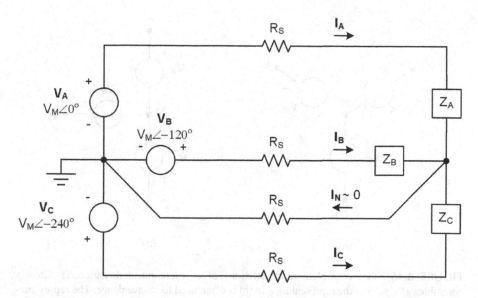

FIGURE 8.47   Delivery of power to three equal load impedances using a three-phase system.

return wire is zero: $I_N = I_A + I_B + I_C \approx 0$. The voltages all have the same magnitude V_M, though they are at different phases, the load impedances all are equal: $Z_A = Z_B = Z_C = Z$, and the load currents are all equal in magnitude though at different phases: $|I_A| = |I_B| = |I_C| = |I|$. In this case, because of the elimination of the power loss in the three return lines, the total power loss in the series resistances is $P_{\text{loss}} = 3|I|^2 R$. Therefore, the power loss in the series wiring is halved compared to the single-phase case, and this is significant. Typically, the power losses in the series resistance amount to 3%, but would be twice this or 6% if single-phase power transmission were used. In the United States, a 3% savings in power is enough to supply all of New Jersey!

In the example considered here, the system was **balanced**, so the loss in the return line was ideally zero. (In a balanced system, the voltage magnitudes are equal, the source phases are spaced by 120°, and the load impedances are equal.) Although real systems are never *exactly* balanced, they tend to be *approximately* balanced, and the benefit in efficiency is nearly the same as for the balanced case.

8.11.1   THREE-PHASE CONFIGURATIONS

There are several possible configurations for a three-phase system because the sources may be connected as a wye (Y) or delta (Δ), and similarly the loads may be connected in a wye or a delta. Figure 8.48 displays two possible ways of drawing a wye-connected source. Although the form in 8.48a more strongly resembles a "wye," the form on the left (also referred to as a "tee") is often used in schematics for convenience.

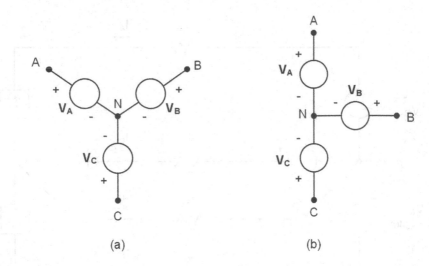

(a)   (b)

FIGURE 8.48 Two equivalent representations of a wye-connected source. (a) Closely resembles a "wye" but the representation in (b) is often used for convenience. The representation in (b) is referred to as a "tee."

Figure 8.49 displays three possible ways of drawing a delta-connected source. Although the form in 8.49a most strongly resembles a "delta," the other two forms are often used for convenience. The representation in 8.49b is sometimes called a "pi."

Four basic configurations for a three-phase system are the Y-Y, Y-Δ, Δ-Y, and Δ-Δ, where the first symbol refers to the source and the second refers to the load. These four configurations are shown in Figure 8.50. It is important to note that a ground (neutral) connection is only available in a wye, and only in the Y-Y configuration is it possible to connect the neutral points of the source and load.

The **positive phase sequence** normally used is A-B-C; so phase A leads phase B and phase B leads phase C. Each phase difference is $2\pi/3$, or 120°, and phase A is our reference. Hence, V_B lags V_A by 120°, and V_C lags V_B by 120°. This is shown in Figure 8.51. The phase angle for V_C may be expressed equivalently as −240° or 120°, but the latter is more commonly used. For the example shown, $V_M = 120\,\text{V}\,(\text{rms})$, and the three **phase voltages** are $V_A = 120\angle 0°\,\text{V}\,(\text{rms})$, $V_B = 120\angle -120°\,\text{V}\,(\text{rms})$, and $V_C = 120\angle 120°\,\text{V}\,(\text{rms})$. V_A, V_B, and V_C are referred to as the "phase voltages" because they appear across one of the circuit elements (in this case, sources). The **line voltages** may be determined by Kirchhoff's law. Thus, $V_{AB} = V_A - V_B = \sqrt{3}V_M\angle 30°$, $V_{BC} = V_B - V_C = \sqrt{3}V_M\angle -90°$, and $V_{CA} = V_C - V_A = \sqrt{3}V_M\angle 150°$, as shown in Figure 8.51. These are called line voltages because they exist between two lines. Hence, in a wye connection of sources, the line voltages are larger in magnitude than the phase voltages by a factor of $\sqrt{3}$.

Figure 8.52 shows the positive phase sequence for a delta-connected source. Here, V_{BC} lags V_{AB} by 120°, and V_{CA} lags V_{BC} by 120°. Phase angles are referenced with respect to V_{AB}. For the example shown, $V_M = 208\,\text{V}\,(\text{rms})$, and the three **phase voltages** are $V_{AB} = 208\angle 0°\,\text{V}\,(\text{rms})$, $V_{BC} = 208\angle -120°\,\text{V}\,(\text{rms})$, and $V_{CA} = 208\angle 120°\,\text{V}\,(\text{rms})$.

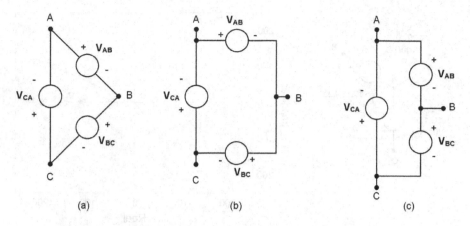

FIGURE 8.49   Three equivalent representations of a delta-connected source. (a) Closely resembles a "delta" but the other two are often used for convenience. The representation in (b) is sometimes referred to as a "pi." (c) is another convenient representation which avoids angled elements.

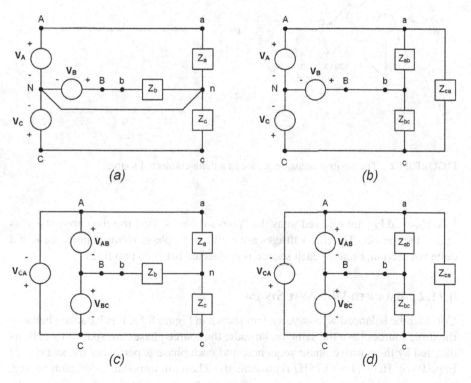

FIGURE 8.50   Four basic configurations for a three-phase system. (a) Y-Y; (b) Y-Δ; (c) Δ-Y; and (d) Δ-Δ.

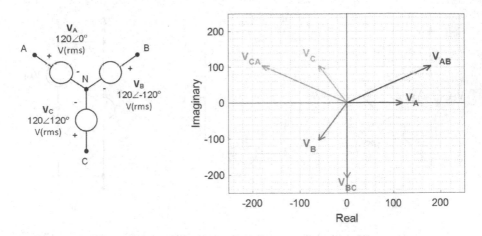

FIGURE 8.51   The positive phase sequence in a wye-connected source.

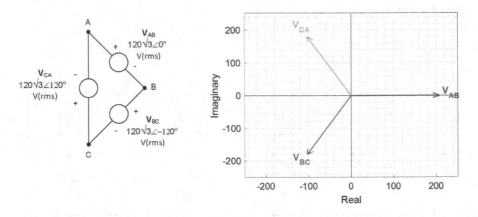

FIGURE 8.52   The positive phase sequence in a delta-connected source.

V_{AB}, V_{BC}, and V_{CA} are referred to as the "phase voltages" because they appear across one of the sources. The **line voltages** are equal to the phase voltages in the case of a delta connection, because each source is connected between two lines.

8.11.2   BALANCED WYE–WYE SYSTEM

Consider the balanced wye–wye system shown in Figure 8.53. It is balanced because the three sources have the same magnitude, the source phases are spaced by 120° as dictated by the positive phase sequence, and each phase experiences the same total impedance. Here, $(1 + j0.75)\,\Omega$ represents the Thevenin impedance for each source, $0.5\,\Omega$ represents the wiring resistance for each line, and $(24 + j6)\,\Omega$ represents the impedance for each load.

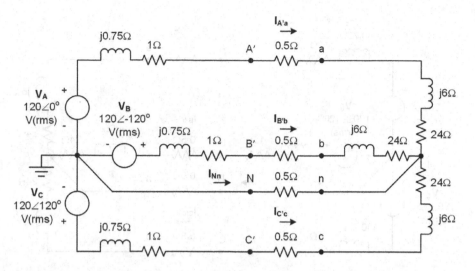

FIGURE 8.53 Balanced wye–wye system.

Because the system is balanced, zero voltage will develop at the neutral point of the load, and zero current will flow in the neutral line. We can show this by use of the node voltage method with the nodes labeled as shown in Figure 8.54. (The mesh current method could also be used, but this would involve three meshes and three equations instead of two essential nodes and one equation in the node voltage method.)

In units of V, A, and Ω, the node voltage equation is

$$N1 \quad \frac{v_n - V_A}{25.5 + j6.75} + \frac{v_n - V_B}{25.5 + j6.75} + \frac{v_n - V_C}{25.5 + j6.75} + \frac{v_n}{0.5} = 0. \qquad (8.174)$$

Collecting like terms,

$$N1 \quad v_n \left[\frac{3}{25.5 + j6.75} + \frac{1}{0.5} \right] = \frac{V_A + V_B + V_C}{25.5 + j6.75}. \qquad (8.175)$$

Solving,

$$v_n = \frac{V_A + V_B + V_C}{25.5 + j6.75} \left[\frac{3}{25.5 + j6.75} + \frac{1}{0.5} \right]^{-1}$$

$$= \frac{\begin{array}{c}(120)V\,(\text{rms}) + (120\cos(-120°) + j120\sin(-120°))V\,(\text{rms}) \\ + (120\cos(120°) + j120\sin(120°))V\,(\text{rms})\end{array}}{25.5 + j6.75} \qquad (8.176)$$

$$\times \left[\frac{3}{25.5 + j6.75} + \frac{1}{0.5} \right]^{-1} = 0.$$

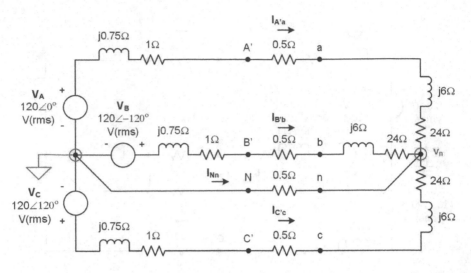

FIGURE 8.54 Node voltage analysis of a balanced wye–wye system.

Therefore, in the balanced wye–wye system, zero voltage develops at the neutral point because the three source voltages add to zero and their contributions are equally weighted in the balanced circuit. Because of this, determination of the line currents is simplified.

$$I_{A'a} = \frac{(120 + j0)\,V\,(rms)}{(25.5 + j6.75)\,\Omega} \tag{8.177}$$

$$= (4.40 - j1.164)\,A\,(rms) = 4.55\angle - 14.8°\,A\,(rms),$$

$$I_{B'b} = \frac{(120\cos(-120°) + j120\sin(-120°))\,V\,(rms)}{(25.5 + j6.75)\,\Omega} \tag{8.178}$$

$$= (-3.21 - j3.23)\,A\,(rms) = 4.55\angle - 134.8°\,A\,(rms),$$

$$I_{C'c} = \frac{(120\cos(120°) + j120\sin(120°))\,V\,(rms)}{(25.5 + j6.75)\,\Omega} \tag{8.179}$$

$$= (-1.191 + j4.39)\,A\,(rms) = 4.55\angle105.2°\,A\,(rms).$$

The phasor voltages and currents are shown in Figure 8.55. It can be seen that the line currents have a similar phase relationship to the phase voltages; that is, I_{Bb} lags I_{Aa} by 120° and I_{Cc} lags I_{Bb} by 120°.

8.11.3 BALANCED DELTA–DELTA SYSTEM

Consider the balanced delta–delta system shown in Figure 8.56. Here, $0.5\,\Omega$ represents the Thevenin impedance for each source, $0.25\,\Omega$ represents the wiring resistance for each line, and $(20 + j4)\,\Omega$ represents the impedance for each load.

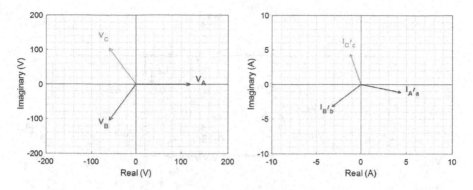

FIGURE 8.55 Phasor diagrams for a balanced wye–wye system.

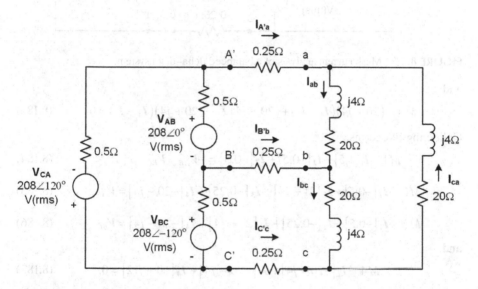

FIGURE 8.56 Balanced delta–delta system.

This system may be solved using the mesh current method with the meshes labeled as shown in Figure 8.57. (The node voltage method could also be used, but this involves six essential nodes and five equations, whereas the mesh current method involves four meshes and four equations.)

In units of V, A, and Ω, the mesh current equations are

$$M1 \quad V_{CA}+0.5I_1+0.5(I_1-I_2)+V_{AB}+0.5(I_1-I_3)+V_{BC}=0, \tag{8.180}$$

$$M2 \quad -V_{AB}+0.5(I_2-I_1)+0.25I_2+(20+j4)(I_2-I_4)+0.25(I_2-I_3)=0, \tag{8.181}$$

$$M3 \quad -V_{BC}+0.5(I_3-I_1)+0.25(I_3-I_2)+(20+j4)(I_3-I_4)+0.25I_3=0, \tag{8.182}$$

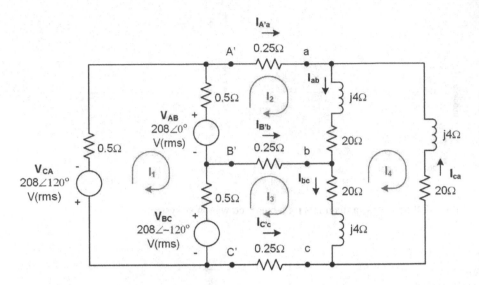

FIGURE 8.57 Mesh current analysis of a balanced delta–delta system.

and

$$M4 \quad (20+j4)(I_4 - I_2)+(20+j4)\,I_4+(20+j4)(I_4 - I_3)= 0. \tag{8.183}$$

Collecting like terms,

$$M1 \quad I_1[1.5]+I_2[-0.5]+I_3[-0.5]=-V_{AB}-V_{BC}-V_{CA}, \tag{8.184}$$

$$M2 \quad I_1[-0.5]+I_2[21+j4]+I_3[-0.25]+I_4[-20-j4]=V_{AB}, \tag{8.185}$$

$$M3 \quad I_1[-0.5]+I_2[-0.25]+I_3[21+j4]+I_4[-20-j4]=V_{BC}, \tag{8.186}$$

and

$$M4 \quad I_2[-20-j4]+I_3[-20-j4]+I_4[60+j12]= 0. \tag{8.187}$$

In matrix form,

$$
\begin{bmatrix}
1.5 & -0.5 & -0.5 & 0 \\
-0.5 & 21+j4 & -0.25 & -20-j4 \\
-0.5 & -0.25 & 21+j4 & -20-j4 \\
0 & -20-j4 & -20-j4 & 60+j12
\end{bmatrix}
\begin{bmatrix}
I_1 \\
I_2 \\
I_3 \\
I_4
\end{bmatrix}
=
\begin{bmatrix}
-V_{AB}-V_{BC}-V_{CA} \\
V_{AB} \\
V_{BC} \\
0
\end{bmatrix}.
\tag{8.188}
$$

Solving,

$$
\begin{bmatrix} I_1 \\ I_2 \\ I_3 \\ I_4 \end{bmatrix} = \begin{bmatrix} (3.19 - j9.08)\,\mathrm{A\,(rms)} \\ (12.63 - j10.86)\,\mathrm{A\,(rms)} \\ (-3.08 - j16.38)\,\mathrm{A\,(rms)} \\ (3.19 - j9.08)\,\mathrm{A\,(rms)} \end{bmatrix}. \tag{8.189}
$$

The line currents are

$$
\begin{bmatrix} I_{A'a} \\ I_{B'b} \\ I_{C'c} \end{bmatrix} = \begin{bmatrix} (12.64 - j10.86)\,\mathrm{A\,(rms)} \\ (-15.72 - j5.52)\,\mathrm{A\,(rms)} \\ (3.08 + j16.37)\,\mathrm{A\,(rms)} \end{bmatrix} = \begin{bmatrix} 16.66\angle -40.7^\circ\,\mathrm{A\,(rms)} \\ 16.66\angle -160.7^\circ\,\mathrm{A\,(rms)} \\ 16.66\angle 79.3^\circ\,\mathrm{A\,(rms)} \end{bmatrix},
$$

$$\tag{8.190}$$

and the phase currents are

$$
\begin{bmatrix} I_{ab} \\ I_{bc} \\ I_{ca} \end{bmatrix} = \begin{bmatrix} (9.45 - j1.779)\,\mathrm{A\,(rms)} \\ (-6.27 - j7.30)\,\mathrm{A\,(rms)} \\ (-3.19 + j9.08)\,\mathrm{A\,(rms)} \end{bmatrix} = \begin{bmatrix} 9.62\angle -10.7^\circ\,\mathrm{A\,(rms)} \\ 9.62\angle -130.7\,\mathrm{A\,(rms)} \\ 9.62\angle 109.3^\circ\,\mathrm{A\,(rms)} \end{bmatrix}. \tag{8.191}
$$

Figure 8.58 shows the phase voltages, line currents, and phase currents on phasor diagrams. It can be seen that the line currents all exhibit the same magnitude and the same phase relationship as the phase voltages; that is, $I_{B'b}$ lags $I_{A'a}$ by 120° and $I_{C'c}$ lags $I_{B'b}$ by 120°. The phase currents all have equal magnitude, but this magnitude is $1/\sqrt{3}$ times the magnitude of the line currents. Also, the phase currents exhibit the same phase relationship as the phase voltages; I_{bc} lags I_{ab} by 120° and I_{ca} lags I_{bc} by 120°.

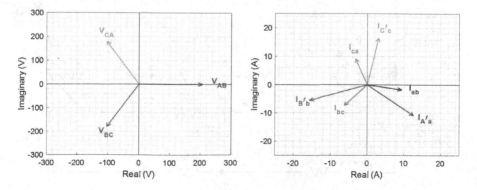

FIGURE 8.58 Phasor diagrams for a balanced delta–delta system.

8.11.4 UNBALANCED THREE-PHASE SYSTEM

Any unbalanced three-phase system may be analyzed using either the node voltage method or the mesh current method. As one example consider the wye–wye system shown in Figure 8.59. This system is unbalanced because the three load impedances are not equal, and therefore a non-zero voltage will develop at the neutral point of the load.

The non-zero voltage at the neutral point of the load may be found by the node voltage method, as shown in Figure 8.60, and once this voltage is known the line currents may be readily determined.

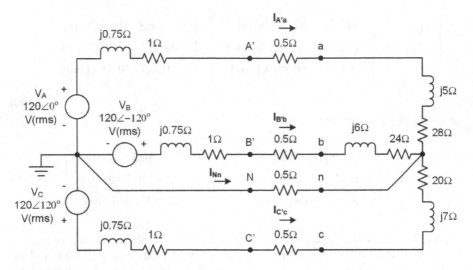

FIGURE 8.59 Unbalanced wye–wye system.

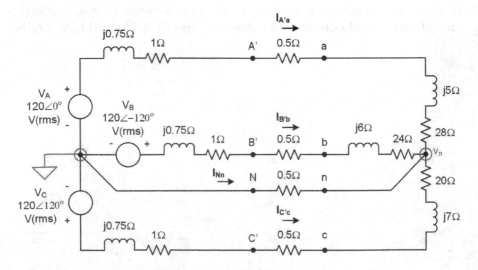

FIGURE 8.60 Node voltage analysis of an unbalanced wye–wye system.

In units of V, A, and Ω, the node voltage equation is

$$N1 \quad \frac{v_n - V_A}{29.5 + j5.75} + \frac{v_n - V_B}{25.5 + j6.75} + \frac{v_n - V_C}{21.5 + j7.75} + \frac{v_n}{0.5} = 0. \tag{8.192}$$

Collecting like terms,

$$N1 \quad v_n \left[\frac{1}{29.5 + j5.75} + \frac{1}{25.5 + j6.75} + \frac{1}{21.5 + j7.75} + \frac{1}{0.5} \right]$$

$$= \frac{V_A}{29.5 + j5.75} + \frac{V_B}{25.5 + j6.75} + \frac{V_C}{21.5 + j7.75}. \tag{8.193}$$

Solving,

$$v_n = (-0.1104 + j0.556)\, V\,(\text{rms}). \tag{8.194}$$

Notice that even though this is an unbalanced system only a small voltage develops at the neutral point of the load because a line has been run between the two neutrals. The line currents are

$$I_{A'a} = \frac{V_A - v_n}{(29.5 + j5.75)\Omega} = (3.92 - j0.783)\, A\,(\text{rms})$$

$$\tag{8.195}$$

$$= 4.00\angle -11.3°\, A\,(\text{rms}),$$

$$I_{B'b} = \frac{V_B - v_n}{(29.5 + j5.75)\Omega} = (-3.21 - j3.25)\, A\,(\text{rms})$$

$$\tag{8.197}$$

$$= 4.56\angle -134.6°\, A\,(\text{rms}),$$

and

$$I_{C'c} = \frac{V_A - v_n}{(25.5 + j6.75)\Omega} = (-0.932 + j5.14)\, A\,(\text{rms})$$

$$\tag{8.198}$$

$$= 5.23\angle 100.3°\, A\,(\text{rms}).$$

These line currents (which are the same as the phase currents for the wye load) are different in magnitude and do not retain the phase relationship of the source voltages because of the imbalance in impedances. This gives rise to a current in the neutral line:

$$I_{Nn} = -I_{A'a} - I_{B'b} - I_{C'c} = (0.221 - j1.113)\, A\,(\text{rms}). \tag{8.199}$$

Note, however, that this is much smaller in magnitude than the line currents.

The phase voltages at the load may be found by use of KVL:

$$V_{an} = v_n + I_{A'a}(28 + j5)\Omega$$

$$\tag{8.200}$$

$$= (113.5 - j1.765)\, V\,(\text{rms}) = 113.5\angle -0.9°\, V\,(\text{rms}),$$

$$V_{bn} = v_n + I_{B'b}(24 + j6)\Omega$$

$$\tag{8.201}$$

$$= (-57.6 - j96.6)\, V\,(\text{rms}) = 112.5\angle -120.8°\, V\,(\text{rms}),$$

and

$$V_{cn} = v_n + I_{C'c}(20 + j7)\,\Omega$$

$$= (-54.7 + j96.9)\,\text{V}\,(\text{rms}) = 111.3\angle119.5°\,\text{V}\,(\text{rms}). \tag{8.202}$$

Similarly, the line voltages are

$$V_{ab} = V_{an} - V_{bn}$$

$$= (171.2 + j94.9)\,\text{V}\,(\text{rms}) = 195.7\angle29.0°\,\text{V}\,(\text{rms}), \tag{8.203}$$

$$V_{bc} = V_{bn} - V_{cn}$$

$$= (-2.87 - j193.6)\,\text{V}\,(\text{rms}) = 193.6\angle-90.8°\,\text{V}\,(\text{rms}), \tag{8.204}$$

and

$$V_{ca} = V_{cn} - V_{an}$$

$$= (-168.3 + j98.7)\,\text{V}\,(\text{rms}) = 195.1\angle149.6°\,\text{V}\,(\text{rms}). \tag{8.205}$$

The phasor diagrams of Figure 8.61 show the line voltages, the phase voltages, and the line currents (same as the phase currents). The expected symmetry of a balanced system is not preserved exactly, but holds approximately.

To review, see Presentation 8.6 in ebook+.

To test your knowledge, try Quiz 8.6 in ebook+.

8.12 MUTUAL INDUCTANCE AND TRANSFORMERS

8.12.1 FUNDAMENTAL CONSIDERATIONS

We saw in Chapter 6 that when a coil of wire produces magnetic flux which links that same coil, there is a **self-inductance**, and by the Faraday law a voltage $L\,di/dt$ is developed when the current changes with time. Here, we will examine the self-inductance

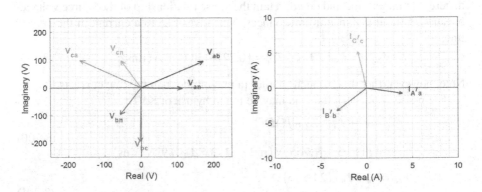

FIGURE 8.61 Phasor diagrams for an unbalanced wye–wye system.

in more detail and then extend those ideas to the situation involving two coils linked by some common magnetic flux (a **transformer** with **mutual inductance** M).

Consider a single coil wrapped on a core of magnetic medium as shown in Figure 8.62.

If the magnetic medium is linear, then the magnetic flux ϕ (in Wb) is given by

$$\phi = AB = A\mu H = A\mu Ni = \mathcal{P}Ni, \tag{8.206}$$

where A is the cross-sectional area (m²), B is the flux density (Wb/m²), μ is the permeability of the medium (Wbm⁻²A⁻¹), N is the number of turns in the coil, i is the current in the coil, and $\mathcal{P}$ is the permeance of the core (Wb/A). By the Faraday law,

$$v = \frac{d\lambda}{dt} = \frac{d(N\phi)}{dt} = N\frac{d}{dt}(\mathcal{P}Ni) = N^2\mathcal{P}\frac{di}{dt} = L\frac{di}{dt}, \tag{8.207}$$

where $\lambda = N\phi$ is the flux linkage (Wb) and L is the self-inductance (H). Now consider the case of two magnetically-couple coils on the same magnetic core as shown in Figure 8.63. A voltage v_1 is applied to the primary (input) coil, resulting in a current i_1. The secondary (output) coil is open-circuited, so no current flows in it, but there is an induced voltage v_2.

The flow of current in the primary gives rise to a total magnetic flux $\phi_1 = \phi_{11} + \phi_{12}$, where ϕ_{12} is the portion of the flux linking both coils and ϕ_{11} is the portion of the flux which links only the primary. The permeance $\mathcal{P}_1 = \mathcal{P}_{11} + \mathcal{P}_{12}$ likewise has two components associated with the two components of flux. By the Faraday law, the primary voltage is

$$v_1 = \frac{d\lambda_1}{dt} = \frac{d(N_1\phi_1)}{dt} = N_1\frac{d}{dt}(\phi_{11} + \phi_{12}) = N_1\frac{d}{dt}(N_1\mathcal{P}_{11}i_1 + N_1\mathcal{P}_{12}i_1)$$

$$= N_1^2(\mathcal{P}_{11} + \mathcal{P}_{12})\frac{di_1}{dt} = N_1^2\mathcal{P}_1\frac{di_1}{dt} = L_1\frac{di_1}{dt}. \tag{8.208}$$

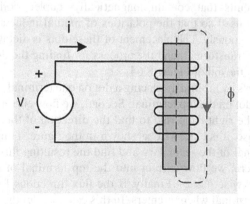

FIGURE 8.62   A coil of wire wrapped on a core of magnetic medium.

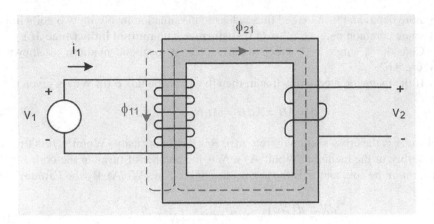

FIGURE 8.63 Two magnetically-coupled coils on a single core.

Similarly, the voltage induced in the secondary is

$$v_2 = \frac{d\lambda_2}{dt} = \frac{d(N_2\phi_{21})}{dt} = N_2 \frac{d}{dt}(N_2 \mathcal{P}_{12} i_1) = N_1 N_2 \mathcal{P}_{12} \frac{di_1}{dt} = M_{12} \frac{di_1}{dt}, \qquad (8.209)$$

where M_{12} is the mutual inductance (H) associated with flux created by the primary and linking the secondary. If $\mathcal{P}_{21} = \mathcal{P}_{12}$ then $M_{21} = M_{12} = M$ and we can use a single value of mutual inductance without subscripts. It can be shown that the secondary coil has its own self-inductance L_2 and that

$$M = k\sqrt{L_1 L_2}, \qquad (8.210)$$

where k is the coefficient of coupling, and $0 \le k \le 1$. Thus, $M \le \sqrt{L_1 L_2}$.

8.12.2 THE DOT CONVENTION FOR POLARITIES

When solving circuits that contain magnetically-coupled coils (**transformers**), a system of dots is used so that the polarities of mutual inductance terms may be determined unambiguously. The placement of these dots is dictated by the physical construction of the transformer, and the process for finding the dot placements may be understood with the aid of Figure 8.64.

The process starts by arbitrarily placing a dot on one terminal of the primary coil; we will place this dot on the top terminal. Second, we flow current i_1 into this dotted terminal and use the right-hand rule to find the direction of the resulting flux ϕ_1 in the core. In this case, it is clockwise as shown in the figure. Third, we flow current i_2 into either terminal of the secondary and find the resulting flux in the core by the right-hand rule. Here, we flow current into the top terminal of the secondary and this results in clockwise flux ϕ_2. Finally, if the flux directions for ϕ_1 and ϕ_2 are the same, we dot the terminal where i_2 enters. In this case, we dot the top terminal of the secondary, so both top terminals are dotted. Now we know that if di_1 / dt is positive,

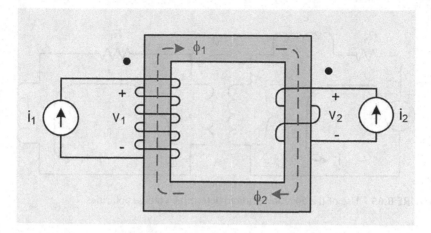

FIGURE 8.64 The dot convention and determination of dot placement.

inducing a positive value of v_1 (positive at the dot) by way of the self-inductance, it will also induce a positive value of v_2 (positive at the dot) by way of the mutual inductance.

To further explore the application of the dot convention in finding polarities of voltage terms in a circuit containing a transformer, refer to Figure 8.65.

The primary voltage is given by

$$v_1 = L_1 \frac{di_1}{dt} - M \frac{di_2}{dt}. \tag{8.211}$$

The sign of the mutual inductance term comes about because if di_2 / dt is positive it induces a self-inductance component of v_2 which is negative at the secondary dot and also produces a mutual inductance component of v_1 which is negative at the primary dot. This is therefore a negative contribution to v_1. By similar reasoning, the secondary voltage is given by

$$v_2 = -L_2 \frac{di_2}{dt} + M \frac{di_1}{dt}. \tag{8.212}$$

The time-domain mesh currents are therefore

$$M1 \quad -v_s + R_s i_1 + L_s \frac{di_1}{dt} + L_1 \frac{di_1}{dt} - M \frac{di_2}{dt} = 0 \tag{8.213}$$

and

$$M2 \quad L_2 \frac{di_2}{dt} - M \frac{di_1}{dt} + R_s i_1 + L_s \frac{di_1}{dt} = 0. \tag{8.214}$$

Often the dots may be omitted from a transformer, and then it can be assumed that they are both on the top terminals (or equivalently, both on the bottom terminals).

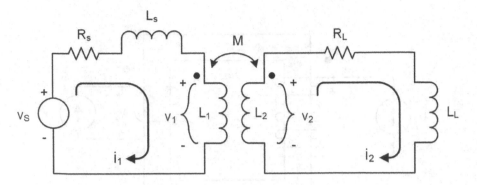

FIGURE 8.65 Use of the dot convention to determine voltage polarities.

8.12.3 THE LINEAR TRANSFORMER IN THE PHASOR DOMAIN

Although the previous section used time-domain mesh equations (differential equations) to illustrate the use of the dot convention, we will normally apply impedance and phasor concepts for the AC steady-state analysis of circuits involving transformers using algebraic equations. Figure 8.66 illustrates a phasor-domain representation of a linear transformer circuit, in which the time-domain self-inductances and mutual inductance have been replaced by $j\omega L_1$, $j\omega L_2$, and $j\omega M$.

For this circuit, the mesh equations are

$$M1 \quad -V_s + Z_s I_1 + R_1 I_1 + j\omega L_1 I_1 - j\omega M I_2 = 0 \qquad (8.215)$$

and

$$M2 \quad -j\omega M I_1 + j\omega L_1 I_2 + R_2 I_2 + Z_L I_2 = 0. \qquad (8.216)$$

Collecting like terms,

$$\begin{bmatrix} Z_s + R_1 + j\omega L_1 & -j\omega M \\ -j\omega M & j\omega L_1 + R_2 + Z_L \end{bmatrix} \begin{bmatrix} I_1 \\ I_2 \end{bmatrix} = \begin{bmatrix} V_s \\ 0 \end{bmatrix}. \qquad (8.217)$$

Solving,

$$I_1 = \frac{\begin{vmatrix} V_s & -j\omega M \\ 0 & j\omega L_1 + R_2 + Z_L \end{vmatrix}}{\begin{vmatrix} Z_s + R_1 + j\omega L_1 & -j\omega M \\ -j\omega M & j\omega L_1 + R_2 + Z_L \end{vmatrix}} = \frac{V_s(j\omega L_1 + R_2 + Z_L)}{(Z_s + R_1 + j\omega L_1)(\omega L_1 + R_2 + Z_L) + \omega^2 M}$$

$$(8.218)$$

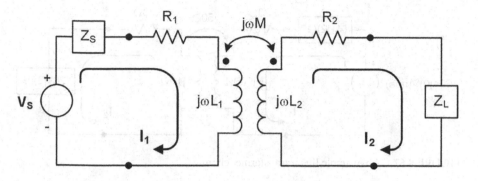

FIGURE 8.66 A linear transformer circuit rendered in the phasor domain.

and

$$I_2 = \frac{\begin{vmatrix} Z_s + R_1 + j\omega L_1 & V_s \\ -j\omega M & 0 \end{vmatrix}}{\begin{vmatrix} Z_s + R_1 + j\omega L_1 & -j\omega M \\ -j\omega M & j\omega L_1 + R_2 + Z_L \end{vmatrix}} = \frac{j\omega M V_s}{(Z_s + R_1 + j\omega L_1)(\omega L_1 + R_2 + Z_L) + \omega^2 M}.$$

(8.219)

As a specific example, consider the linear transformer circuit of Figure 8.67.
The mesh equations, in units of A, V, and Ω, are

$$M1 \quad -50 + (4 + j2) I_1 + 30 I_1 + j900 I_1 - j50 I_2 = 0 \tag{8.220}$$

and

$$M2 \quad -j50 I_1 + j10 I_2 + 3 I_2 + (20 + j3) I_2 = 0. \tag{8.221}$$

Collecting like terms,

$$\begin{bmatrix} 34 + j900 & -j50 \\ -j50 & 23 + j13 \end{bmatrix} \begin{bmatrix} I_1 \\ I_2 \end{bmatrix} = \begin{bmatrix} 50 \\ 0 \end{bmatrix}. \tag{8.222}$$

Solving,

$$I_1 = (7.8 + j57.5)\,\text{mA}\,(\text{rms}) \tag{8.223}$$

and

$$I_2 = (102.1 - j40.6)\,\text{mA}\,(\text{rms}). \tag{8.224}$$

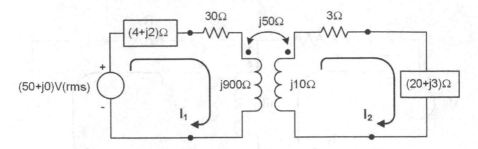

FIGURE 8.67   An example linear transformer circuit.

8.12.4   The Ideal Transformer

Often an ideal transformer model may be used to analyze transformer circuits with satisfactory accuracy. The three attributes of an ideal transformer are the following: (i) the coupling coefficient is unity, so that $M = \sqrt{L_1 L_2}$; (ii) the self-inductances are infinite; and (iii) the coil losses due to resistance are negligible. Therefore for an ideal transformer such as the one shown in Figure 8.68, there is no need to specify the self or mutual inductances, and instead we specify the turns ratio, which may be expressed in the form $N_1 : N_2$ or as a voltage ratio $V_1 : V_2$.

Using the reference polarities and dot placement in Figure 8.68, the basic relationships for the ideal transformer are

$$\frac{V_1}{V_2} = \frac{N_1}{N_2} \tag{8.225}$$

and

$$N_1 I_1 = -N_2 I_2. \tag{8.226}$$

Combining these, we see that

$$\underbrace{|V_1||I_1|}_{\text{primary volt amperes}} = \underbrace{|V_2||I_2|}_{\text{secondary volt amperes}}. \tag{8.227}$$

Therefore, a transformer which steps up the voltage will step down the current, and vice versa.

Consider the general case of a source with series impedance Z_S driving a load with impedance Z_L through an ideal transformer with a turns ratio $N_1 : N_2$, as shown in Figure 8.69.

The mesh equations are

$$M1 \quad -V_S + Z_S I_1 + V_1 = 0 \tag{8.228}$$

and

$$M2 \quad -V_2 + Z_2 I_2 = 0. \tag{8.229}$$

But

$$V_1 = \left(\frac{N_1}{N_2}\right) V_2 = \left(\frac{N_1}{N_2}\right) Z_L I_2 = \left(\frac{N_1}{N_2}\right) Z_L \left(\frac{N_1}{N_2}\right) I_1 = \left(\frac{N_1}{N_2}\right)^2 Z_L I_1. \tag{8.230}$$

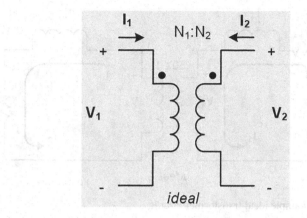

FIGURE 8.68   Ideal transformer.

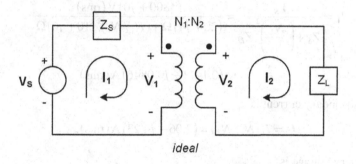

FIGURE 8.69   Use of an ideal transformer to drive a load impedance.

Substituting this into the M1 equation,

$$M1 \quad -V_S + Z_S I_1 + \left(\frac{N_1}{N_2}\right) Z_L \left(\frac{N_1}{N_2}\right) I_1 = 0, \tag{8.231}$$

and solving,

$$I_1 = \frac{V_S}{Z_S + \left(\dfrac{N_1}{N_2}\right)^2 Z_L}. \tag{8.232}$$

Therefore, the apparent load impedance seen looking into the primary side of the transformer is $(N_1 / N_2)^2 Z_L$, and therefore an interesting function of the transformer is that of impedance matching. An example would be a case where maximum power transfer is desired but practical constraints make it impossible to match the load to the Thevenin impedance of the source.

As a specific example, consider the ideal transformer circuit in Figure 8.70.

The primary current is

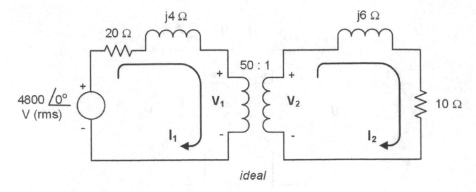

FIGURE 8.70 An example ideal transformer circuit.

$$I_1 = \frac{V_S}{Z_S + \left(\dfrac{N_1}{N_2}\right)^2 Z_L} = \frac{(4800 + j0)\,\text{V}\,(\text{rms})}{(20 + j4)\,\Omega + (N_1/N_2)^2\,(10 + j6)\,\Omega} \tag{8.233}$$

$$= (0.1411 - j0.0846)\,\text{A}\,(\text{rms}),$$

and the secondary current is

$$I_2 = I_1(N_1/N_2) = (7.06 - j4.23)\,\text{A}\,(\text{rms}). \tag{8.234}$$

The primary voltage is

$$V_1 = V_S - I_1 Z_S = (4797 + j0.113)\,\text{V}\,(\text{rms}), \tag{8.235}$$

and the secondary voltage is

$$V_2 = V_1(N_2/N_1) = (95.9 + j0.023)\,\text{V}\,(\text{rms}). \tag{8.236}$$

The complex power for the source is

$$S_S = -V_S I_1^* = (-677.3 - j406.2)\,\text{VA}, \tag{8.237}$$

so the source develops 677.3 W average power.

The complex power for the load is

$$S_L = V_2 I_2^* = (-676.8 - j406.0)\,\text{VA}, \tag{8.238}$$

so 99.9% of the average power developed by the source is delivered to the load. This high efficiency is possible despite the relatively large magnitude of source impedance because of the action of the transformer.

To review, see Presentation 8.7 in ebook+.

To test your knowledge, try Quiz 8.7 in ebook+.

8.13   SUMMARY

In a circuit driven by one or more sinusoidal sources with a particular frequency, we may use AC steady-state analysis to determine the voltages and currents. The voltages and currents are all sinusoidal with the same frequency as the source(s) so we only need to find their magnitudes and phases. In order to avoid the need to solve differential equations in the time domain, we transform to the phasor domain. Passive components are replaced with complex impedances; the impedance of a resistor is R, the impedance of an inductor is $j\omega L$, and the impedance of a capacitor is $1/(j\omega C)$. The real part of an impedance Z is a resistance R and the imaginary part is j times a reactance X, so that in general $Z = R + jX$. Impedances combine in series and parallel like resistances. Voltages and currents are treated as complex phasors. The phasor has a magnitude and a phase, which corresponds to the magnitude and phase in the time domain. A phasor may be expressed in polar or rectangular form and may be plotted visually on the complex plane in a phasor diagram. In the phasor domain, Ohm's law and Kirchhoff's laws apply. Therefore, we may use the node voltage method, the mesh current method, the current-divider rule, the voltage-divider rule, and Thevenin's and Norton's theorems in the phasor domain for AC steady-state analysis.

The instantaneous power for an element in a circuit with sinusoidal excitation is sinusoidal with a frequency twice the frequency of the source(s). Of more interest than the instantaneous power are the average power P and the reactive power Q. The complex power S embodies both quantities: $S = P + jQ$. To find the complex power for an element, we multiply the phasor voltage by the conjugate of the current: $S = VI^*$. For this calculation, both the current and the voltage should be in root mean square (rms) form, and the passive sign convention should be applied to determine whether a minus sign should be applied in the complex power equation.

For maximum power transfer from a two-terminal network to a load impedance, the load impedance should be set to the conjugate of the Thevenin impedance for the two-terminal network. If there are constraints on $Z_L = R_L + jX_L$ which make this impossible, then the load reactance should be set as close as possible to $-X_{\text{Th}}$, and then the load resistance should be set as close as possible to $\sqrt{R_{\text{Th}}^2 + (X_L + X_{\text{Th}})^2}$.

Three-phase circuits and systems are commonly used for power transmission and distribution. They make use of three sinusoidal sources at 120° intervals, which decreases transmission losses significantly and improves efficiency. Sources and loads may be connected in wye or delta configurations, leading to four basic combinations. Other configurations are possible if transformers are connected to the source or the load. Three-phase systems may be analyzed using mesh or nodal analysis.

Transformers utilize two or more coils which are magnetically coupled to step up or step down a voltage using the mutual inductance between the coils. In a linear transformer, the mutual inductance is used in the time or phasor domain to write mesh equations. For an ideal transformer, only the turns ratio of the transformer is used to find approximate currents and voltages. In either case, mesh analysis is preferred and nodal analysis may be cumbersome.

To evaluate your mastery of chapter eight, solve Example Exam 8.1 in ebook+ or Example Exam 8.2 in ebook+.

PROBLEMS

Problem 8.1. Find the time-domain currents $i_L(t)$, $i_R(t)$, and $i_C(t)$ of the circuit in Figure P8.1, and plot them together as functions of time with the time interval from zero to 50 ms. Assume the circuit operates under sinusoidal steady-state conditions.

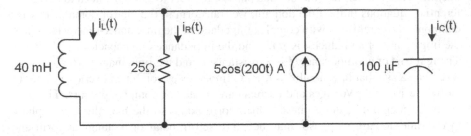

FIGURE P8.1   Parallel circuit with sinusoidal excitation

Problem 8.2. Find the time-domain voltages $v_A(t)$, $v_B(t)$, and $v_C(t)$ in the diagram of Figure P8.2. Determine the associated phasor-domain voltages V_A, V_B, and V_C and plot these on a phasor diagram. Assume the circuit operates under sinusoidal steady-state conditions.

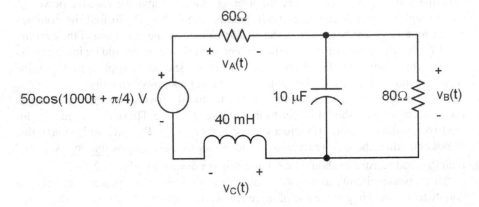

FIGURE P8.2   Circuit with sinusoidal excitation and two meshes.

Problem 8.3. Find the time-domain currents $i_X(t)$, $i_Y(t)$, and $i_Z(t)$ in the diagram of Figure P8.3. Determine the associated phasor-domain voltages I_X, I_Y, and I_Z and plot these on a phasor diagram. Assume the circuit operates under sinusoidal steady-state conditions.

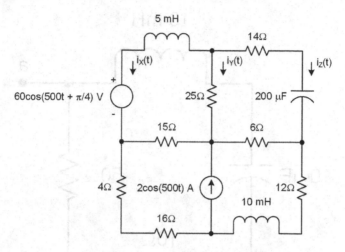

FIGURE P8.3 Circuit with sinusoidal excitation and four meshes.

Problem 8.4. For the network in Figure P8.4, calculate the complex impedance Z_{ab} with respect to the terminals a and b at a radial frequency of $10,000$ rad/s. Also find the complex admittance Y_{ab} at this same frequency.

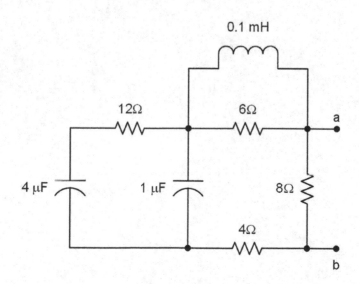

FIGURE P8.4 Two-terminal network involving three reactive elements and three resistances.

Problem 8.5. For the network in Figure P8.5, find the complex impedance Z_{ab} as a function of radial frequency. Plot the real and imaginary parts of this impedance, both as functions of the radial frequency, for the frequency range from 10^2 rad/s to 10^5 rad/s.

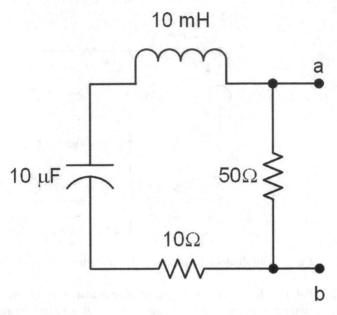

FIGURE P8.5   Two-terminal network involving two reactive elements and two resistances.

Problem 8.6. For the network in Figure P8.6, find the complex impedance Z_{ab} as a function of the capacitance C, assuming $\omega = 5000 \, \text{rad/s}$. Plot the real and imaginary parts of this impedance, both as a function of the capacitance C, choosing an appropriate range of this capacitance. Is there a value of the capacitance C which makes the impedance real? If so, what is the value of capacitance and the resulting impedance?

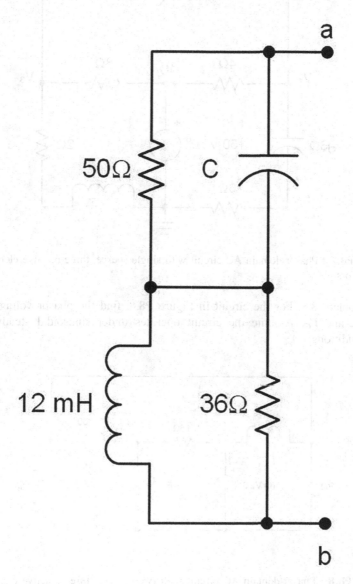

FIGURE P8.6   Two-terminal network involving reactive elements and resistances in two meshes.

Problem 8.7. For the circuit in Figure P8.7, determine the phasor voltages V_1, V_2, and V_3 with respect to ground. Assume the circuit operates under sinusoidal steady-state conditions.

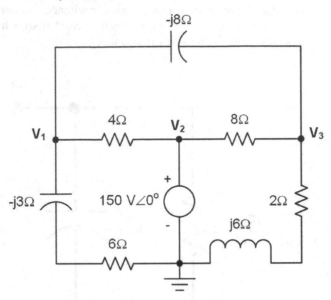

FIGURE P8.7 Phasor-domain AC circuit with single source, three reactive elements, and four resistors.

Problem 8.8. For the circuit in Figure P8.8, find the phasor voltages V_A, V_B, and V_C. Assume the circuit operates under sinusoidal steady-state conditions.

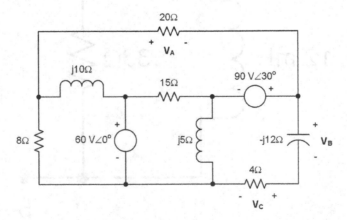

FIGURE P8.8 Phasor-domain AC circuit with two sources, three reactive elements, and four resistors.

Problem 8.9. Find the phasor currents I_A, I_B, and I_C of the circuit shown in Figure P8.9. Assume the circuit operates under sinusoidal steady-state conditions.

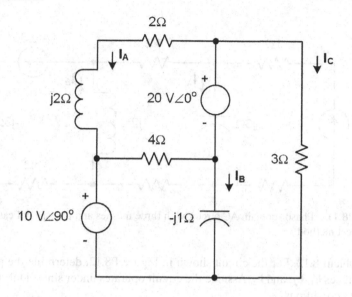

FIGURE P8.9 Phasor-domain AC circuit with two sources, two reactive elements, and three resistors.

Problem 8.10. For the circuit shown in Figure P8.10, determine the phasor voltages V_A, V_B, and V_C. Assume the circuit operates under sinusoidal steady-state conditions.

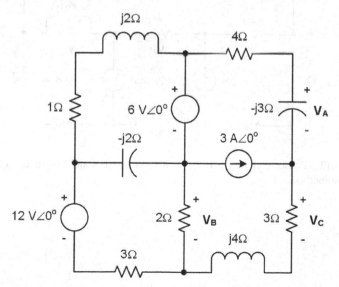

FIGURE P8.10 Phasor-domain AC circuit with five essential nodes and two special cases for the node voltage method.

Problem 8.11. Find the phasor currents I_A and I_B of the circuit shown in Figure P8.11. Assume the circuit operates under sinusoidal steady-state conditions.

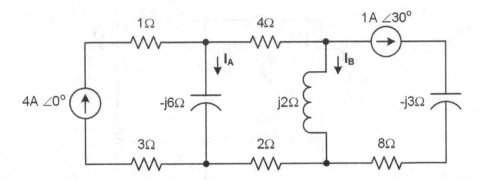

FIGURE P8.11   Phasor-domain AC circuit with three meshes and two special cases for the mesh current method.

Problem 8.12. For the circuit shown in Figure P8.12, determine the phasor voltages V_1, V_2, and V_3. Assume the circuit operates under sinusoidal steady-state conditions.

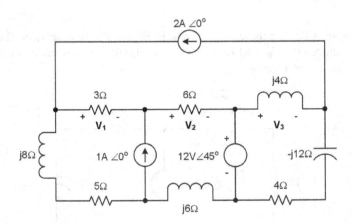

FIGURE P8.12   Phasor-domain AC circuit with four meshes and two special cases for the mesh current method.

Problem 8.13. Find the phasor currents I_X, I_Y, and I_Z of the circuit shown in Figure P8.13. Assume the circuit operates under sinusoidal steady-state conditions.

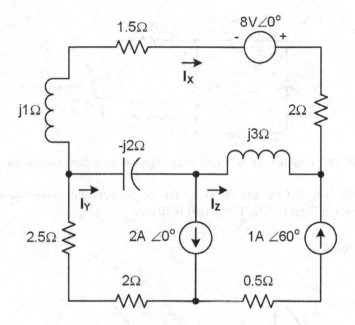

FIGURE P8.13 Phasor-domain AC circuit with a supermesh containing a known mesh current.

Problem 8.14. Determine the phasor currents I_A, I_B, and I_C of the circuit shown in Figure P8.14. Assume the circuit operates under sinusoidal steady-state conditions.

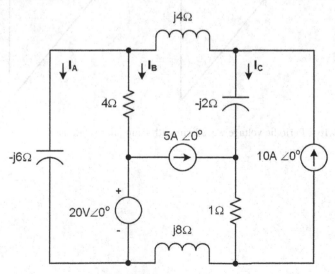

FIGURE P8.14 Phasor-domain AC circuit with four meshes, including a supermesh and a known mesh current.

Problem 8.15. Use superposition to find the time-domain currents $i_1(t)$ and $i_2(t)$ of the circuit in Figure P8.15. Assume the circuit operates under sinusoidal steady-state conditions.

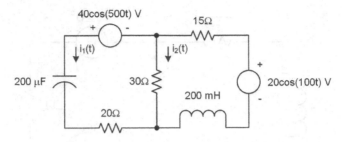

FIGURE P8.15   Time-domain AC circuit operating with two different frequencies.

Problem 8.16. Calculate the rms value of the periodic voltage waveform shown in Figure P8.16. The period is 10 ms.

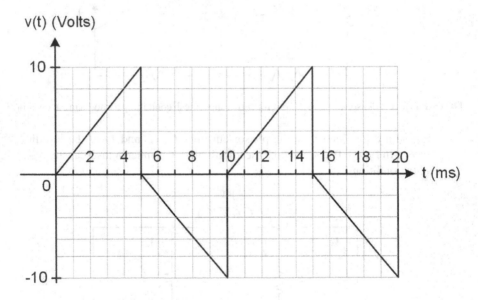

FIGURE P8.16   Periodic voltage waveform with triangular sections.

Problem 8.17. Find the rms value of the periodic voltage waveform of Figure P8.17. The period is 2 ms.

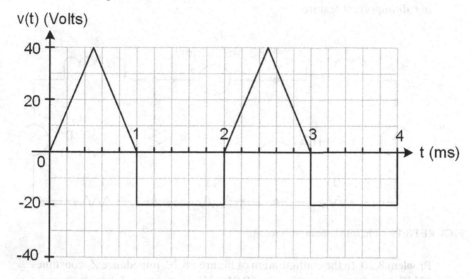

FIGURE P8.17   Periodic voltage waveform having triangular and rectangular sections.

Problem 8.18. Determine the complex power for each of the sources of the circuit in Figure P8.18. Assume the circuit operates under sinusoidal steady-state conditions.

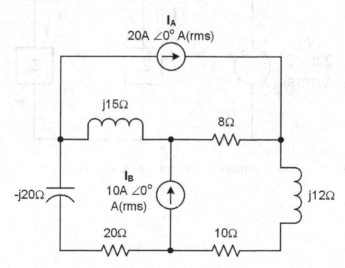

FIGURE P8.18   Phasor-domain AC circuit.

Problem 8.19. For the circuit in Figure P8.19, determine the complex power for each of the sources. Construct the power triangle for each source, labeling all important features.

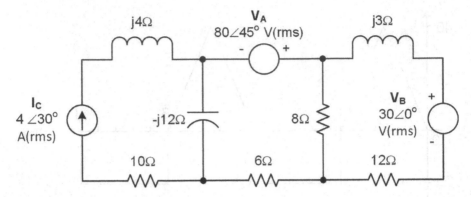

FIGURE P8.19   Phasor-domain AC circuit.

Problem 8.20. In the configuration of Figure P8.20, impedance Z_1 consumes 15 kW at a lagging power factor of 0.94 while impedance Z_2 has an apparent power of 10 kVA at a leading power factor of 0.90. Find the impedances Z_1 and Z_2, the phasor currents I_1 and I_2, and the complex power for the source.

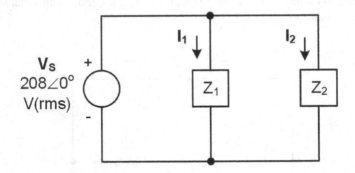

FIGURE P8.20   Parallel impedances connected to a voltage source.

Problem 8.21. For the circuit in Figure P8.21, impedance Z_1 has an apparent power of 40 kVA at a lagging power factor of 0.85 while impedance Z_2 has an apparent power of 30 kVA at a leading power factor of 0.92. Find the impedances Z_1 and Z_2, the phasor currents I_1 and I_2. Determine the value of the impedance Z_{corr} which will correct the source power factor to unity. (If there is more than one possible value of Z_{corr}, it is desirable to minimize power loss.)

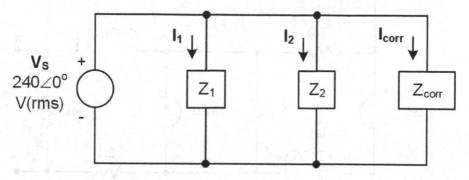

FIGURE P8.21   Three parallel impedances connected to a voltage source.

Problem 8.22. For the circuit in Figure P8.22, impedance Z_1 has an apparent power of 200 kVA at a leading power factor of 0.93 while impedance Z_2 has an apparent power of 300 kVA at a lagging power factor of 0.96. Find the impedances Z_1 and Z_2, the phasor current I_s, the phasor voltages and V_2, and the complex power for the source.

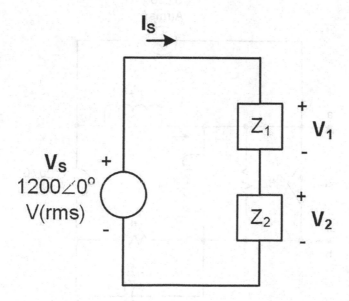

FIGURE P8.22   Series impedances connected to a voltage source.

Problem 8.23. For the network in Figure P8.23, find the Thevenin equivalent with respect to the terminals *a* and *b*.

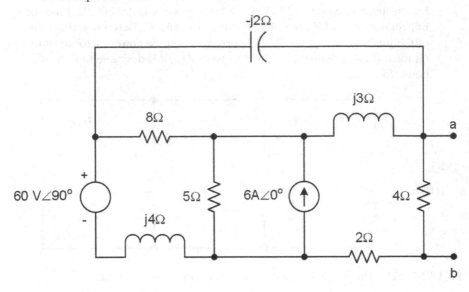

FIGURE P8.23   A two-terminal network containing only independent sources and impedances.

Problem 8.24. For the network in Figure P8.24, find the Norton equivalent with respect to the terminals *a* and *b*.

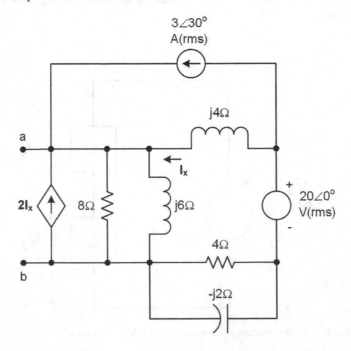

FIGURE P8.24   A two-terminal network containing mixed sources and impedances.

Problem 8.25. For the circuit in Figure P8.25, determine the load impedance Z_L which will dissipate maximum power. Assuming the load impedance is set to this value, find the phasor load voltage V_L, the phasor load current I_L, and the complex power S_L for the load.

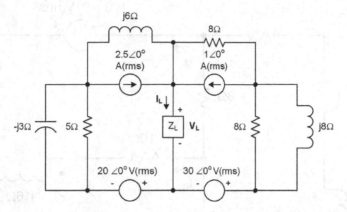

FIGURE P8.25 Two-terminal network with an unspecified load impedance connected.

Problem 8.26. For the circuit in Figure P8.26, plot the average power in the load resistor as a function of the load resistance R_L. Use a logarithmic scale for resistance with a range of resistance values $0.1\,\Omega \le R_L \le 1000\,\Omega$. Find the value of load resistance which will dissipate maximum average power, and for this value of resistance, find the phasor load voltage V_L, the phasor load current I_L, and the load power P_L.

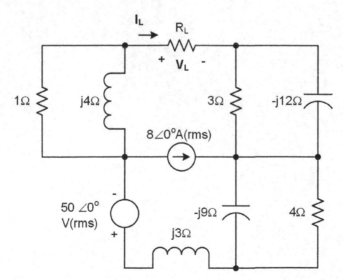

FIGURE P8.26 Two-terminal network with an unspecified load resistance connected.

Problem 8.27. For the circuit in Figure P8.27, calculate the phasor currents I_A, I_B, and I_C, and determine the complex power for each of the voltage sources.

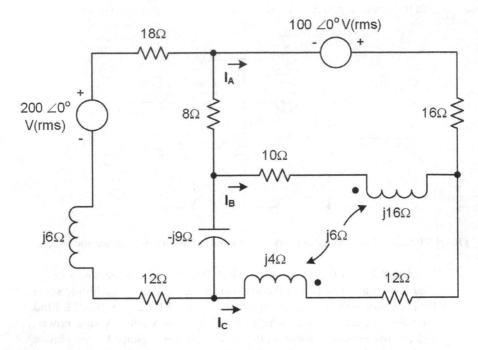

FIGURE P8.27 Circuit involving mutual inductance.

Problem 8.28. For the circuit in Figure P8.28, determine the phasor currents I_A, I_B, and I_C.

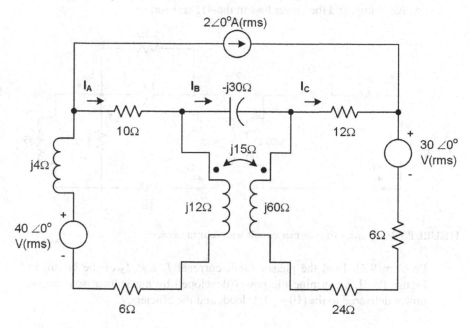

FIGURE P8.28   Circuit involving a linear transformer.

Problem 8.29. Find the phasor mesh currents I_1 and I_2 in the linear transformer circuit of Figure P8.29. Determine the average power developed by the voltage source and the average power delivered to the $2\,\Omega$ resistor.

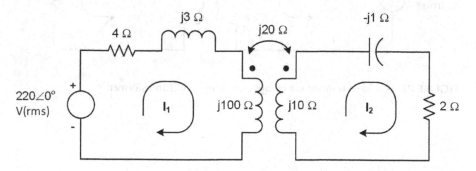

FIGURE P8.29   Linear transformer circuit with two meshes.

Problem 8.30. Find the phasor currents I_1, I_2, I_3, and I_4 in the linear transformer circuit of Figure P8.30. Determine the power developed by the voltage source and the power loss in the 4 Ω resistor.

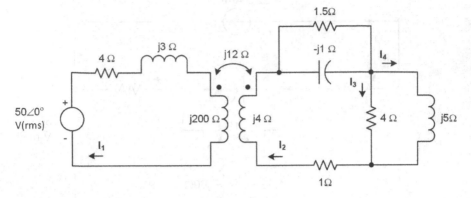

FIGURE P8.30   Linear transformer circuit with four meshes.

Problem 8.31. Find the phasor mesh currents I_1 and I_2 of the circuit in Figure P8.31. Determine the power developed by the voltage source, the power delivered to the $(10 + j2)$ Ω load, and the efficiency.

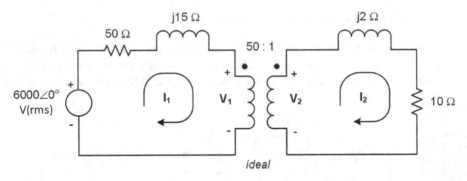

FIGURE P8.31   Ideal transformer circuit containing a 50:1 transformer.

Problem 8.32. Find the phasor mesh currents I_1 and I_2 of the circuit in Figure P8.32. Find the power developed by the voltage source and the power delivered to the $(6+j2)\,\Omega$ load. Calculate the power factor of the $(6+j2)\,\Omega$ load, and compare this to the power factor "seen" by the voltage source.

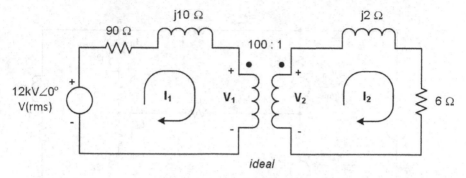

FIGURE P8.32   Ideal transformer circuit containing a 100:1 transformer.

Problem 8.33. For the balanced wye–wye system of Figure P8.33, find the line currents $I_{A'a}$, $I_{B'b}$, and $I_{C'c}$, and plot these on a phasor diagram. Determine the total power developed and the power delivered to the wye-connected load.

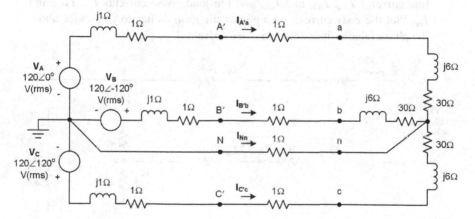

FIGURE P8.33   Balanced wye–wye system.

Problem 8.34. For the balanced delta–delta system of Figure P8.34, find the line currents $I_{A'a}$, $I_{B'b}$, and $I_{C'c}$, and the load phase currents I_{ab}, I_{Bbc}, and I_{ca}. Determine the total power developed and the power delivered to the delta-connected load.

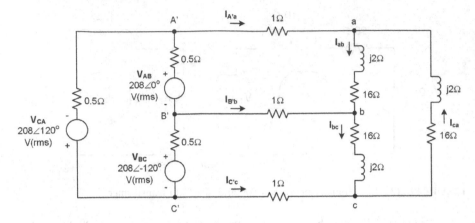

FIGURE P8.34   Balanced delta–delta system.

Problem 8.35. For the balanced wye–delta system of Figure P8.35, find the line currents $I_{A'a}$, $I_{B'b}$, and $I_{C'c}$, and the load phase currents I_{ab}, I_{Bbc}, and I_{ca}. Plot these six currents on a phasor diagram. What do you notice about the phase relationships between these currents?

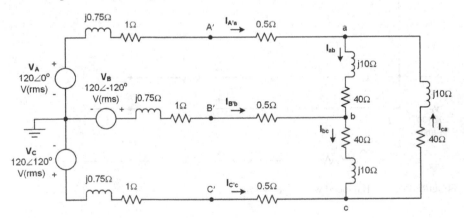

FIGURE P8.35   Balanced wye–delta system.

Problem 8.36. For the unbalanced wye–wye system of Figure P8.33, find the line currents $I_{A'a}$, $I_{B'b}$, and $I_{C'c}$, and the neutral current I_{Nn}. Plot these on a phasor diagram. By how much do the line current phase angle differences depart from the 120° value of the balanced system?

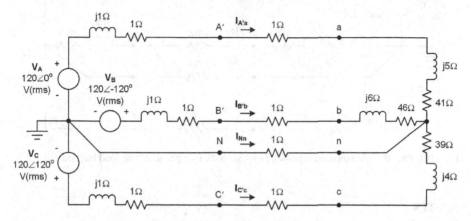

FIGURE P8.36   Unbalanced wye–wye system.

Problem 8.37. For the unbalanced wye–delta system of Figure P8.37, find the line currents $I_{A'a}$, $I_{B'b}$, and $I_{C'c}$, and the load phase currents I_{ab}, I_{bc}, and I_{ca}. Determine the complex power for each voltage source, and the complex power for each of the complex loads.

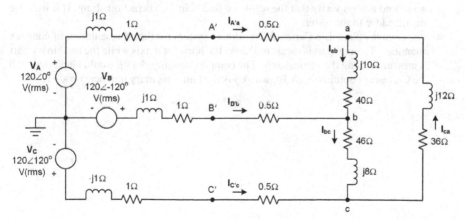

FIGURE P8.37   Unbalanced wye–delta system.

Problem 8.38. The wye–wye system shown in Figure P8.38 was balanced until the c phase burned out on the load side. Find the line currents $I_{A'a}$ and $I_{B'b}$, and the neutral current I_{Nn}. How does the magnitude of the neutral current compare to the magnitude of the other two currents?

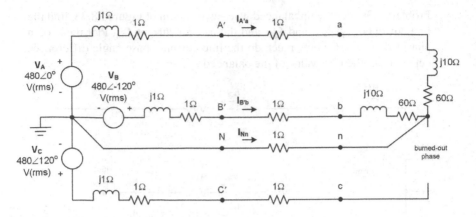

FIGURE P8.38   Unbalanced wye–wye system with one phase of the load burned out.

NOTES

1 Although mathematicians use i to denote the imaginary unit, we as electrical engineers will use j to avoid confusion with our notation for an electrical current.

2 Euler's relationship states that $e^{j\theta} = cos\theta + jsin\theta$.

3 When converting from a rectangular to a polar form, we must realize that $tan^{-1}(-B/-A)$ is not the same as $tan^{-1}(B/A)$, because whereas the latter is in the first quadrant the former is in the third quadrant. For example, $tan^{-1}(-1/-1) = 3\pi/2$ whereas $tan^{-1}(1/1) = \pi/2$. Many calculators do not recognize this difference. Hence, we should always verify that the result we find is in the correct quadrant. If it isn't, we should add π to the result.

4 The complex plane is a Cartesian coordinate system for the representation of complex quantities. The real part is measured along the horizontal axis while the imaginary part is measured along the vertical axis. The complex quantity $A + jB$ would be plotted with the Cartesian coordinates (A,B), where j is the unit imaginary number, $j = \sqrt{-1}$.

9 Frequency Response

9.1 INTRODUCTION

In the previous chapter, we considered the ac steady-state analysis of circuits operating with sinusoidal excitation at a single frequency. When there were multiple sources, they were all considered to operate at the same frequency so we could treat the impedances of capacitors and inductors as constants. If we had multiple sinusoidal sources operating at different frequencies, we could apply the same concepts with each source and use superposition to find the overall time-domain voltages and currents in the circuit. Often, however, we have signals that are non-sinusoidal or non-periodic. Such signals can be considered to be built from sinusoidal components, so the response of a circuit to such general signals can be understood by finding its response to sinusoids as a function of frequency. In this chapter, we will therefore explore the frequency response of circuits containing reactive components such as capacitors and inductors.

To determine the frequency response of a two-port network with an input and an output, we first find the transfer function H, which is the ratio of the phasor output voltage to the phasor input voltage: $H = V_{out} / V_{in}$. In general, this will be a complex quantity that varies with frequency. The amplitude response, or gain, of the circuit is the magnitude of the transfer function as a function of frequency:

$$\text{gain} = |H| = |V_{out} / V_{in}|. \tag{9.1}$$

Sometimes, the amplitude response is plotted in decibel units:

$$\text{gain in dB} = 20\log_{10}|H| = 20\log_{10}|V_{out} / V_{in}|. \tag{9.2}$$

The phase response is the phase of the transfer function as a function of frequency:

$$\text{phase} = \angle H = \angle(V_{out} / V_{in}) = \angle V_{out} - \angle V_{in}. \tag{9.3}$$

Taken together, the amplitude response and phase response specify the frequency response of a circuit.

Generally, circuits will exhibit frequency-dependent behavior whether or not it is desirable, because of the presence of reactive components. However, we may design circuits to achieve specific frequency-dependent behavior, and such circuits are called **filters**.

Broadly, filters may be designed to achieve four types of behavior. A **low-pass filter** allows low-frequency components to pass through while blocking high-frequency components of the signal. There is a transition from the low-frequency

DOI: 10.1201/9781003408529-9

range (the **passband**) to the high-frequency range (the **stopband**) at the **cutoff fre-quency**. A **high-pass filter** allows high frequencies to pass while blocking low fre-quencies, again with a transition at the cutoff frequency. The **bandpass** filter passes a band of frequencies while blocking both higher and lower frequencies. It therefore has two cutoff frequencies. A **bandstop** (or notch) filter blocks only a band of fre-quencies but passes lower and higher frequencies. Like the bandpass filter, it has two cutoff frequencies.

The cutoff frequency for a filter is defined to be the **half-power frequency**; in other words, the cutoff frequency is the frequency at which the power delivered to a resistive load has dropped to half its value in the passband. (If there are two or more cutoff frequencies, the half-power criterion applies to each.) The power delivered to a resistive load is proportional to the square of the magnitude of the output voltage. Hence,

$$\frac{1}{2} = \frac{P_{\text{cutoff}}}{P_{\text{passband}}} = \frac{|V_{\text{out}}|^2_{\text{cutoff}}}{|V_{\text{out}}|^2_{\text{passband}}} = \frac{|H|^2_{\text{cutoff}}}{|H|^2_{\text{passband}}} \tag{9.4}$$

and

$$\frac{|H|_{\text{cutoff}}}{|H|_{\text{passband}}} = \frac{1}{\sqrt{2}}, \tag{9.5}$$

so the cutoff frequency is the frequency at which the gain has dropped to $1/\sqrt{2}$ times the passband gain (~70% of the passband gain). Moreover,

$$20\log_{10}\left(\frac{|H|_{\text{cutoff}}}{|H|_{\text{passband}}}\right) = 20\log_{10}\left(\frac{1}{\sqrt{2}}\right) = -3\,\text{dB}, \tag{9.6}$$

so

$$20\log_{10}\left(|H|_{\text{cutoff}}\right) - 20\log_{10}\left(|H|_{\text{passband}}\right)$$

$$= 20\log_{10}\left(\frac{1}{\sqrt{2}}\right) = -3\,\text{dB}, \tag{9.7}$$

and the cutoff frequency is the frequency at which the gain in dB has dropped by $3\,\text{dB}$ relative to the passband. (It is the $-3\,\text{dB}$ frequency.)

Filters may be realized in passive circuitry (using only passive components such as resistors, inductors, and capacitors) or in active circuitry (using active devices such as operational amplifiers and transistors as well as passive devices). In the following sections, we will investigate a few example filter circuits, both passive and active types, to illustrate the general principles of frequency response analysis. However, the same general approach may be applied to other filters as well, even if they involve higher order and much greater complexity.

To review, see Presentation 9.1 in ebook+.

To test your knowledge, try Quiz 9.1 in ebook+.

9.2 PASSIVE FILTERS

A passive filter is constructed using only passive components (resistors, inductors, and capacitors). An example is the first-order resistor-capacitor (RC) low-pass filter shown in Figure 9.1.

The transfer function for this filter may be found by application of the voltage-divider rule. If we use the compact notation $s = j\omega$, the impedance of the capacitor is $1/(sC)$, and the transfer function is

$$H = \frac{V_{out}}{V_{in}} = \frac{1/(sC)}{R + 1/(sC)} = \frac{1}{1 + sRC}. \tag{9.8}$$

The amplitude response is given by

$$|H| = \frac{1}{|1 + sRC|} = \frac{1}{|1 + j\omega RC|} = \frac{1}{|1 + j2\pi fRC|} = \frac{1}{\sqrt{1 + (2\pi fRC)^2}}. \tag{9.9}$$

The passband gain is unity, so at the cutoff frequency

$$|H| = \frac{1}{\sqrt{1 + (2\pi f_c RC)^2}} = \frac{1}{\sqrt{2}}. \tag{9.10}$$

Solving for the cutoff frequency,

$$f_c = \frac{1}{2\pi RC}. \tag{9.11}$$

The phase response is given by

$$\angle H = \angle \left(\frac{1}{1 + sRC} \right) = -\angle(1 + sRC) = -\angle(1 + j\omega RC)$$

$$= -\angle(1 + j2\pi fRC) = -\tan^{-1}(2\pi fRC). \tag{9.12}$$

Figure 9.2 shows the amplitude and phase response for an RC first-order low-pass filter with $R = 1000\,\Omega$ and $C = 100\,\text{nF}$. For this combination $f_c = 1.59\,\text{kHz}$, and we can

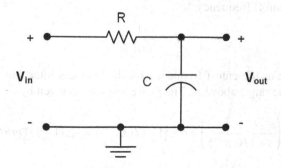

FIGURE 9.1 First-order RC low-pass filter.

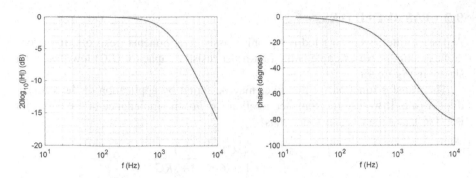

FIGURE 9.2 Amplitude and phase response for an RC first-order low-pass filter.

see that this is the $-3\,\mathrm{dB}$ frequency. The passband gain is unity ($0\,\mathrm{dB}$) and the stop-band slope is $-6\,\mathrm{dB/octave}$ or $-20\,\mathrm{dB/decade}$. (An octave is a factor of two change in frequency while a decade is a factor of ten change in frequency.)

We can also construct a first-order RC high-pass filter, by swapping the positions of the capacitor and resistor from the previous example. This is shown in Figure 9.3.

The transfer function is

$$H = \frac{V_{out}}{V_{in}} = \frac{R}{R + 1/(sC)} = \frac{1}{1 + 1/(sRC)}. \tag{9.13}$$

The amplitude response is given by

$$|H| = \frac{1}{|1 + 1/(sRC)|} = \frac{1}{|1 + 1/(j\omega RC)|} = \frac{1}{|1 - j(1/(2\pi f RC))|} = \frac{1}{\sqrt{1 + 1/(2\pi f RC)^2}}. \tag{9.14}$$

The passband gain is unity, so at the cutoff frequency

$$|H| = \frac{1}{\sqrt{1 + 1/(2\pi f RC)^2}} = \frac{1}{\sqrt{2}}. \tag{9.15}$$

Solving for the cutoff frequency,

$$f_c = \frac{1}{2\pi RC}. \tag{9.16}$$

This is the same as the cutoff frequency for the low-pass filter, but in this case the passband is in the range above f_c. The phase response is given by

$$\angle H = \angle\left(\frac{1}{1 + 1/(sRC)}\right) = -\angle(1 + 1/(sRC)) = -\angle(1 - j/(\omega RC))$$

$$= -\angle(1 - j/(2\pi f RC)) = -\tan^{-1}(-1/(2\pi f RC)). \tag{9.17}$$

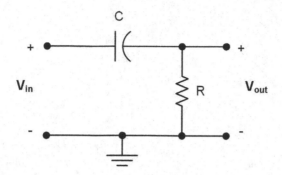

FIGURE 9.3   First-order RC high-pass filter.

Figure 9.4 shows the amplitude and phase response for an RC first-order high-pass filter with $R = 1000\,\Omega$ and $C = 100\,nF$, with $f_c = 1.59\,kHz$. As for the low-pass filter, the passband gain is unity ($0\,dB$) but the stopband slope is positive ($6\,dB$/octave or $20\,dB$/decade).

Passive filters of higher order may be constructed as well. For example, the circuit of Figure 9.5 is a second-order low-pass filter.

The transfer function for this filter may be found by use of the node voltage method, with the essential nodes labeled in Figure 9.6.

The node voltage equation is

$$\frac{V_1 - V_{in}}{R} + \frac{V_1}{1/(sC)} + \frac{V_1}{R + 1/(sC)} = 0. \tag{9.18}$$

Collecting like terms,

$$V_1 \left[\frac{1}{R} + \frac{1}{1/(sC)} + \frac{1}{R + 1/(sC)} \right] = \frac{V_{in}}{R}. \tag{9.19}$$

Multiplying through by R,

$$V_1 \left[1 + \frac{R}{1/(sC)} + \frac{R}{R + 1/(sC)} \right] = V_{in}. \tag{9.20}$$

Simplifying,

$$V_1 \left[1 + sRC + \frac{sRC}{1 + sRC} \right] = V_{in}, \tag{9.21}$$

$$V_1 \left[\frac{(1 + sRC)^2 + sRC}{1 + sRC} \right] = V_{in}, \tag{9.22}$$

and

$$V_1 = V_{in} \left[\frac{1 + sRC}{(1 + sRC)^2 + sRC} \right]. \tag{9.23}$$

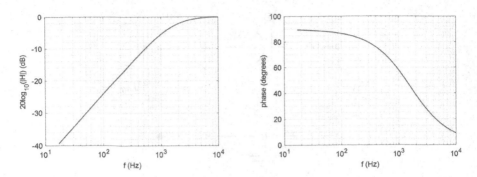

FIGURE 9.4   Amplitude and phase response for an RC first-order high-pass filter.

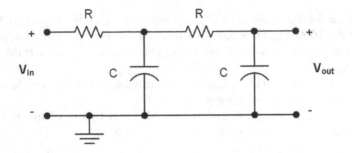

FIGURE 9.5   Second-order low-pass filter.

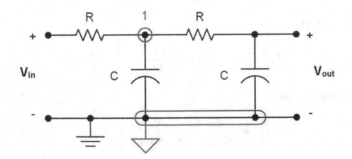

FIGURE 9.6   Second-order low-pass filter labeled for nodal analysis.

By the voltage-divider rule,

$$
V_{out} = V_1 \left[\frac{\dfrac{1}{sC}}{R + \dfrac{1}{sC}} \right] = V_1 \left[\frac{1}{1+sRC} \right] = V_{in} \left[\frac{1+sRC}{(1+sRC)^2 + sRC} \right] \left[\frac{1}{1+sRC} \right],
$$

$$
= V_{in} \left[\frac{1}{(1+sRC)^2 + sRC} \right] = V_{in} \left[\frac{1}{s^2 R^2 C^2 + 3sRC + 1} \right].
$$

(9.24)

The transfer function is therefore

$$H = \frac{V_{\text{out}}}{V_{\text{in}}} = \left[\frac{1}{s^2R^2C^2 + 3sRC + 1}\right] = \frac{1}{\left[1 - (2\pi fRC)^2\right] + j6\pi fRC}. \quad (9.25)$$

Figure 9.7 shows the amplitude and phase response for the second-order filter circuit with $R = 1000\,\Omega$ and $C = 100\,\text{nF}$. Although the corner frequency (the frequency where the two straight-line pieces of the amplitude response appear to meet) is $1.59\,\text{kHz}$, this corner frequency is not the same as the cutoff frequency. This is because the response at the corner frequency is $-6\,\text{dB}$, and the actual cutoff frequency is approximately $600\,\text{Hz}$ and we can see that this is the $-3\,\text{dB}$ frequency. The passband gain is unity ($0\,\text{dB}$) and the stopband slope is twice that for the first-order filter ($-12\,\text{dB/octave}$ or $-40\,\text{dB/decade}$).

To review, see Presentation 9.2 in ebook+

To test your knowledge, try Quiz 9.2 in ebook+. To put your knowledge to practice, try Laboratory Exercise 9.1 in ebook+.

9.3 ACTIVE FILTERS

Active filters are commonly built using operational amplifiers, although they may also be designed with discrete transistors. In either case, these filter circuits are most easily analyzed using the node voltage method.

9.3.1 SALLEN AND KEY LOW-PASS FILTER

As an example of an active filter, consider the Sallen and Key low-pass filter shown in Figure 9.8.

When writing the node voltage equations, we note that the voltage at the non-inverting input of the op amp is equal to V_{out} because of the virtual short. Thus, for node one,

$$N1 \quad \frac{V_1 - V_{\text{in}}}{R} + \frac{V_1 - V_{\text{out}}}{R} + \frac{V_1 - V_{\text{out}}}{1/(sC)} = 0. \quad (9.26)$$

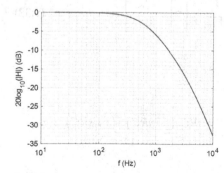
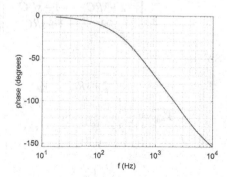

FIGURE 9.7 Amplitude and phase response for a second-order low-pass filter.

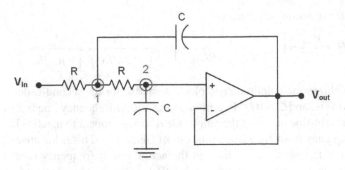

FIGURE 9.8   Sallen and Key low-pass filter.

For the non-inverting input node,

$$N2 \quad \frac{V_{out} - V_1}{R} + \frac{V_{out}}{1/(sC)} = 0. \tag{9.27}$$

Collecting like terms,

$$N1 \quad V_1 \left[\frac{1}{R} + \frac{1}{R} + \frac{1}{1/(sC)} \right] + V_{out} \left[-\frac{1}{R} - \frac{1}{1/(sC)} \right] = \frac{V_{in}}{R} \tag{9.28}$$

and

$$N2 \quad V_1 \left[-\frac{1}{R} \right] + V_{out} \left[\frac{1}{R} + \frac{1}{1/(sC)} \right] = 0. \tag{9.29}$$

Simplifying,

$$N1 \quad V_1 \left[2 + sRC \right] + V_{out} \left[-1 - sRC \right] = V_{in} \tag{9.30}$$

and

$$N2 \quad V_1 \left[-1 \right] + V_{out} \left[1 + sRC \right] = 0. \tag{9.31}$$

In matrix form,

$$\begin{bmatrix} 2 + sRC & -1 - sRC \\ -1 & 1 + sRC \end{bmatrix} \begin{bmatrix} V_1 \\ V_{out} \end{bmatrix} = \begin{bmatrix} V_{in} \\ 0 \end{bmatrix}. \tag{9.32}$$

Solving,

$$V_{out} = \frac{\begin{vmatrix} 2 + sRC & V_{in} \\ -1 & 0 \end{vmatrix}}{\begin{vmatrix} 2 + sRC & -1 - sRC \\ -1 & 1 + sRC \end{vmatrix}} = \frac{V_{in}}{(2 + sRC)(1 + sRC) - (1 + sRC)}$$

$$= \frac{V_{in}}{s^2 R^2 C^2 + 2sRC + 1}. \tag{9.33}$$

The transfer function is

$$H = \frac{1}{s^2 R^2 C^2 + 2sRC + 1}. \tag{9.34}$$

Figure 9.9 shows the amplitude and phase response for this filter, with $R = 10\,\text{k}\Omega$, $C = 2.2\,\text{nF}$, and a corner frequency of $1/(2\pi RC) = 7.23\,\text{kHz}$.

9.3.2 SALLEN AND KEY HIGH-PASS FILTER

As a second example, the Sallen and Key high-pass filter is shown in Figure 9.10. For node one,

$$N1 \quad \frac{V_1 - V_{in}}{1/(sC)} + \frac{V_1 - V_{out}}{1/(sC)} + \frac{V_1 - V_{out}}{R} = 0. \tag{9.35}$$

For the non-inverting input node,

$$N2 \quad \frac{V_{out} - V_1}{1/(sC)} + \frac{V_{out}}{R} = 0. \tag{9.36}$$

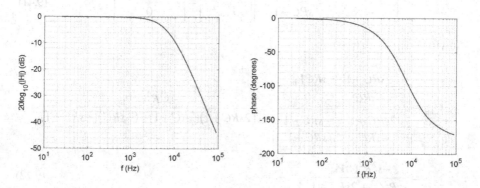

FIGURE 9.9 Amplitude and phase response for a Sallen and Key LPF with $R = 10\,\text{k}\Omega$, $C = 2.2\,\text{nF}$, and a corner frequency of $1/(2\pi RC) = 7.23\,\text{kHz}$.

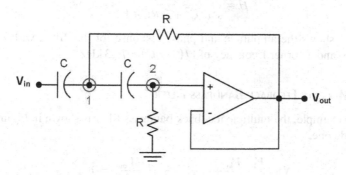

FIGURE 9.10 Sallen and Key high-pass filter.

Collecting like terms,

$$N1 \quad V_1\left[\frac{1}{1/(sC)}+\frac{1}{1/(sC)}+\frac{1}{R}\right]+V_{\text{out}}\left[-\frac{1}{1/(sC)}-\frac{1}{R}\right]=\frac{V_{\text{in}}}{1/(sC)} \tag{9.37}$$

and

$$V_1\left[-\frac{1}{1/(sC)}\right]+V_{\text{out}}\left[\frac{1}{1/(sC)}+\frac{1}{R}\right]=0. \tag{9.38}$$

Simplifying,

$$N1 \quad V_1[2sRC+1]+V_{\text{out}}[-sRC-1]=sRCV_{\text{in}} \tag{9.39}$$

and

$$N2 \quad V_1[-sRC]+V_{\text{out}}[sRC+1]=0. \tag{9.40}$$

In matrix form,

$$\begin{bmatrix} 2sRC+1 & -sRC-1 \\ -sRC & sRC+1 \end{bmatrix}\begin{bmatrix} V_1 \\ V_{\text{out}} \end{bmatrix}=\begin{bmatrix} sRCV_{\text{in}} \\ 0 \end{bmatrix}. \tag{9.41}$$

Solving,

$$V_{\text{out}} = \frac{\begin{vmatrix} 2sRC+1 & sRCV_{\text{in}} \\ -sRC & 0 \end{vmatrix}}{\begin{vmatrix} 2sRC+1 & -sRC-1 \\ -sRC & sRC+1 \end{vmatrix}} = \frac{-s^2R^2C^2V_{\text{in}}}{(2sRC+1)(sRC+1)-(-sRC)(-sRC-1)}$$

$$= \frac{-s^2R^2C^2V_{\text{in}}}{s^2R^2C^2+2sRC+1}. \tag{9.42}$$

The transfer function is

$$H = \frac{-s^2R^2C^2}{s^2R^2C^2+2sRC+1}. \tag{9.43}$$

Figure 9.11 shows the amplitude and phase response for this filter, with $R=10\,\text{k}\Omega$, $C=2.2\,\text{nF}$, and a corner frequency of $1/(2\pi RC)=7.23\,\text{kHz}$.

9.3.3 MULTIPLE FEEDBACK BANDPASS FILTER

As a third example, the multiple feedback bandpass filter is shown in Figure 9.12.
    For node one,

$$N1 \quad \frac{V_1-V_{\text{in}}}{R_1}+\frac{V_1}{1/(sC)}+\frac{V_1-V_{\text{out}}}{1/(sC)}=0. \tag{9.44}$$

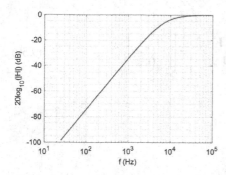
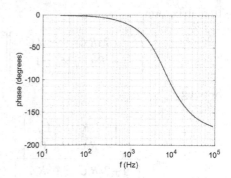

FIGURE 9.11 Amplitude and phase response for a Sallen and Key high-pass filter (HPF) with $R = 10\,k\Omega$, $C = 2.2\,nF$, and a corner frequency of $1/(2\pi RC) = 7.23\,kHz$.

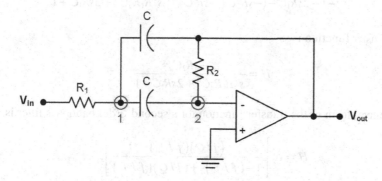

FIGURE 9.12 Multiple feedback bandpass filter.

For the inverting input node,

$$N2 \quad \frac{-V_1}{1/(sC)} + \frac{-V_{out}}{R_2} = 0. \tag{9.45}$$

Collecting like terms,

$$N1 \quad V_1\left[\frac{1}{R_1} + \frac{1}{1/(sC)} + \frac{1}{1/(sC)}\right] + V_{out}\left[-\frac{1}{1/(sC)}\right] = \frac{V_{in}}{R_1} \tag{9.46}$$

and

$$N2 \quad V_1\left[-\frac{1}{1/(sC)}\right] + V_{out}\left[-\frac{1}{R_2}\right] = 0. \tag{9.47}$$

Simplifying,

$$N1 \quad V_1[1 + 2sR_1C] + V_{out}[-sR_1C] = V_{in} \tag{9.48}$$

and

$$V_1[-sR_2C] + V_{out}[-1] = 0. \tag{9.49}$$

In matrix form,

$$\begin{bmatrix} 1+2sR_1C & -sR_1C \\ -sR_2C & -1 \end{bmatrix} \begin{bmatrix} V_1 \\ V_{out} \end{bmatrix} = \begin{bmatrix} V_{in} \\ 0 \end{bmatrix}. \tag{9.50}$$

Solving,

$$V_{out} = \frac{\begin{vmatrix} 1+2sR_1C & V_{in} \\ -sR_2C & 0 \end{vmatrix}}{\begin{vmatrix} 1+2sR_1C & -sR_1C \\ -sR_2C & -1 \end{vmatrix}} \tag{9.51}$$

$$= \frac{-sR_2CV_{in}}{-1-2sR_1C-(-sR_2C)(-sR_1C)} = \frac{-sR_2CV_{in}}{s^2R_1R_2C^2+2sR_1C+1}.$$

The transfer function is

$$H = \frac{-sR_2C}{s^2R_1R_2C^2+2sR_1C+1}. \tag{9.52}$$

The general form of the transfer function for a second-order bandpass filter is

$$H = A_r\left[\frac{(j/Q)(f/f_0)}{1-(f/f_0)^2+(j/Q)(f/f_0)}\right], \tag{9.53}$$

where f_0 is the center frequency for the passband, A_r is the resonant gain at the center frequency, and Q is the quality factor. One interpretation of the quality factor is the ratio of the center frequency to the bandwidth, where the bandwidth is the difference between the two cutoff frequencies. By comparing these two equations for the transfer function, we find that for the multiple feedback bandpass filter,

$$f_0 = \frac{1}{2\pi C\sqrt{R_1R_2}}, \tag{9.54}$$

$$A_r = -\frac{R_2}{2R_1}, \tag{9.55}$$

and

$$Q = \frac{1}{2}\sqrt{\frac{R_2}{R_1}}. \tag{9.56}$$

Figure 9.13 shows the amplitude and phase response for two designs of this filter. For the first design, $C = 10\,nF$, $R_1 = 1\,k\Omega$, and $R_2 = 4\,k\Omega$, giving a center frequency of 7.96 kHz, a quality factor of 1.0, and a bandwidth of 7.96 kHz. For the second design,

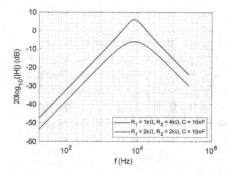 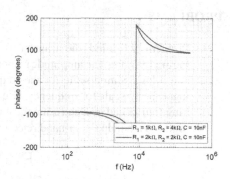

FIGURE 9.13 Amplitude and phase response for multiple feedback bandpass filters with two designs. For the first design, $C = 10\,\text{nF}$, $R_1 = 1\,\text{k}\Omega$, and $R_2 = 4\,\text{k}\Omega$, giving a center frequency of $7.96\,\text{kHz}$, a quality factor of 1.0, and a bandwidth of $7.96\,\text{kHz}$. For the second design, $C = 10\,\text{nF}$, $R_1 = 2\,\text{k}\Omega$, and $R_2 = 2\,\text{k}\Omega$, giving a center frequency of $7.96\,\text{kHz}$, a quality factor of 0.50, and a bandwidth of $15.92\,\text{kHz}$.

$C = 10\,\text{nF}$, $R_1 = 2\,\text{k}\Omega$, and $R_2 = 2\,\text{k}\Omega$, giving a center frequency of $7.96\,\text{kHz}$, a quality factor of 0.50, and a bandwidth of $15.92\,\text{kHz}$.

To review, see Presentation 9.3 in ebook+.

To test your knowledge, try Quiz 9.3. To put your knowledge to practice, try Laboratory Exercises 9.2 and 9.3 in ebook+.

9.4 SUMMARY

In circuits with reactive components (capacitors and inductors), the phasor voltages and currents vary with frequency. **Filters** are circuits designed to be frequency selective in a desired fashion. Broadly speaking, there are four basic types of filters: **low pass, high pass, bandpass**, and **bandstop**. Filters may be further characterized as **passive** or **active**; a passive filter uses only passive components (resistors, capacitors, inductors) whereas an active filter also contains active components such as transistors or operational amplifiers. The **transfer function** H for a filter is the phasor ratio of the output and input voltages: $H = V_{out} / V_{in}$. Generally, the transfer function for a circuit may be found by the use of the node voltage method. The **amplitude response** is the magnitude of H as a function of frequency. The **phase response** is the phase angle of H as a function of frequency. Together, the amplitude response and phase response make up the **frequency response** of the filter. Often, the amplitude response is plotted in **decibel (dB)** units, where the amplitude response in dB is $20\log_{10}|H|$. The cutoff frequency for a filter is the half-power or -3dB frequency. If a filter has more than one cutoff frequency, this criterion applies to each.

To evaluate your comprehensive knowledge of Chapters 1–9, solve Example Exam 9.1 in ebook+ or Example Exam 9.2 in ebook+.

PROBLEMS

Problem 9.1. Find the transfer function for the filter circuit shown in Figure P9.1. Plot the amplitude response using a decibel scale for the response and a logarithmic scale for the frequency. Plot the phase response using a logarithmic scale for frequency. Use an appropriate range of frequency spanning four decades. Determine the nature of the filter response (i.e., low pass, high pass, bandpass, or band reject) and determine the cutoff frequency(ies).

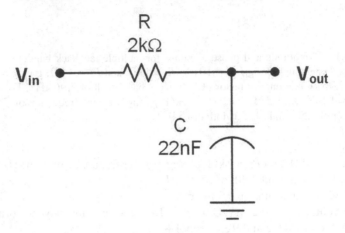

FIGURE P9.1   First-order RC low-pass filter with $R = 2k\Omega$ and $C = 22$ nF.

Problem 9.2. Find the transfer function for the filter circuit shown in Figure P9.2. Plot the amplitude response using a decibel scale for the response and a logarithmic scale for the frequency. Plot the phase response using a logarithmic scale for frequency. Use an appropriate range of frequency spanning four decades. Determine the nature of the filter response (i.e., low pass, high pass, bandpass, or band reject) and determine the cutoff frequency(ies).

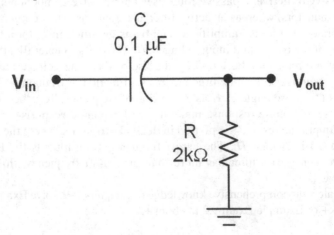

FIGURE P9.2   First-order RC high-pass filter with $R = 2k\Omega$ and $C = 0.1$ μF.

Problem 9.3. Find the transfer function for the filter circuit shown in Figure P9.3. Plot the amplitude response using a decibel scale for the response and a logarithmic scale for the frequency. Plot the phase response using a logarithmic scale for frequency. Use an appropriate range of frequency spanning four decades. Determine the nature of the filter response (i.e., low pass, high pass, bandpass, or band reject) and determine the cutoff frequency(ies).

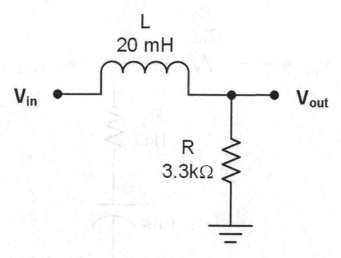

FIGURE P9.3 First-order resistor-inductor (RL) low-pass filter with $R = 3.3$ kΩ and $L = 20$ mH.

Problem 9.4. Find the transfer function for the filter circuit shown in Figure P9.4. Plot the amplitude response using a decibel scale for the response and a logarithmic scale for the frequency. Plot the phase response using a logarithmic scale for frequency. Use an appropriate range of frequency spanning four decades. Determine the nature of the filter response (i.e., low pass, high pass, bandpass, or band reject) and determine the cutoff frequency(ies).

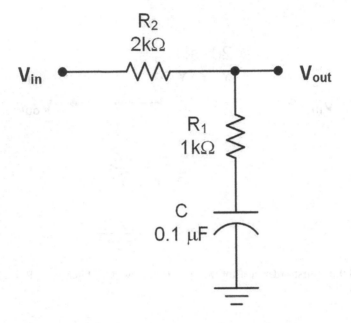

FIGURE P9.4   First-order RC high-pass filter including two resistors.

Problem 9.5. Find the transfer function for the filter circuit shown in Figure P9.5. Plot the amplitude response using a decibel scale for the response and a logarithmic scale for the frequency. Plot the phase response using a logarithmic scale for frequency. Use an appropriate range of frequency spanning four decades. Determine the nature of the filter response (i.e., low pass, high pass, bandpass, or band reject) and determine the cutoff frequency(ies).

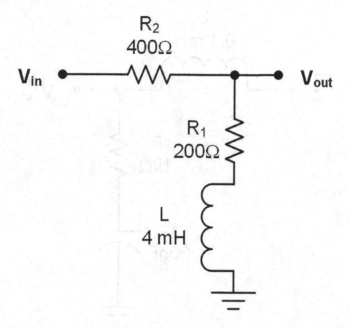

FIGURE P9.5   First-order RL high-pass filter including two resistors.

Problem 9.6. Find the transfer function for the filter circuit shown in Figure
P9.6. Plot the amplitude response using a decibel scale for the response
and a logarithmic scale for the frequency. Plot the phase response using
a logarithmic scale for frequency. Use an appropriate range of frequency
spanning four decades. Determine the nature of the filter response (i.e.,
low pass, high pass, bandpass, or band reject) and determine the cutoff
frequency(ies).

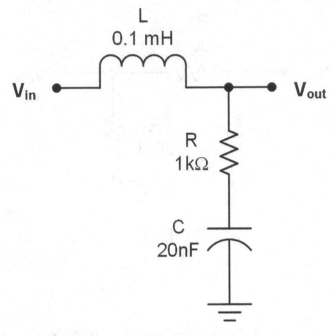

FIGURE P9.6 Second-order resistor-inductor-capacitor (RLC) low-pass filter with R = 1
kΩ, L = 0.1 mH, and C = 20 nF.

Problem 9.7. Find the transfer function for the filter circuit shown in Figure P9.7. Plot the amplitude response using a decibel scale for the response and a logarithmic scale for the frequency. Plot the phase response using a logarithmic scale for frequency. Use an appropriate range of frequency spanning four decades. Determine the nature of the filter response (i.e., low pass, high pass, bandpass, or band reject) and determine the cutoff frequency(ies).

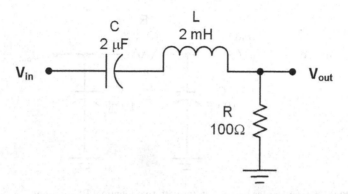

FIGURE P9.7   Second-order RLC filter with R = 100 Ω, L = 2 mH, and C = 2 μF.

Problem 9.8. Find the transfer function for the filter circuit shown in Figure P9.8. Plot the amplitude response using a decibel scale for the response and a logarithmic scale for the frequency. Plot the phase response using a logarithmic scale for frequency. Use an appropriate range of frequency spanning four decades. Determine the nature of the filter response (i.e., low pass, high pass, bandpass, or band reject) and determine the cutoff frequency(ies).

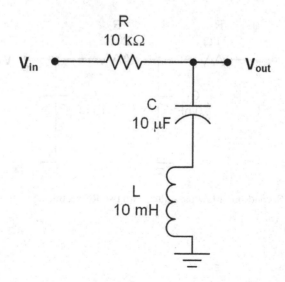

FIGURE P9.8   Second-order RLC filter with R = 10 kΩ, L = 10 mH, and C = 10 μF.

Problem 9.9. Find the transfer function for the filter circuit shown in Figure P9.9. Plot the amplitude response using a decibel scale for the response and a logarithmic scale for the frequency. Plot the phase response using a logarithmic scale for frequency. Use an appropriate range of frequency spanning four decades. Determine the nature of the filter response (i.e., low pass, high pass, bandpass, or band reject) and determine the cutoff frequency(ies).

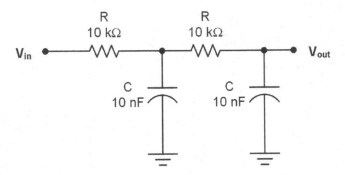

FIGURE P9.9   Second-order low-pass filter with two identical RC sections.

Problem 9.10. Find the transfer function for the filter circuit shown in Figure P9.10. Plot the amplitude response using a decibel scale for the response and a logarithmic scale for the frequency. Plot the phase response using a logarithmic scale for frequency. Use an appropriate range of frequency spanning four decades. Determine the nature of the filter response (i.e., low pass, high pass, bandpass, or band reject) and determine the cutoff frequency(ies).

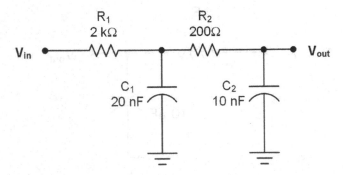

FIGURE P9.10   Second-order low-pass filter with two RC sections.

Problem 9.11. Find the transfer function for the filter circuit shown in Figure P9.11. Plot the amplitude response using a decibel scale for the response and a logarithmic scale for the frequency. Plot the phase response using a logarithmic scale for frequency. Use an appropriate range of frequency spanning four decades. Determine the nature of the filter response (i.e., low pass, high pass, bandpass, or band reject) and determine the cutoff frequency(ies).

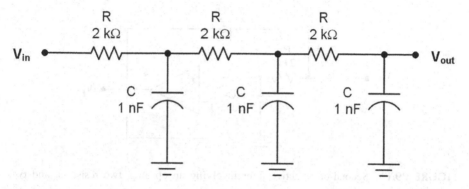

FIGURE P9.11   Third-order low-pass filter with three identical RC sections.

Problem 9.12. Find the transfer function for the filter circuit shown in Figure P9.12. Plot the amplitude response using a decibel scale for the response and a logarithmic scale for the frequency. Plot the phase response using a logarithmic scale for frequency. Use an appropriate range of frequency spanning four decades. Determine the nature of the filter response (i.e., low pass, high pass, bandpass, or band reject) and determine the cutoff frequency(ies).

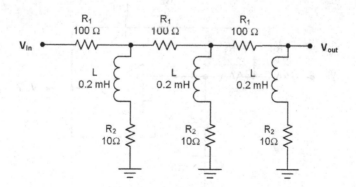

FIGURE P9.12   Third-order filter with three RL sections.

Problem 9.13. Find the transfer function for the filter circuit shown in Figure P9.13. Plot the amplitude response using a decibel scale for the response and a logarithmic scale for the frequency. Plot the phase response using a logarithmic scale for frequency. Use an appropriate range of frequency spanning four decades. Determine the nature of the filter response (i.e., low pass, high pass, bandpass, or band reject) and determine the cutoff frequency(ies).

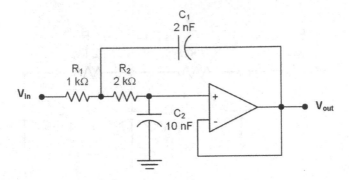

FIGURE P9.13 Second-order active filter involving an op amp, two resistors, and two capacitors.

Problem 9.14. Find the transfer function for the filter circuit shown in Figure P9.14. Plot the amplitude response using a decibel scale for the response and a logarithmic scale for the frequency. Plot the phase response using a logarithmic scale for frequency. Use an appropriate range of frequency spanning four decades. Determine the nature of the filter response (i.e., low pass, high pass, bandpass, or band reject) and determine the cutoff frequency(ies).

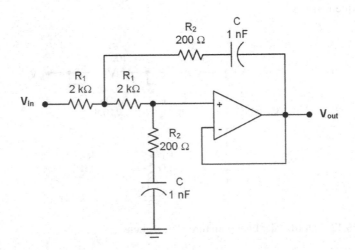

FIGURE P9.14 Second-order active filter involving an op amp, four resistors, and two capacitors.

Problem 9.15. Find the transfer function for the filter circuit shown in Figure P9.15. Plot the amplitude response using a decibel scale for the response and a logarithmic scale for the frequency. Plot the phase response using a logarithmic scale for frequency. Use an appropriate range of frequency spanning four decades. Determine the nature of the filter response (i.e., low pass, high pass, bandpass, or band reject) and determine the cutoff frequency(ies).

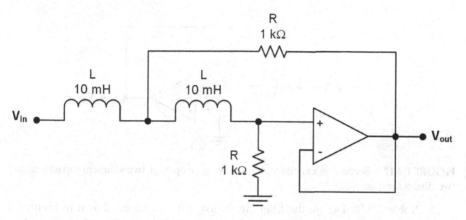

FIGURE P9.15   Second-order active filter involving an op amp, two resistors, and two inductors.

Problem 9.16. Find the transfer function for the filter circuit shown in Figure P9.16. Plot the amplitude response using a decibel scale for the response and a logarithmic scale for the frequency. Plot the phase response using a logarithmic scale for frequency. Use an appropriate range of frequency spanning four decades. Determine the nature of the filter response (i.e., low pass, high pass, bandpass, or band reject) and determine the cutoff frequency(ies).

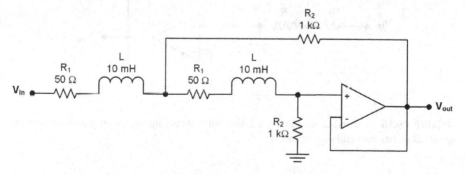

FIGURE P9.16   Second-order active filter involving an op amp, four resistors, and two capacitors.

Problem 9.17. Find the transfer function for the filter circuit shown in Figure P9.17. Plot the amplitude response using a decibel scale for the response and a logarithmic scale for the frequency. Plot the phase response using a logarithmic scale for frequency. Use an appropriate range of frequency spanning four decades. Determine the nature of the filter response (i.e., low pass, high pass, bandpass, or band reject) and determine the cutoff frequency(ies).

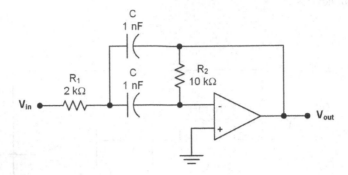

FIGURE P9.17   Second-order active filter involving an op amp, two identical capacitors, and two distinct resistors.

Problem 9.18. Design the filter circuit using the topology shown in Figure P9.18 to achieve a nominal cutoff frequency of 15 kHz ±10%. (The nominal cutoff frequency is the value calculated without taking component tolerances into account. Note that because this is a second-order filter the corner frequency will differ slightly from the cutoff frequency.) Use standard 5% tolerance resistor values and standard 20% tolerance capacitor values. Verify the amplitude response and cutoff frequency.

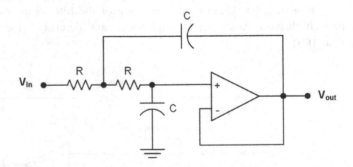

FIGURE P9.18   Second-order low-pass active filter involving an op amp and unspecified values of resistances and capacitances.

Problem 9.19. Design the filter circuit using the topology shown in Figure P9.19 to achieve a nominal quality factor of 2.0 and a nominal center frequency of 5 kHz ± 10%. (The nominal values are those calculated without taking component tolerances into account.) Use standard 5% tolerance resistor values and standard 10% tolerance capacitor values. Plot the amplitude response, verify the center frequency, and determine the bandwidth for the filter.

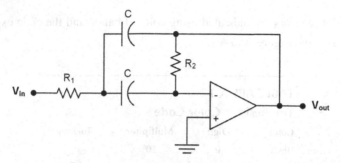

FIGURE P9.19 Second-order bandpass active filter involving an op amp and unspecified values of resistances and capacitances.

Problem 9.20. Design the filter circuit using the topology shown in Figure P9.20 to achieve a nominal quality factor of 5.0 and a nominal center frequency of 10 kHz ± 10%. (The nominal values are those calculated without taking component tolerances into account.) Use standard 5% tolerance resistor values and standard 20% tolerance capacitor values. Plot the amplitude response, verify the center frequency, and determine the bandwidth for the filter.

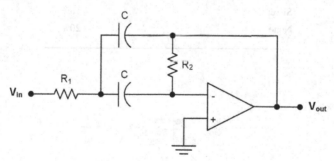

FIGURE P9.20 Second-order bandpass active filter involving two capacitors and two distinct resistors, all of which have unspecified values.

APPENDIX A
Resistor Color Code

The values of resistors are indicated using colored bands and the code explained in Table App A.1 and Figure App A.1.

TABLE APP A.1
The Resistor Color Code

Color	Digit	Multiplier	Tolerance
Black	0	10^0	
Brown	1	10^1	1%
Red	2	10^2	2%
Orange	3	10^3	
Yellow	4	10^4	
Green	5	10^5	0.5%
Blue	6	10^6	0.25%
Violet	7	10^7	0.1%
Grey	8	10^8	
White	9	10^9	
Gold			5%
Silver			10%
None			20%

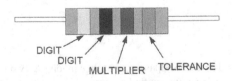

DIGIT
DIGIT
MULTIPLIER   TOLERANCE

FIGURE APP A.1   Explanation of the bands used for the resistor color code.

APPENDIX B
Standard Values of 5% Resistors

The commercial values of resistors with 5% tolerance have been standardized to those given in Table App B.1. These values repeat every decade. Therefore, 15 Ω, 150 Ω, and 1.5 kΩ are all standard values. The adjacent values differ by approximately 10% so the ranges will meet or overlap slightly.

TABLE APP B.1
Standard Values of Resistors with 5% Tolerance

1.0	1.1	1.2	1.3	1.5	1.6
1.8	2.0	2.2	2.4	2.7	3.0
3.3	3.6	3.9	4.3	4.7	5.1
5.6	6.2	6.8	7.5	8.2	9.1

The values given repeat every decade.

APPENDIX C
Standard Values of 10% Capacitors

The commercial values of capacitors with 10% tolerance have been standardized to those given in Table App C.1. The values repeat every decade. Therefore, 22 pF, 220 nF, and 2.2 μF are all standard values. The adjacent values differ by approximately 20% so the ranges will meet or overlap slightly.

TABLE APP C.1
Standard Values for Capacitors with 10% Tolerance

1.0	1.2	1.5	1.8	2.2	2.7
3.3	3.9	4.7	5.6	6.8	8.2

The values repeat every decade.

APPENDIX D
Ceramic Capacitors

Capacitor values are generally given in pF (1 pF= 10^{-12} F). A ceramic capacitor will typically give its value using three numbers; the first two represent digits and the third is a multiplier. For example, "223" designates 22,000 pF or 22 nF as shown in Figure App D.1.

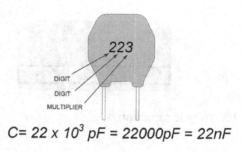

$$C= 22 \times 10^3\ pF = 22000pF = 22nF$$

FIGURE APP D.1   A ceramic capacitor with an explanation of the capacitance code.

APPENDIX E
Electrolytic Capacitors

Electrolytic capacitors generally give their capacitance in μF as in the example shown in Figure App E.1. Also given on the case is the voltage rating and the polarity (for polarized capacitors). **It is important to not reverse the polarity across an electrolytic capacitor; it may evolve hydrogen gas, explode, and cause personal injury.**

FIGURE APP E.1   Electrolytic capacitor.

APPENDIX F
Complex Numbers

In order to solve AC circuits algebraically, without the need for differential equations, we will make extensive use of complex numbers and complex mathematics. It should be understood that this represents a system of shortcuts and that we are using complex quantities to represent real circuit elements, real currents, and real voltages. Nonetheless, working with complex quantities is a critical part of sinusoidal steady-state analysis so we will review their mathematics.

A complex number $A + jB$ involves a real component A and an imaginary component, jB, where j is the imaginary unit,[1] $j = \sqrt{-1}$. This complex number may be shown on the complex plane using Cartesian (rectangular) coordinates or polar coordinates of the form $R\angle\theta$, as shown in Figure App F.1.

Using Euler's relationship,[2] we can convert from rectangular coordinates to polar coordinates[3]:

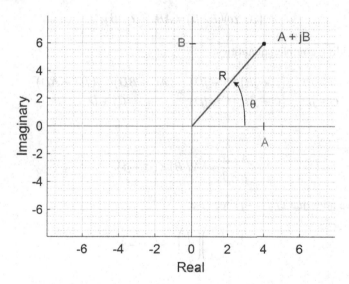

FIGURE APP F.1   A complex quantity $A + jB$ shown in the complex plane.

$$\underbrace{A + jB}_{\text{rectangular coordinates}} \rightarrow \underbrace{\left(A^2 + B^2\right)^{1/2} e^{j\tan^{-1}(B/A)}}_{\text{polar coordinates}} = Re^{j\theta}. \qquad \text{(F.1)}$$

We can also convert from polar coordinates to rectangular coordinates:

$$\underbrace{Re^{j\theta}}_{\text{polar coordinates}} \rightarrow \underbrace{R\cos\theta + jR\sin\theta = A + jB.}_{\text{rectangular coordinates}} \qquad (F.2)$$

Often, we will express complex numbers in polar form using the shorthand notation $R\angle\theta$; it should be recognized that this means the same thing as $Re^{j\theta}$.

When adding complex quantities,

$$(A + jB) + (C + jD) = (A + C) + j(B + D) \qquad (F.3)$$

and

$$A\angle\alpha + B\angle\beta = (A\cos\alpha + B\cos\beta) + j(A\sin\alpha + B\sin\beta). \qquad (F.4)$$

When multiplying complex quantities,

$$(A + jB)(C + jD) = (AC - BD) + j(BC + AD) \qquad (F.5)$$

and

$$(A\angle\alpha)(B\angle\beta) = AB\angle(\alpha + \beta). \qquad (F.6)$$

When dividing complex quantities,

$$\frac{A + jB}{C + jD} = \frac{(A + jB)}{(C + jD)}\frac{(C - jD)}{(C - jD)} = \frac{(AC + BD) + j(BC - AD)}{C^2 + D^2} \qquad (F.7)$$

and

$$\frac{A\angle\alpha}{B\angle\beta} = (A / B)\angle(\alpha - \beta). \qquad (F.8)$$

Finally, we note that $1/j = -j$, because

$$\frac{1}{j} = \left(\frac{1}{j}\right)\left(\frac{j}{j}\right) = \frac{j}{-1} = -j. \qquad (F.9)$$

NOTES

1 Although mathematicians use i to denote the imaginary unit, we as electrical engineers will use j to avoid confusion with our notation for an electrical current.
2 Euler's relationship states that $e^{j\theta} = \cos\theta + j\sin\theta$.
3 When converting from a rectangular to a polar form, we must realize that $tan^{-1}(-B/-A)$ is not the same as $tan^{-1}(B/A)$, because whereas the latter is in the first quadrant the former is in the third quadrant. For example, $tan^{-1}(-1/-1) = 3\pi/2$ whereas $tan^{-1}(1/1) = \pi/2$. Many calculators do not recognize this difference. Hence, we should always verify that the result we find is in the correct quadrant. If it isn't, we should add π to the result.

APPENDIX G
Cramer's Method

When using the node-voltage method or the mesh-current method, we need to solve systems of simultaneous equations. One way to do this is by Cramer's method. Cramer's method is a matrix approach which is convenient when solving either symbolically, as when finding a transfer function, or numerically, as when finding specific voltages. Other matrix approaches may be used as well, and in general, these are more convenient than substitution or subtraction when solving three or more equations. These other approaches are discussed in detail in books on linear algebra. Here though, we focus on Cramer's method, which is used throughout this book for solving node and mesh equations.

Suppose we need to solve a system of two equations involving two unknowns x_1 and x_2:

$$A_{11}x_1 + A_{12}x_2 = B_1 \tag{G.1}$$

and

$$A_{21}x_1 + A_{22}x_2 = B_2. \tag{G.2}$$

These two equations may be written in matrix form as

$$\begin{bmatrix} A_{11} & A_{12} \\ A_{21} & A_{22} \end{bmatrix} \begin{bmatrix} x_1 \\ x_2 \end{bmatrix} = \begin{vmatrix} B_1 \\ B_2 \end{vmatrix}. \tag{G.3}$$

To solve for each of the unknowns, we calculate a ratio of determinants. When finding the first unknown x_1, the numerator is the determinant of a modified matrix involving the replacement of the *first* column by the matrix B, and the denominator is the determinant of the original matrix:

$$x_1 = \frac{\begin{vmatrix} B_1 & A_{12} \\ B_2 & A_{22} \end{vmatrix}}{\begin{vmatrix} A_{11} & A_{12} \\ A_{21} & A_{22} \end{vmatrix}} = \frac{B_1 A_{22} - B_2 A_{12}}{A_{11}A_{22} - A_{21}A_{12}}. \tag{G.4}$$

When finding the second unknown x_2, the numerator is the determinant of a modified matrix involving the replacement of the *second* column by the matrix B, and the denominator is the determinant of the original matrix:

423

$$x_2 = \frac{\begin{vmatrix} A_{11} & B_1 \\ A_{21} & B_2 \end{vmatrix}}{\begin{vmatrix} A_{11} & A_{12} \\ A_{21} & A_{22} \end{vmatrix}} = \frac{B_1 A_{22} - B_2 A_{12}}{A_{11} A_{22} - A_{21} A_{12}}. \tag{G.5}$$

We can extend this to a system of three or more equations as well. For example, in the case of three equations in three unknowns x_1, x_2, and x_3, the system of equations could be written as

$$A_{11}x_1 + A_{12}x_2 + A_{13}x_3 = B_1, \tag{G.6}$$

$$A_{21}x_1 + A_{22}x_2 + A_{23}x_3 = B_2, \tag{G.7}$$

and

$$A_{31}x_1 + A_{32}x_2 + A_{33}x_3 = B_3. \tag{G.8}$$

In matrix form,

$$\begin{bmatrix} A_{11} & A_{12} & A_{13} \\ A_{21} & A_{22} & A_{23} \\ A_{31} & A_{32} & A_{33} \end{bmatrix} \begin{bmatrix} x_1 \\ x_2 \\ x_3 \end{bmatrix} = \begin{vmatrix} B_1 \\ B_2 \\ B_3 \end{vmatrix}. \tag{G.9}$$

To solve for the nth unknown, we take the ratio of two determinants; the numerator is the determinant of a modified matrix in which the nth column has been replaced by matrix B and the denominator is the determinant of the original matrix. For example, to find x_1,

$$x_1 = \frac{\begin{vmatrix} B_1 & A_{12} & A_{13} \\ B_2 & A_{22} & A_{23} \\ B_3 & A_{32} & A_{33} \end{vmatrix}}{\begin{vmatrix} A_{11} & A_{12} & A_{13} \\ A_{21} & A_{22} & A_{23} \\ A_{31} & A_{32} & A_{33} \end{vmatrix}} = \frac{\begin{matrix} B_1 A_{22} A_{33} + A_{12} A_{23} B_3 + A_{13} B_2 A_{32} \\ -A_{13} A_{22} B_3 - A_{12} B_2 A_{33} - B_1 A_{23} A_{32} \end{matrix}}{\begin{matrix} A_{11} A_{22} A_{33} + A_{12} A_{23} A_{31} + A_{13} A_{21} A_{32} \\ -A_{13} A_{22} A_{31} - A_{12} A_{21} A_{33} - A_{11} A_{23} A_{32} \end{matrix}}. \tag{G.10}$$

The method of finding a determinant is explained in Figures App G.1 and App G.2. For a 2×2 matrix as shown in Figure App G.2, we simply subtract the NE-SW diagonal product from the NW-SE diagonal product, and a numerical example is given on the right. For a matrix of dimension 3×3 or greater, we first concatenate the matrix with itself before finding the diagonal products, as shown in Figure App G.2. Although a 3×3 example is shown, this same general procedure applies to matrices of higher dimension.

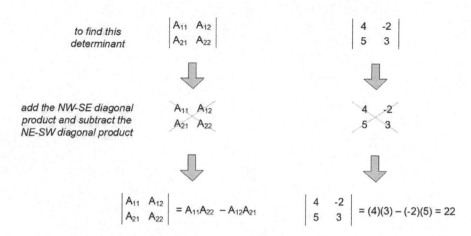

FIGURE APP G.1 Finding the determinant of a 2 × 2 matrix.

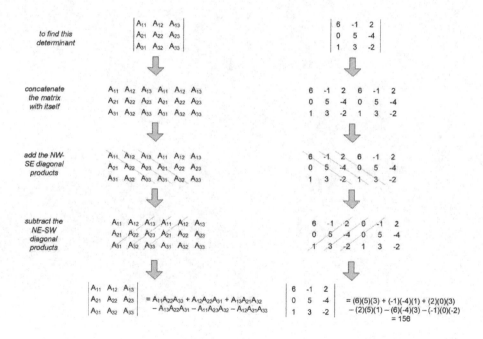

FIGURE APP G.2 Finding the determinant of a 3 × 3 matrix.

Index

Printed in the United States
by Baker & Taylor Publisher Services